The Process Improvement Strategy

Steps   **Sample Tools**

Understand the process   • Flowchart

Collect data on key   • Checksheet
input, process, and   • Data sheet
output measures   • Survey

Assess process stability • Run chart (time plot)
                         • Control chart

Eliminate special-   • See Problem-Solving Strategy
cause variation

Evaluate process   • Histogram
capability   • Standards
                         • Capability analysis

Analyze common-cause   • Pareto chart
variation   • Statistical inference
                         • Stratification

Study cause-and-effect   • Experimental design   • Scatter plot
relationships   • Interrelationship digraph   • Box plot
                         • Cause-and-effect diagram
                         • Model building

Plan and implement
changes

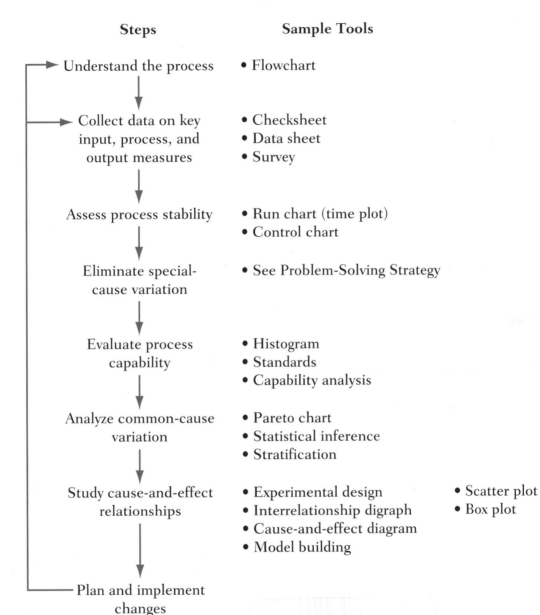

D0605289

STATISTICAL THINKING
Improving Business Performance

Roger W. Hoerl

Ronald D. Snee

DUXBURY

THOMSON LEARNING

Australia ■ Canada ■ Mexico ■ Singapore ■ Spain ■ United Kingdom ■ United States

DUXBURY

THOMSON LEARNING ™

Sponsoring Editor: *Curt Hinrichs*
Marketing Team: *Tom Ziolkowski,
    Samantha Cabaluna*
Editorial Assistant: *Nathan Day*
Production Editor: *Keith Faivre*
Production Service: *Scratchgravel Publishing
    Services*
Manuscript Editors: *Kay Mikel, Carol Lombardi*
Permissions Editor: *Connie Dowcett*

Interior Design: *Anne Draus*
Cover Design: *Christine Garrigan*
Cover Illustration: *Todd Daman*
Interior Illustration: *Richard Sheppard*
Print Buyer: *Vena Dyer*
Typesetting: *Scratchgravel Publishing Services*
Printing and Binding: *R. R. Donnelley & Sons,
    Crawfordsville*

For more information about this or any other Duxbury product, contact:
DUXBURY
511 Forest Lodge Road
Pacific Grove, CA 93950 USA
www.duxbury.com
1-800-423-0563 (Thomson Learning Academic Resource Center)

Printed in United States of America

10   9   8   7   6   5   4   3   2   1

Library of Congress Cataloging-in-Publication Data
Hoerl, Roger, [date]–
    Statistical thinking: improving business performance / Roger Hoerl, Ronald Snee.
        p. cm.
    Includes bibliographical references and index.
    ISBN 0-534-38158-8 (alk. paper)
        1. Commercial statistics.   I. Snee, Ronald, [date]–   II. Title.

HF1017. H58 2001
658.4'033—dc21
                                                          00-04828

To the memory of Arthur E. Hoerl, Horace P. Andrews, and Ellis R. Ott—
great teachers from whom we learned much about
the theory and use of statistical thinking

Contents

PART III Formal Statistical Methods   217

■ Introduction to Minitab   219

■ Introduction to JMP   224

8 Applications of Statistical Inference Tools   327

9 The Underlying Theory of Statistical Inference 374

Preface

This book offers a new approach to introductory business statistics. It is designed to provide business students with the ability to understand the strategic value of data and statistics within the context of real problems and to make informed judgments and improvements with data analysis. Our approach has been influenced by recent research and by recommendations from many outstanding teachers and authors; we combine the most relevant material both in terms of content and pedagogy. We have tried to write a book that will not only make students think about statistics in a practical way but also will improve their attitudes about the subject. Our perspective on statistics has developed from our own experience in both academic and professional settings.

Much about the business world has changed in recent years, largely due to the developments and influence of information technology and global competition. Indeed, many businesses today find themselves "drowning" in data, yet many professionals lack the ability to harness the data for competitive advantage. Statistical thinking* and methods are the keys to unleashing the powerful information contained in data. Global competition has grown even more pervasive through the Internet, requiring businesses to improve rapidly and continuously. Clearly, the need to utilize statistical thinking for tangible business improvements is greater now than ever before. We believe that introductory statistics plays a primary role in preparing students for this modern business reality.

---

*We use the definition of statistical thinking published in the *Glossary of Terms for Statistical Quality Control* (1996): "Statistical thinking is a philosophy of learning and action based on the following fundamental principles:
- All work occurs in a system of interconnected processes,
- Variation exists in all processes, and
- Understanding and reducing variation are keys to success."

Students cannot focus on larger problem-solving strategies if the statistics course focuses only on the nuts and bolts of calculations. We have therefore adopted an approach that relegates calculation to the computer. Conversely, we take great care to accurately develop the concepts to ensure that readers understand the proper application of statistical methods. To that end, this book provides students with the strategic and analytical skills they will need to deal with, and appropriately respond to, data.

Purpose and Objectives

We believe that the purpose of an introductory course for business students should be to develop their capability to apply statistical thinking to improve business processes. Therefore, our objectives include developing the students' *knowledge* to describe the role of statistical thinking and methods for problem solving and process improvement, including the need for understanding, quantifying, and reducing variation. In addition, we aim to develop *skills* to improve business performance by identifying and understanding business processes, collecting appropriate data for a specified purpose, recognizing limitations in existing data, graphically analyzing data using basic tools, deriving actionable conclusions from data analyses, and understanding the limitations of statistical analyses. Finally, it is also our objective that students will develop the *attitude* that statistical thinking and methods can help them do a better job in their chosen careers. With these objectives we emphasize statistical thinking rather than statistical methods per se.

Organization

We have adopted a "teach by example" approach, using real, sequential case studies from the business world to motivate and illustrate the topics. We have organized the material to first present the "big picture" (i.e., overall approaches to business improvement) so that students will understand why they need statistical thinking and methods and what these tools can do for them. Then we offer more detailed instruction in the individual tools. Research indicates that students learn better when they see the big picture first.

The organization of this text may appear quite different from other introductory statistics textbooks—for good reason. Educational and behavioral research (e.g., Forester 1990) has shown that people generally learn most effectively when instruction proceeds from tangible examples to abstract theory—from the big picture, or "whole," to details, or "parts"—and from a conceptual understanding to ability to perform specific tasks. Consequently, we begin with a discussion of why statistical thinking is necessary and helpful to develop conceptual understanding. We then provide real, sequential case studies that illustrate how to integrate several statistical tools into an overall improvement methodology to apply the tools. Based on these case studies, we then provide an overall approach, or model, for applying statistical thinking. Only then do we present individual tools in detail and show how they fit into the big picture. Students can then develop the skills necessary to apply the tools to real business problems. In this approach learning is viewed as a process. We have taught courses, in both academia and business settings, using this approach and have found it to be very effective.

Pedagogy

In our experience, students learn best by doing. We therefore strive to involve students directly in actually applying the material to real problems so that they can learn and develop competency. In addition to numerous exercises, we encourage use of course projects. It is generally recognized that learners must actually practice a new skill to develop proficiency. Lecturing is an important part of the learning process, but it is not sufficient. This text specifically refers to a sequential course project in each chapter. In addition, the instructor's manual gives more detailed advice on how to implement experiential learning, and interactive scenarios will be available on the course Web site.

Role of the Computer

Because this text emphasizes conceptual understanding over formulas and calculations, the role of the computer is an important one. Because of its prominence in business circles and ease of use, Excel is used in most cases to illustrate the basic graphs and statistical methods. However, Excel is not especially suited for formal statistical methods such as regression and design of experiments, so we have utilized Minitab primarily for these methods. Rather than teach individual Excel or Minitab commands for each technique reviewed, we provide introductions to Excel (prior to Chapter 5) and Minitab and JMP (prior to Chapter 6). These will help students get up to speed quickly on these software packages. In addition, the discussion of formal statistical methods includes computer output, with emphasis on proper interpretation.

Instructor and Student Support

A link to the Web site to accompany the text is available at http://www.duxbury. com; select Online Book Companions, then click on the cover icon for this book. The Web site includes data files, relevant PowerPoint presentations, corrections, and comments. In addition, we offer Excel "scenarios" that provide another opportunity for experiential learning. These Excel spreadsheets include a brief description of the situation and list several columns of data. Functions relating the y's to the x's have been programmed into the spreadsheet but are hidden from the user. Students can therefore enter values of the x's, and Excel will calculate y values, which will include random variation and, in some cases, drifts over time, outliers, increases in variation, and so on. These scenarios can be used sequentially to practice basic graphs and summary statistics, two-way comparisons and graphs, regression and design of experiments, and so forth. Users can change the underlying functions however they wish. The Web site will be expanded and updated continuously, and we hope you find it to be a useful resource.

We also include several interactive computer exercises in the instructor's manual. These can help illustrate key concepts throughout the text. See *StatConcepts: A Visual Tour of Statistical Ideas* (1997) by Newton and Harvill for similar examples.

Use of the Text

This text was designed for a one-semester introductory statistics course for undergraduate business students or MBA's and also for professionals within a business environment. Since the need for improvement is generic, the text can also be used for introductory courses in other contexts, such as general introductory or engineering statistics. If two semesters are allotted for introductory statistics, we recommend using this text to develop understanding and skills in statistical thinking to set the appropriate context in the first term. The second term can then go into more detail on relevant statistical methods, such as experimental design, regression analysis, econometrics, and time series.

If it is not possible to cover all ten chapters, we strongly recommend that Part I (Chapters 1–3) be covered. While this part addresses topics not typically covered in introductory statistics, it teaches the most critical concepts of statistical thinking and lays the groundwork for the concepts and methods that follow. Chapters 4 and 5 present overall improvement strategies and basic graphical tools. It is amazing how much students can accomplish with just these basics! Chapters 6 and 7 cover the core methodologies of model building and experimental design. Chapters 8–10 cover the more traditional topics of statistical inference methods (8) and theory (9), as well as an overall summary and potential next steps for students (10).

Depending upon the course emphasis, the text may be used in several ways:

Emphasis	*Chapters Covered*
Concepts, Methods, and Theory	1–10 (recommended)
Concepts and Methods (No Theory)	1–7 or 1–8
Basic Improvement Tools	1–5
Methods and Theory	4–9 (not recommended)
Inference	8, 9, then 6, 7 (not recommended)

Acknowledgments

We would like to acknowledge the numerous individuals who provided insights, suggestions, and constructive criticism in the development of this text. These include Sal Agnihothri, Binghamton University; Rekha Agrawal, GE Corporate Audit Staff; Frank Alt, University of Maryland; Paul Baum, California State University, Northridge; Gayle Bryce, Brigham Young University; John Chiu, University of Washington; James C. Ford, Ford Consulting Associates/Drexel University; Malcom Getz, Vanderbilt University; David K. Hildebrand, University of Pennsylvania; Bob Hogg, University of Iowa; John Houlihan, Boston University; Fred

Hulme, Baylor University; Glenn W. Milligan, Ohio State University; Mike Parzen, University of Chicago; Harry V. Roberts, University of Chicago; Edwin M. Saniga, University of Delaware; Al Schainblatt, San Francisco State University; Hirokuni Tamura, University of Washington; and Nancy Weida, Bucknell University.

We would also like to express our sincere gratitude to Curt Hinrichs and the other members of the publication team at Duxbury.

Special appreciation goes to our spouses, Senecca and Marjorie, who were most helpful and understanding for longer than could be reasonably expected.

References

Forester, A. D. (1990). "An Examination of the Parallels Between Deming's Model for Transforming Industry and Current Trends in Education." *Small College Creativity,* 2(2), 43–66.

Newton, H. J., & Harvill, J. L. (1997). *StatConcepts: A Visual Tour of Statistical Ideas.* Pacific Grove, CA: Duxbury Press.

Statistics Division, ASQC. (1996). *Glossary and Tables for Statistical Quality Control.* Milwaukee, WI: Quality Press.

PART I

Statistical Thinking Concepts

1

The Need for Business Improvement

If you don't keep doing it better—your competition will.

Anonymous

1.1 Overview

In today's global marketplace success—even survival —hinges on an organization's ability to improve everything it does. In this chapter we demonstrate why corporations need to improve how they run their businesses and how the use of statistical thinking can improve business operations. Statistical thinking can be applied to both business operations and methods of management.

The main learning objective of Chapter 1 is to better understand the effect of global competition on business and other organizations in our society and how this impact is forcing us to improve. You will become familiar with the various approaches to improvement and how statistical thinking plays a role in each of these methods. This will enable you to see how the broad use of statistical thinking can help businesses and other organizations improve.

We begin with a short case study. Generalizing from the case study, we then discuss today's business realities, the need to improve, and the recognition that improving how we work is part of the job. The need to improve while we accomplish our work is illustrated with an overall model for business improvement. We then briefly review some new management approaches. Common themes that run through these approaches are identified, and the role of statistical thinking in these themes, and hence in the improvement effort, is noted.

1.2 Today's Business Realities and the Need to Improve

Consider the following business scenario. A large publication corporation, Kowalski and Sons, is having trouble with their monthly billing process. They have

3

discovered that it takes about 17 days to send bills out to customers. But there is a lot of variation from billing cycle to billing cycle, with some bills taking much longer than 17 days. Management's expectation is that the billing should be done in less than 10 days with minimal variation. This target is important from both the company's and the customers' point of view. A shorter cycle time for the bills would improve the company's cash flow, and it would allow customers to enter the billing information in their accounting systems promptly so they can close their monthly books sooner. The current situation results in numerous "late" payments, for which Kowalski and their customers often blame each other. Customers complain that other publishers are not as tardy in sending out bills.

Does this sound like a bad situation? Actually, this is a typical situation in many businesses. In fact, when one of the authors consulted on this problem and began to dig deeper, the situation became worse! Assessing the process revealed that three different departments were involved in billing. Each department worked separately, and no one understood the process from beginning to end. When problems occurred, there was a lot of finger pointing: "The problem is not with us, it's with them. If they would clean up their act, the billing process would be OK." Similarly, there were no standard operating procedures—that is, formal, agreed-upon methods of doing the job. Everybody did it their own way. This resulted in a lot of "fire fighting" to keep the bills going out—heroic efforts requiring long hours and shifting priorities.

The one clear advantage was that a quantitative measure to monitor performance did exist, the number of days required to send bills out. Without a clear measure of success, it is difficult—if not impossible—to effectively manage and improve a process.

Traditional business leaders faced with this situation might attempt to assign blame so the persons responsible could be reprimanded. The approach we recommend is just the opposite. Here is how we approached this problem: A systems map was created for the overall process, along with a flowchart of the critical process steps. The systems map identified the responsible departments and the information or materials that flowed back and forth between the groups. The flowchart was used to construct a production schedule for the monthly billing cycle. This schedule showed what had to be done each month by each group along with a timetable for doing so.

Next, critical subprocesses were identified and cycle time measurements were monitored for each of these critical subprocesses as well as for the overall process. These measurements highlighted key problem areas. Cross-functional teams were formed to troubleshoot the process daily and to review the billing process at the end of the cycle. These teams identified problems and suggested procedures for creating and implementing solutions.

Efforts were also made to document the process and the procedures used in its operation. This documentation helped reduce variation in the process and was central to training new employees. A process owner was also assigned. The process owner's job was to care for the "health" of the process by seeing that the various aspects of the process management system were used and improved to handle the changing conditions the process would experience.

Use of this statistical thinking approach significantly improved the billing process. Over a 5-month period, the monthly billing cycle time was reduced from an average of 17 days to about 9.5 days, with less variation. This resulted in annual savings of more than $2.5 million, more satisfied customers, and a less stressful work environment for employees.

The use of statistics in business has grown over the years as a result of political, social, technological, and economic forces that have affected our world economy. Each new force has created a new need for statistics that typically results in new concepts, methods, tools, and applications. For example, World War II created the need for statistical quality control: munitions needed to be manufactured consistently to very tight tolerances. The need for statistical design of experiments resulted from the demand for major increases in farm production in the early 1900s, which required experimentation with new farming techniques. This movement was accelerated both by the former Soviet Union's launch of the *Sputnik* satellite and by the increasing focus on research and development in the chemical and process industries during the 1950s and 1960s.

The U.S. Food, Drug, and Cosmetics Act and the U.S. Environmental Protection Act resulted in increased use of statistics in the pharmaceutical industry and in environmental studies in the 1970s. The advent of the computer also made statistical calculations easier and available to a broader range of people. The 1980s brought a new economic force—global competition—which has created the need to make major changes in how we run our businesses. The need for change is driven by increasing customer demands for more responsive companies and for higher quality products and services at lower costs.

Evidence of the effects of the global marketplace on the U.S. economy can be seen in the balance of trade and average wages (adjusted for inflation) shown in Figures 1.1 and 1.2. These plots indicate a robust U.S. economy in the 1950s and 1960s, but things clearly changed in the 1970s and 1980s. Global competition became a serious challenge to the U.S. economy. Figure 1.1 shows that the trade balance of goods and services (exports minus imports) was positive until 1971, when it turned negative. Despite some positive upturns, it remained significantly negative in the 1980s and 1990s. In Figure 1.2 we see that average hourly earnings adjusted for inflation increased until 1973 and decreased after that date so that earnings in the 1990s were at the same levels as those of the 1960s. This indicates a declining standard of living for the country as a whole.

Global competition has had an impact on the U.S. economy in other ways as well. Companies find it difficult to compete, which results in layoffs, downsizing, mergers, and bankruptcies. Many of the 1960 Fortune 500 companies are not in business today. The General Electric Company is the only surviving corporation from the original members of the Dow Jones Industrial Average in 1896. In the new millennium the Internet and e-commerce are the driving forces. Many "traditional" businesses are being replaced by electronic competitors with a radically different business model, such as Dell's direct marketing of personal computers.

The changes taking place in U.S. business have ripple effects throughout society, including government, education, health care, and nonprofit organizations. For example, difficult economic times often result in reduced contributions to

FIGURE 1.1
U.S. Balance of Trade, 1946–1993

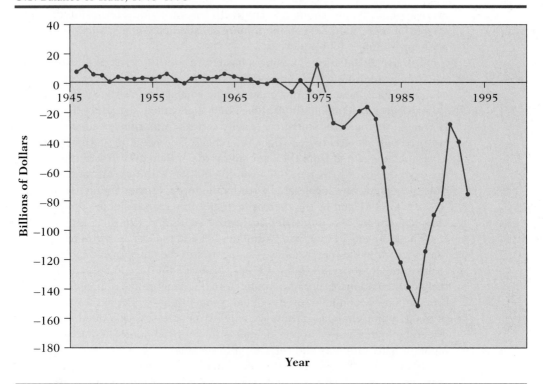

FIGURE 1.2   U.S. Hourly Earnings, 1959–1994 (1982 dollars seasonally adjusted)

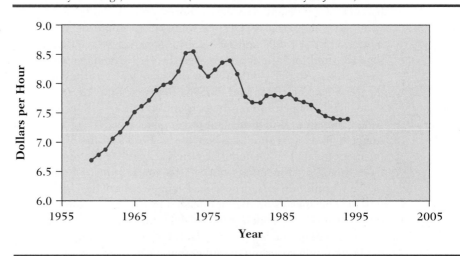

artistic, charitable, and religious groups. Poor business earnings and declining real wages reduce tax revenues to governments, and high unemployment demands greater expenditures by these same governments. Organizations are continually being asked to do more with less, to work in different ways, to be more responsive and caring, to provide better service, and so on. Those organizations that can't keep up are left behind.

The increase in competition is in large part due to progress in the rest of the world, as opposed to changes in the United States alone. After World War II, the United States dominated the world's manufacturing capacity, being the only world economic power that did not suffer significant destruction during the war. The significant prewar economies of Germany and Japan were in shambles, and those of the United Kingdom, France, Italy, and many others suffered a great deal of damage. Over the years since 1945, these countries have regained their competitive edge, and developing countries are becoming players in the world market. The obvious result of these changes is that a healthy U.S. economy, abundant jobs, high wages, and the comfortable lifestyle desired by most Americans cannot be taken for granted, they must be fought for and earned! So what should we do?

1.3   We Now Have Two Jobs: A Model for Business Improvement

We used to have only one job—to do our work. We came to work, did our job, provided a product or a service, and our work was complete. There was no need to change how we did things because there was little competition. No one else was doing things differently. To survive and prosper in this new economic era, we have to make some changes. Now we must accept a second job—improving how we do our work.

Having two jobs means that we each must work to improve our personal knowledge and skills and how we do our jobs as well as get our daily work done. Managers must lead, plan, and manage how the organization can improve its performance as well as operate its day-to-day processes effectively and efficiently. This was illustrated in the billing scenario in Section 1.2, when Kowalski and Sons needed to improve the billing process to keep its current customers.

Organized team sports provide an excellent analogy to business because team sports operate in a competitive environment, have well-defined rules, use teamwork to succeed, and have clear measures of success (winning scores) that are monitored regularly. (We will present a statistical thinking case study involving a soccer team in Chapter 2.) The dual focus on "doing" and "improving" activities can be seen clearly in sports. For example, the "doing" work of baseball is playing the game itself. Professional baseball teams play 160 regular season games per year. But the work activities of baseball go way beyond showing up for each game, playing nine innings, and going home. The "improving" work of baseball is building individual and team skills.

The improvement cycle begins with spring training, where players get in shape and hone their skills. Players work on improving their hitting, running, and pitching. Pitchers work on controlling the curve ball, learning to throw a knuckle ball, and developing pitches they didn't have before. Hitters work on hitting the curve ball or fast ball and other aspects of hitting. This work on improvement goes on all year: before the game, after the game, in the bullpen, viewing videotapes of pitching and hitting, and so on. In the off-season improvement activities involve weight training to build strength and speed or playing winter baseball. Coaches frequently state that star performers are not necessarily the most naturally talented but typically are those who work the hardest at improving their game.

Figure 1.3 shows that the amount of time and effort we spend on improving how we work will increase in the future. We will also be doing more work in the future, as depicted by the larger pie on the chart on the right side of the figure. Increasing the rate of improvement is key. If the competition is also improving, the organizations that succeed will be those with the fastest rate of improvement. It is likely that Kowalski and Sons' competitors are also improving; hence, they cannot view the improvements to the billing process as a one-time event but must make improvement part of the job. Companies must continually improve or go out of business.

Government, health care, and nonprofit organizations also operate in this competitive environment. For example, states compete with one another and with overseas foreign countries for investment from business and industry, which creates new jobs. States that can offer businesses the best-educated workforce and the best infrastructure (transportation, communication, and so on) at the lowest cost (taxes and regulations) tend to get new investments and jobs. The goal for all types of organizations must therefore be to improve faster than their competition.

Figure 1.4 depicts an overall model for business improvement. The doing activity is represented by the Business Process shown at the top. A series of activi-

FIGURE 1.3   We Have Two Jobs: Doing and Improving

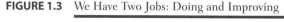

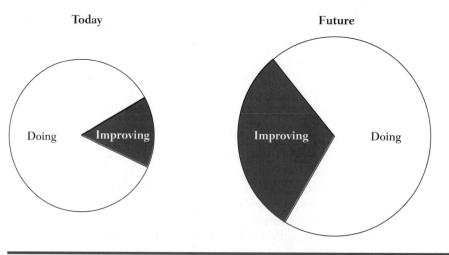

FIGURE 1.4   Improvement Model

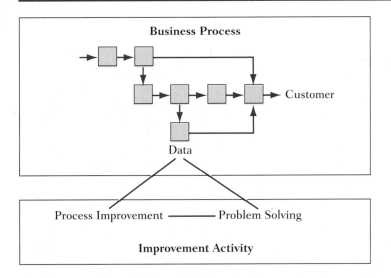

ties, each with its own inputs and outputs, are done as a sequence of steps to produce the desired output for the customer. For example, Kowalski and Sons went through several processing steps to send out their monthly bills. The purpose, or aim, of the process is to provide a product or service of value to the customer. Note that a customer need not be someone outside the organization that purchases a product or service. A *customer* is anyone who uses the output of the process, whether within or outside the organization. Internal customers, the employees, are the key customers of the payroll process.

The improving activity is shown at the bottom of the figure. There are many different approaches to improvement, but we will focus on two types of improvement: process improvement and problem solving. *Process improvement* is a series of activities aimed at fundamentally improving the performance of the process. Here are some typical process improvement activities:

- Flowcharting the process to understand it better
- Collecting data to assess the current performance
- Identifying areas where the process could be fundamentally improved
- Changing the process to implement improvement ideas
- Checking the impact of improvement efforts
- Making the improvements part of the standard way of doing business

Problem solving addresses specific problems that are not part of the normal behavior of the process. These issues are often discovered in the process improvement analysis and can be resolved without fundamentally changing the process. For example, if only one customer's bill was a problem at Kowalski and Sons, they would investigate what happened to that particular bill rather than change the whole billing process. Problem solving usually involves significantly less time and cost investment than that required for true process improvement. Here are the problem-solving steps:

- Document the scope of the problem.
- Identify the root causes.
- Select, implement, and standardize corrections.

A company may go through the process improvement and problem-solving cycle many times in the course of improving a process. Problem-solving strategies and tools will be discussed in greater detail in Chapters 4 and 5. Kowalski and Sons used the process improvement model, which can require significant time and effort. If the process needs to be completely redesigned from scratch, the re-design activity is often called *reengineering*. (Reengineering is briefly outlined in Appendix H.)

Data are the connector or link between the doing and improving activities. Data fuel process improvement and problem-solving activities and increase their effectiveness. Data help us document process performance, identify problems, and evaluate the impact of proposed solutions. This was certainly the case at Kowalski and Sons. But *data* is not synonymous with *information*. For example, we presented average times for bills to be sent out, but the actual time varies from bill to bill. How should we interpret this variation? Customers don't care about average time; they only care about their bill. Therefore, we need both theoretical understanding and practical experience to properly translate these data into actionable information. A thorough conceptual understanding of statistical thinking provides us with the theoretical understanding we need, and a personal project (see Chapter 2) will help provide the experience. First, let's look at why statistical thinking is required.

1.4 New Management Approaches Require Statistical Thinking

New demands to improve have created the need for new management approaches, and a wide range of approaches on how to change have been proposed. Among these approaches are the following:

- Reengineering
- Total quality management
- Learning organizations
- Self-managed work teams
- Benchmarking
- Six Sigma

In addition to this list are the philosophies proposed by Peter Drucker, Stephen Covey, W. Edwards Deming, Joseph Juran, Tom Peters, Peter Senge, and many others. As you can see, management has many choices in today's business climate. Let's look at a few of these approaches in more detail.

Reengineering is "the fundamental rethinking and radical redesign of business processes to achieve dramatic improvements in critical, contemporary measures of performance, such as cost, quality, service, and speed" (Hammer & Champy, 1993, p. 32). The approach is to "start with a clean sheet of paper" and redesign critical processes and the business as a whole, if needed, to become and remain competitive. The key distinction of this approach is to replace rather than improve key business processes, often utilizing information technology.

Total quality management (TQM), as generally practiced, is a broader, less radical approach to business improvement (Berry, 1991). The basic elements of TQM are to focus on improving the quality of all aspects of the business to better satisfy the needs of customers. This involves cooperative efforts from all employees, from the CEO to those sweeping the floor, and typically stresses data-based decisions and use of statistical tools to reduce process variation.

Learning organizations create change and improvement by learning how to work in more effective and efficient ways (see Senge, 1990). This includes both individual learning and learning by the organization. Learning how to view the organization as a system of interconnected processes is key to this approach. The focus is on improving the system as a whole rather than looking at problems of individual departments. This approach requires an open mind and routine gathering of data, both quantitative and qualitative, from which to learn.

Self-managed work teams were created in response to the need to reduce layers of management and to empower the workforce. In self-managed work teams employees work as a team without direct supervision from management, using principles and guidelines developed jointly with management. A key rationale for this approach is a belief that those who work with the process every day understand it best and therefore should make most of the day-to-day decisions (see Scholtes, Streibel, & Joiner, 1996).

Benchmarking is the process of improvement that finds the best practices in other organizations and adapts those practices to make improvements (Camp, 1989). The best practices are often identified in outside industries. Examples could include the billing process, approaches to new product development, compensation plans, organizational structure, and so on. This approach avoids the problem of "reinventing the wheel." Internal benchmarking, identifying and using the best practices of one department in others, also helps reduce variation from department to department.

Six Sigma is a business improvement approach that seeks to find and eliminate causes of mistakes or defects in business processes (Breyfogle, 1999; Harry & Schroeder, 2000). Six Sigma is a statistical term that roughly translates to only 3.4 defects per million opportunities. The Six Sigma approach emphasizes understanding and documenting the business process, developing metrics and hard data, and reducing variation. This approach uses a breakthrough strategy that consists of four process improvement phases: measure, analyze, improve, and control. The goal is to improve the process in such a way that customer satisfaction increases and there is a positive impact on the bottom line. The Six Sigma approach was originally pioneered in 1987 by Motorola, which focused primarily on manufacturing, and was later applied by other companies including Allied Signal and General Electric, which broadened the approach to include general business activities such as financial services. Use of the Six Sigma approach is growing rapidly in the United States and around the world. The Six Sigma breakthrough strategy is discussed in greater detail in Chapter 4, and elaboration of the tools used is provided in Appendix D.

Each of these approaches and philosophies is useful, and the best aspects of each can be integrated with the management approach an organization is currently using. The result is a new management approach that helps the organization better serve the needs of its customers and compete effectively in the marketplace. Three common themes run through these management approaches:

- Viewing work as a process
- Using data to guide decisions
- Responding wisely to variation

These three items are part of the body of knowledge known as statistical thinking. This body of knowledge and its associated skills are essential to the successful management and improvement of any business. Statistical thinking is a philosophy of learning and action based on these fundamental principles (Statistics Division of the American Society for Quality Control, 1996):

- All work occurs in a system of interconnected processes.
- Variation exists in all processes.
- Understanding and reducing variation are keys to success.

These principles work together to create the power of statistical thinking. The steps in implementing statistical thinking are shown in Figure 1.5. We begin by recognizing that all work is a process and all processes are variable. We must analyze the process variation to develop knowledge of the process. You cannot improve a process that you do not understand. Note that these core principles are similar to the common themes of recent management improvement efforts presented earlier. With knowledge of the process, we're in a position to take action to improve that process.

From a statistical point of view improvement activity—both fundamental process improvement and problem solving—can be viewed as working on either of two process characteristics: (1) reducing variation through tighter control of the process or (2) improving the overall level (average value) by changing the process target, which may also result in reduced variation. For example, the primary objective of Kowalski and Sons' billing efforts was to reduce the average time to get bills out. They also wanted to reduce the variation from bill to bill. The end result of using statistical thinking is business performance that satisfies the stakeholders: customers, employees, the community in which the business operates, and the shareholders.

The terms *average* and *variation* are critical to applying statistical thinking. For most processes the average is the central value around which the process varies. Variation results when two or more measures of the process are different, which is the rule rather than the exception. Figure 1.6 illustrates these concepts by plotting 15 consecutive process measurements. Although any units could be used here, as an example, let's use monthly gross sales. The process is centered between $10,000 and $11,000, with an average value of $10,600. The average

FIGURE 1.5
Steps in Implementing Statistical Thinking

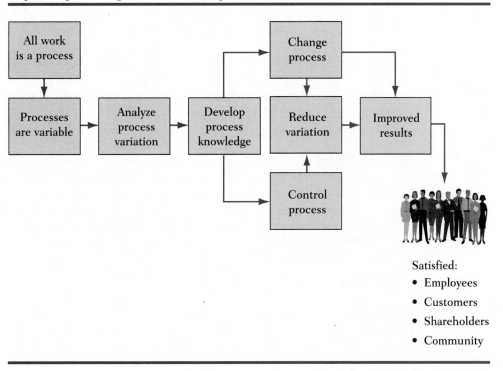

FIGURE 1.6 Process Average and Variation

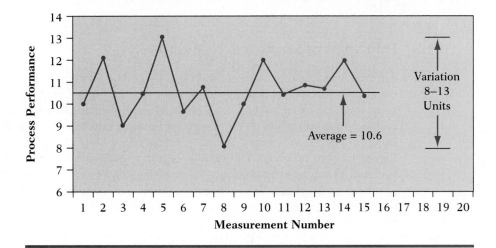

Standard Deviation   The standard deviation is an overall measure of how far individual data points vary about the average. Most data points will fall within 1 standard deviation of the average. The standard deviation can be thought of as the typical (average) deviation. It is calculated by taking the deviation of each data point from the average, squaring these deviations (so they are all positive), averaging the squared deviations, and then taking the square root to go back to the original units; that is, the standard deviation $= \sqrt{\text{sum of squared deviations} \div n}$, where n is the number of data points. For example, to calculate the standard deviation of 1, 3, 5, 7, and 9, we first calculate the average, $(1 + 3 + 5 + 7 + 9) \div 5 = 5$, then we subtract the deviations of each value from 5, $1 - 5 = -4$, $3 - 5 = -2$, $5 - 5 = 0$, $7 - 5 = 2$, and $9 - 5 = 4$. Next we square the deviations to get 16, 4, 0, 4, and 16. The average of these values is 8, and the standard deviation is 2.83. Note that in many cases we divide the sum of the squared deviations by $n - 1$ rather than by n when calculating the standard deviation. The reasons for this are discussed in Chapter 9.

value is computed by adding the values to get the total and then dividing by the number of values—that is, $159 \div 15 = 10.6$ in this case. The observed variation in the process is from about $8000 to $13,000, resulting in a range of about $5000. Another common measure of variation is called the *standard deviation,* which can be thought of as the "typical" deviation of individual data points from the average. (See the box on standard deviation.)

With an understanding of the meaning of statistical thinking, we can now discuss the principles underlying statistical thinking.

1.5   Principles of Statistical Thinking

The first principle of statistical thinking is that *all work occurs in a system of interconnected processes.* This principle provides the context for understanding the organization, improvement potential, and sources of variation mentioned in the second and third principles. A process is one or more connected activities in which *inputs* are transformed into *outputs* for a specific purpose. This is illustrated in Figure 1.7. For example, mailing bills requires that records are kept on charges (inputs). These records must be processed (aggregated for a month, reduced by payments made, checked for accuracy and applicable discounts, and so on), often with the aid of computer systems, into a monthly bill. Any discrepancies or errors must be resolved, often through a manual process. These bills must then be physically printed and stuffed into the appropriately addressed envelopes (or sent to appropriate electronic addresses) and delivered to the desired mailing system (U.S. mail, Federal Express, UPS, etc.). The bill a customer receives is the output

FIGURE 1.7   SIPOC: A Process View of Work

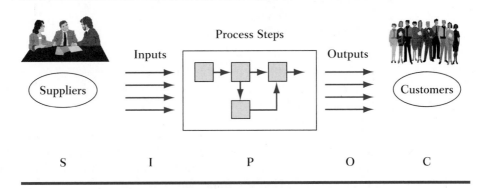

of the process. For a simpler example, the series of activities one goes through to get to work or school in the morning (getting out of bed, bathing, dressing, eating, and driving or riding) can be thought of as a process. Any activity in which a change of state takes place is a process, as depicted in Figure 1.7.

We encourage people to "blame the process, not the people" when working on improvement. Joseph M. Juran pointed out that the source of most problems is in the process we use to do our work. He discovered the "85/15 rule," which states that 85% of the problems are in the process and the remaining 15% are due to the people who operate the process. W. Edwards Deming states that the true figure is more like 96/4 (Joiner, 1994, pp. 33–34). We may debate the correct figure, but it is clear that the vast majority of problems are in the process.

Businesses and other organizations are made up of a collection of processes. Business processes interconnect and interact to form a system that typically provides a product or service for a customer (the person who receives the output of the system). Viewing business as a collection of processes is discussed in detail in Chapter 3. Here are some typical business processes:

- Customer service
- Business planning
- New product development
- Employee recruiting and orientation
- Company mail delivery
- Manufacturing
- Equipment procurement
- Patent application
- Accounting
- Laboratory measurement

The second principle of statistical thinking is that *variation exists in all processes.* This provides the focus for improvement work. Variation is the key. If there were no variation,

- Processes would run better.
- Products would have the desired quality.
- Service would be more consistent.
- Managers would manage better.

Focusing on variation is a key strategy to improve performance. Of course, in certain situations variation is desirable, such as wanting variety in menu items in a restaurant, valuing diversity among team members, and so on. Intended or desirable variation is valuable and should be promoted. It is unintended variation that we will focus on here.

Variation is a fact of life. Variation is all around us. It is present in everything we do, in all the processes we operate, and in all the systems we create. Variation results when two or more things, which we may think are exactly the same, turn out to be different. Here are some examples:

- Restaurant service time varies from day to day or even from customer to customer.
- Tires wear at different rates.
- Tomatoes of the same variety vary in weight.
- Shirts of the same size fit differently.
- Cars of the same model perform differently.

To understand and improve a process we must take variation into account. Variation creates the need for statistical thinking. If there were no variation, there would be little need to study and use statistical thinking.

The third principle of statistical thinking is that *understanding and reducing variation are keys to success*. The focus is on unintended variation and how it is analyzed to improve performance. First we must identify, characterize, and quantify variation to understand both the variation and the process that produced it. With this knowledge we work to change the process (for example, operate it more consistently) to reduce its variation.

The average performance of any process (for example, average time in days to get bills out, average waiting time in minutes to be served in a restaurant, or average pounds of waste of a printing process) is a function of various factors involved in the operation of the process. When we understand the variation in the output of the process, we can determine which factors within the process influence the average performance. We can then attempt to modify these factors to move the average to a more desirable level.

Businesses, as well as customers, are interested in the variation of the process output around its average value. Typically, consistency of product and service is a key customer requirement. For example, customers of a shipping company do not want shipments to arrive sometimes in a day and other times in a week. Such variation would make it difficult to plan when to actually ship goods. Similarly, in accounts payable, we do not want to pay bills too late because we may lose discounted terms (often a 2% discount for prompt payment). If we pay too early, however, we lose interest on the cash. If we can reduce the variation in the process, we can consistently pay right at the due date and receive discounted terms and minimize loss of interest. This results in many improvement efforts that focus on reducing variation, for example, reducing restaurant waiting time from 0–10 minutes to 0–5 minutes, reducing the variation in real estate sales from 20–40 units/month to 25–30 units/month, or reducing bill mailing time from 10 ± 4 days to 10 ± 1 days. Low variation is important to customers, and they will sometimes accept less desirable "average performance" to obtain better consistency.

Two types of variation that we may need to reduce are "special cause" and "common cause." *Special-cause variation* is outside the normal or typical variation a process exhibits. Normal or typical variation is called *common-cause variation*. The distinction between these types of variation will be discussed in more detail in Chapter 2. The result of special-cause variation may be unpredictable or unexpected values that are too high or too low for the customer. Some examples include waiting for service for 30 minutes when 10 minutes or less is typical performance for a particular restaurant, a real estate sales office selling 10 units this month when typical sales are 20–40 units, or a printing press running at 10% waste this month when typical waste varies from 1% to 3%.

Because special-cause variation is atypical, we can often eliminate the root causes without fundamentally changing the process. Using a problem-solving approach, we identify what was different in the process when it produced the unusual result. For example, in the printing press scenario they may have purchased their raw paper stock from a new supplier that month. A hypothetical example is shown in Figure 1.8. The two points that clearly stick out from the rest are due to special causes. There should be a specific, identifiable reason these waste values were so high. Even if we eliminate the causes for these points through problem solving, however, we are left with the normal level of waste due to common causes. To make further improvement, we need to fundamentally change the system; that is, we need to improve the overall process.

FIGURE 1.8 Special-Cause and Common-Cause Variation

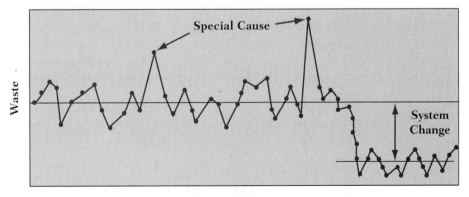

Reducing the inherent common-cause variation typically requires studying the process as a whole because there are no unusual results to investigate. In other words, there is no single, identifiable reason waste was high one day and low the next. Using a process improvement approach, we study the normal variation and try to discover the input and process factors that are the largest sources of variation. Knowledge and understanding of common-cause variation is best obtained using a statistical approach and planned experiments. Historical process data analysis can also be helpful but is not as effective as planned experiments. Once the primary sources of variation are identified, we can determine how to modify the process to eliminate or mitigate their effects. Using the printing press example, we could experiment to find optimal combinations of paper stock, ink, and printing conditions, which would result in the fundamental improvement of the process.

1.6   Applications of Statistical Thinking

Statistical thinking can be used in all parts of an organization and in all job functions. For example, the job of the manager is to lead the organization in a common direction. The manager must gain agreement on the desired direction and get employees to align all their activities with this common direction. The result of this reduced variation in activity is less wasted effort, less rework, reduced costs, and faster improvement of organizational performance. An engineer in a factory may want to reduce the variation in a product characteristic to meet customers' specifications a higher percentage of the time. This may require using a better process control procedure or developing a deeper understanding of the process so that fundamental changes can be made to better operate the process.

A business analyst may wish to find ways to change the "closing" process (calculating final profit and loss for a specific time period, such as quarterly or annually) so that, on average, the accounting books are "closed" in 2 days rather than the current 15-day average, thereby reducing overtime work required and increasing customer satisfaction (customers would include corporate management and the financial community). This objective will likely require a better understanding of how the closing process works and what persons and groups are involved. Armed with this knowledge, we can find ways to improve the process to meet business needs.

A restaurant manager may want to understand why it is taking a long time to serve some customers. Data on customer service time is needed to document current performance and to determine whether newly introduced procedures are working. Properly collected data are also useful in identifying causes of poor performance and possible solutions.

All of the problems identified in these examples can be addressed using statistical thinking. As noted earlier, statistical thinking also applies to various kinds of work in all organizations and functions, including the following:

Type of Organization	*Type of Function*
Manufacturing	Marketing
Financial services	Sales
Education	Manufacturing
Government	Research and development
Health care	Engineering
Retail sales	Human resources
Transportation	Information systems
Software	Purchasing
Restaurants	Finance

Statistical thinking is also useful in the management of organizations. We noted earlier that the job of the manager is to move the organization in a common direction. To do this the manager must gain agreement among the employees regarding the direction the organization should go. This will reduce variation in employee actions.

Similarly, many organizations have reduced the number of their suppliers. This action is based on the fact that a large number of suppliers increases the variation in incoming supplies, which in turn increases the variation in the output of the organization. Customers want consistent products and services. When a company decreases the number of suppliers, the company generally decreases the variation in their product or service.

1.7 Summary

1. Global competition has forced U.S. companies to change.
2. Improvement is needed for an organization to survive.
3. We now have two jobs: to do our work and to improve on how we do our work.
4. The U.S. system of management is changing, and many new approaches have been proposed.
5. Statistical thinking is an integral part of the common themes that run through these new approaches.
6. Statistical thinking is based on three principles: all work occurs in a system of interconnected processes, variation exists in all processes, and understanding and reducing variation are keys to success.
7. Broad use of statistical thinking can help an organization improve operations and its management system.

EXERCISES

1. Update the databases on the U.S. balance of trade and the U.S. average hourly earnings (Figures 1.1 and 1.2). Construct a plot of these measures of the U.S. Economy versus time and comment on the trends you see based on the latest information.

2. Describe how global competition and its effects on the U.S. economy have affected you and your family personally. Consider the effects of global competition on all aspects of society (education, health care, government, churches, and so on). Be sure to consider how these effects link together. For example, how do effects on business affect churches and government?

3. Describe a work task, hobby, or other activity you regularly do and list sequentially the various things you do to complete this activity. How complex is your list and how many steps are required to complete the activity?

4. Describe an activity (work or play) and explain how you would go about improving this activity. Differentiate between *doing* and *improving* activities.

5. Understanding variation is at the core of statistical thinking. Describe how variation has affected you in your daily life.

6. Using Figure 1.5 as a model, explain how statistical thinking can be used to improve the situation you described in Exercise 3.

7. Briefly discuss the impact the Internet is having on the world business climate. How might this radical change provide new opportunities for applications of statistical thinking?

8. Using one of the management methods discussed in Section 1.4, describe how that approach can improve organizational performance and how statistical thinking is an integral part of the approach. Use the Internet or other sources to find background information on and examples of the chosen management method.

9. Identify specific ways the Internet can be used to obtain information and data to improve business processes. (*Hint:* Note the elements of business processes and the various types of improvements—such as reducing errors and decreasing cycle time—and approaches to improvement.)

10. Respond to the scenario that follows. What would you do to improve this learning situation?

Statistical Thinking in Academia

(Doug Zahn, Florida State University)

Professor Zachary Davis has been teaching a large-lecture, 250-student, introductory statistics course at Maryland State University for more than 10 years, experimenting over the years with a number of different strategies for improving the course. To his delight, the department has hired a new statistics professor, Dr. Leon Cornwall, who is well respected for his ability in statistical thinking.

"Welcome to MSU, Dr. Cornwall. I'm looking forward to hearing how you will apply statistical thinking to the course we will be teaching together."

"Great! And please call me Leon. What is the situation with respect to the course?"

"Well, for starters, call me Zack. The situation is challenging. We teach in a stark, 250-seat auditorium with 100 seats on the main floor and 150 in the balcony. Students in the balcony are far from me and have a tendency to talk among themselves. What with the acoustics, this makes it even tougher for all the students to hear."

"Sounds challenging. What sort of results have you seen lately?"

"Some results are amazingly persistent, even in the face of lots of work. About 20% of the students fail the course each term; the average score on the final exam hovers around 60%; after 3 or 4 weeks, only about 50% of the students come to lectures; only 50–70% attend the recitation sessions now that there are no weekly quizzes; the teaching assistants are often international students with English-language problems and are from cultures that have much different relationships between faculty and students; some of the student project teams self-destruct, some overcome their troubles, and some love the course project. The students come from many places, including high school, community colleges, and prerequisite courses here at MSU. They have a variety of skills. All students vary in majors and topics that interest them. About half of them are working now."

"Whew! Anything else?"

"Oh, yes, I'm just getting started. There are adversarial relationships all over the place. Faculty teaching the course don't think the faculty teaching prerequisite courses or the students are doing their jobs; faculty teaching courses for which this is a prerequisite are always asking for one more topic to be added to the course; employers complain that our graduates can't express themselves orally or in writing, can't work in teams, and can't think critically; students just want to know what will be on the test and why I am wasting time if I try to do anything other than lecture to them; parents and the legislature wonder why our graduates can't get good jobs. Students resist shifting from passive spectators to active participants in the course. I guess this isn't too big of a surprise, since this is the system they have succeeded in up to now."

"Anything else?"

"Sure, teachers' assistants are untrained and often regard time spent on the course as detracting from their research time. You come highly recommended for your statistical thinking skills. I don't know much about that, but I understand that it means looking at the system as a whole. To me, that means everything, from the prerequisite courses to placement of our students in employment positions. You've got your work cut out for you! Where shall we start?"

REFERENCES

Berry, T. H. (1991). *Managing the total quality transformation.* New York: McGraw-Hill.

Breyfogle, F. W. III. (1999). *Implementing Six Sigma: Smarter solutions using statistical methods.* New York: Wiley-Interscience.

Camp, R. C. (1989). *Benchmarking: The search for industry best practices that lead to superior performance.* Milwaukee, WI: Quality Press.

Hammer, M., & Champy, J. (1993). *Reengineering the corporation.* New York: Harper Business.

Harry, M., & Schroeder, R. (2000). *Six Sigma: The breakthrough management strategy revolutionizing the world's top corporations.* New York: Currency.

Joiner, B. L. (1994). *Fourth generation management.* New York: McGraw-Hill.

Senge, P. (1990). *The fifth discipline: The art and practice of the learning organization.* New York: Doubleday/Currency.

Scholtes, P., Streibel, B., & Joiner, B. (1996). *The team handbook* (2nd ed.). Madison, WI: Oriel, Inc.

Statistics Division of the American Society for Quality Control. (1996). *Glossary and tables for statistical quality control* (3rd ed.). Milwaukee WI: Quality Press.

2

The Overall Statistical Thinking Approach

Statistical thinking will one day be as necessary for efficient citizenship as the ability to read or write.

H. G. Wells

2.1 Overview

The purpose of this chapter is to illustrate how various statistical thinking tools fit together to form an overall approach to business improvement. This overall approach is the common theme throughout the text. Each new tool introduced in later chapters will be positioned as a component of this approach.

The main learning objective is a good conceptual understanding of the statistical thinking approach and how it applies in practice. Other learning objectives include understanding of the following:

- The existence of variation in business processes
- The synergy between data (empiricism) and subject matter theory
- The dynamic nature of business processes
- The sequential nature of statistical thinking

These are all critically important aspects of applying statistical thinking effectively to real business processes.

We begin by reviewing two case studies that illustrate various concepts of statistical thinking. The first case study involves a traditional business function (sales) and demonstrates how statistical thinking can be applied. The second case study is an application by a high school soccer coach. Although coaching can be considered a business, the main reason for including this case study is to show how statistical thinking applies to everyday situations. This case study may provide a useful example when you carry out your own project (Section 2.9). The section following the case studies shows the similarities of the two approaches and describes an overall model for statistical thinking. The last three sections elaborate on key elements of this approach: variation, the synergy of data and theory, and the dynamic nature of business processes.

2.2   CASE STUDY: The Effect of Advertising on Sales

This case study is a condensed version of one given in Ackoff (1978). It is a classic application of statistical thinking to advertising. August Busch Jr., CEO of Anheuser-Busch, Inc., contacted Robert Ackoff and his associates to evaluate a decision relative to Busch's advertising expenditures. The vice president of marketing had asked Busch for a significant increase in funding to advertise the Budweiser brand more extensively in 12 marketing areas. The request was justified by claiming that it would produce a significant increase in sales. Busch told Ackoff and his associates that he received similar requests every year and usually approved them, but he never felt he knew whether he got what he paid for. His specific request was that Ackoff estimate the actual increased sales resulting from the increase in advertising funds.

The First Experiment

The original proposal was to allow the marketing department to select any 6 of the 12 marketing areas to increase advertising, and use the other 6 as a control group with which the results of the test group could be compared. Allowing the marketing department to select the 6 areas could bias the results because they could pick the 6 most likely to experience a sales increase, but it was believed that this method would help overcome resistance from the marketing department. Furthermore, because equations for forecasting monthly sales had previously been developed, the plan was to measure deviation from the forecast rather than from raw sales. This would allow the researchers to take into account the effects of other factors on sales, such as seasonality.

This first study was conducted over 6 months and resulted in 72 data points, 6 for each of the 12 areas. The 6 highest advertising regions did produce higher sales (relative to forecast) than did the control group. However, given the variation in the data, there was still some doubt as to whether this effect was real and repeatable. In any case the apparent increase was not large enough to justify the increased expenditures in advertising. Busch was encouraged by these findings and asked the researchers to investigate what amount should be spent on advertising to maximize profitability. He wanted to proceed with caution, however, because he believed much of Budweiser's success was due to the effectiveness with which its quality was communicated through advertising. He therefore authorized experimentation in up to 15 marketing areas, provided they did not include any of the major markets.

The Second Experiment

Many companies fix advertising as a percentage of forecasted sales, and this created a problem for researchers. If forecasted sales increase or decrease, so do advertising dollars. When comparing historical data for advertising with sales, there will generally be a strong correlation between the two. This is not necessarily because advertising produces sales but rather because sales forecasts drive advertising. To overcome this problem, the researchers decided to consciously manipulate advertising without regard for sales forecasts.

Having little theory to go on, the researchers hypothesized a *stimulus–response* relationship between advertising (stimulus) and sales (response), as depicted in Figure 2.1. This suggests that a small amount of advertising has little impact on sales, but once advertising gets past a certain threshold, it begins to produce a significant response from consumers. This effect flattens out as consumers become saturated with advertising. Any further advertising produces *supersaturation*—consumers get fed up with the advertising and react negatively.

To test this theoretical relationship, the researchers suggested that advertising be varied over four levels, including a control of no change, a 50% reduction, a 50% increase, and a 100% increase. Four levels were needed to test whether this was the correct shape of the stimulus–response curve. Nine marketing areas per advertising level were needed to have confidence in the results because of the expected level of variability in sales. Unfortunately, this plan was rejected by Busch because it required too many marketing areas. A compromise plan was to use two levels of advertising and a control of no change. In addition, the reduction in advertising was changed from 50% to 25% because of marketing's concern that a 50% reduction might do permanent damage to the brand. The final plan used a 25% reduction, no change, and a 50% increase, with each level tested over 9 areas. This was not the researchers' first choice, but they accepted

FIGURE 2.1   Stimulus–Response Curve

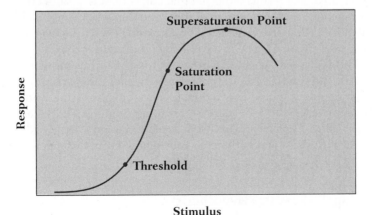

FIGURE 2.2   Results of Second Experiment

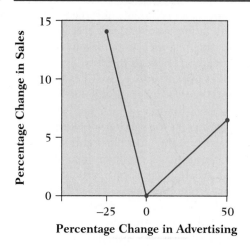

it because it included experimenting in 18 areas, as opposed to the original restriction of 15 areas.

Within the 9 areas for each level of advertising, two other important marketing variables were explicitly varied: the amount spent on sales effort (on the sales
force) and the amount spent on point-of-sales materials (displays, signs, and so
forth). Pricing could also have been varied, but this was rejected by marketing and
sales. As you will see later in the text, it is possible to simultaneously vary several
factors using experimental techniques and still be able to interpret the results at
the end of the experiment. This provides significant efficiency advantages and
speeds our pace of learning.

The experiment was conducted over 12 months, resulting in 12 data points
for each area. The results indicated that current levels of sales effort and point-of-
sales expenditures were close to optimal and that advertising levels behaved independently of these two variables. The surprising result, however, was that the pattern of sales versus change in advertising produced a V shape, as shown in Figure
2.2. This is the only possible pattern that is not consistent with the original hypothesis, and it puzzled both the researchers and the marketing department. Although no one disputed that a 50% increase in advertising could produce a 7%
increase in sales, the 14% increase resulting from a 25% *decrease* in advertising
was shocking. The researchers were asked to design another experiment to validate these findings and perhaps to convince skeptics that sales could actually be
increased with a decrease in advertising.

Refining the Research Hypothesis

Because the findings of this experiment were inconsistent with the original hypothesis, the researchers gave careful thought to how their current understanding
of the stimulus–response relationship needed to be modified to explain the re-

FIGURE 2.3   Response Function of Segmented Populations

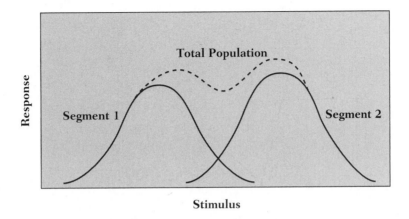

sults. They hypothesized that if there were distinct market segments, each with its own curve, the combined curve might have a V shape. This is illustrated in Figure 2.3. This theory matched marketing's experience with beer drinkers. Marketing suggested that there should be three distinct segments: heavy, moderate, and infrequent beer drinkers. It was further suggested that heavy users would be particularly sensitive to advertising and infrequent users least sensitive.

To test this new hypothesis, the researchers used previous findings that suggested that beer consumption correlated positively with discretionary income within typical income ranges. Data on average discretionary income for each marketing area were readily available, and these data were compared to the deviations from sales forecast in each area. A positive correlation was again observed, lending credence to the hypothesis of distinct market segmentation.

To accurately estimate the aggregate stimulus–response curve across all three segments, the researchers obtained data for other levels of advertising. Levels of 100% reduction (no advertising), 50% reduction, 25% reduction, no change, 50% increase, 100% increase, and 200% increase were used. Data were collected for 12 months. Results from the levels used in the previous experiment were consistent with these results, providing greater confidence in the experimental design. The deviation from sales forecast results are shown in Figure 2.4.

Research Outcomes

Two surprising outcomes were noted. First, there were only two "humps" rather than the three that were expected, one for each segment. It was suspected that a third hump was hidden on the right side of the curve, possibly around 150% increase, but this level was not actually tested. Even more surprising was the fact that complete elimination of advertising produced sales at about the same level as sales in the control group. Although most people did not believe this result, a few attributed it to the long history, strength, and exposure of the brand. Distributors

FIGURE 2.4   Results of Third Experiment

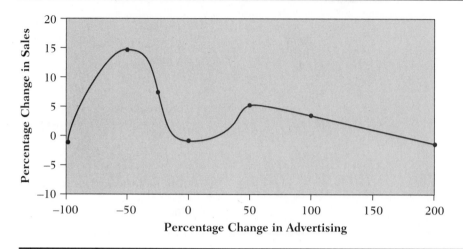

confirmed the theory of supersaturation for the areas that received a 200% increase in advertising. They noted that they were receiving negative feedback to the overexposure of the brand. The researchers were intrigued by these results and suggested further tests to evaluate elimination of advertising. The study was continued to determine how long it would take for sales to begin to drop off. The researchers also wanted to determine how long it would take to recapture lost sales. At this same time additional research was initiated on the relative effectiveness of different media.

Because of some cash flow issues within the company, management suggested reductions in advertising expenses. Based on the estimated stimulus–response curve, the researchers suggested a 15% reduction in nonstrategic market areas and predicted a 5% increase in sales. Even though an increase in sales from reduced advertising seemed counterintuitive to management, they agreed to the plan. Surprisingly (to some), sales did increase by about 5% for 6 months. Management then decided to double the number of reduced advertising market areas and to increase the advertising reduction to 25%. As successful results were achieved, this process was extended to other areas.

Another contributing factor to these savings was the discovery that it took over a year and a half for sales to decrease in those areas where advertising was eliminated. From that point on a small decline in sales was noted each month. Once estimates of the rate of decline were obtained, advertising was reinitiated and the markets returned to normal sales levels in about 6 months. These results led to speculation that a more economic advertising strategy might be to pulse advertising, using an on-off pattern. The experimental results suggested that this would produce the same amount of sales at significantly reduced advertising expenditures.

Two types of pulsing were evaluated. In the first type all advertising media were discontinued. In the second type only one medium was used at a time and media were rotated. The media included billboards, magazines, newspapers, radio, and television. Several forms of the first type were tested, along with variation in levels of expenditure. The results indicated that one type of pulsing was best for low expenditure, and the other was best for high expenditures. In statistics this type of phenomenon, wherein the best pulsing pattern depends on the level of expenditure, is called an "interaction." It was later determined that the best pulsing pattern also depended on median income levels and growth rate in the area. Subsequent testing revealed a negligible difference between the two types of pulsing, but the second type of media pulsing was easier to administer.

Summary

These learnings were gradually implemented and continued as the results matched predictions. Further evaluation of different media showed that television was slightly superior to the others and that billboards were significantly inferior. It was hypothesized that billboards could do little more than convey a product name and slogan—that is, remind consumers that the product still exists. It was estimated that urban consumers saw the Budweiser name about 10 times a day, so there was no need to remind them that it still existed. On the basis of this finding all billboard advertising was discontinued, which saved 20% of the total advertising budget.

The impact of implementing these changes was that overall advertising expenses for Budweiser were eventually reduced by well over 50%. During this same time period, the market share of Budweiser increased from 8% to about 13%.

2.3   CASE STUDY: Improvement of a Soccer Team's Performance

This case study is based on Hau (1990) and has been elaborated on through conversations with him about additional facts and issues that were not included in the original version.

Background

This experiment took place when Hau was a soccer coach for a Monona Grove, Wisconsin high school in 1988/1989. Hau was a graduate student in the Department of Statistics at the University of Wisconsin and felt that some of the tools he was learning about could be directly applied to improving the performance of his soccer teams (boys' and girls' teams). The commentary will follow the girls' team, but some data from the boys' team have been interspersed where they are more illustrative.

Overall Approach

Hau summarized the overall approach he took when he became coach as follows:

1. *Study the current situation.* In the first week of the soccer season short drills were designed to find out the character and the skill level of each player.

2. *Define quality performance.* Having found out where the team stood at the time, coaches, players, players' parents, and the school athletic director worked together to define quality performance so that everybody had a common goal. Quality performance was defined as the number of games won.

3. *Analyze performance.* The coaches and players together analyzed how to break down the overall performance (winning) into smaller subgoals. Areas the team needed to work on were then identified.

4. *Focus on the main problems.* Data gathered from games were used to focus practice.
    -   After each game coaches and players together analyzed improvements in play, the root causes of every loss, and why scoring chances were missed. Drills were then designed to improve skills in those areas.
    -   Data from the games were gathered and analyzed to identify the weaknesses of the team. (Pareto analysis was used to identify the major causes of an issue.)
    -   An experiment was designed to determine the best players to use for penalty kicks.
    -   A modified seven-step method was applied to improve players' heading skill. (The seven-step method is a formal approach to problem solving.)
    -   An experiment was designed to find out where the team should attack to score more goals.

5. *Monitor and evaluate progress.* Data from games as well as practice sessions were gathered continuously to measure progress and to use as a base for changing the focus in practice.

Getting Started

In his initial observation of the team in the first 2 weeks, Hau's impression was that the girls were well disciplined, motivated, and eager to cooperate. Unfortunately, he also found that they had a relatively low skill level, lacked knowledge about the game itself, and were not physically strong. The season lasted only 10 weeks, and he realized he needed to see improvement by the playoffs at the end of the season.

Everyone had agreed that the measure of success was the number of games won. What was not widely understood by the team, the parents, or the team's supporters were the key factors that determined winning and losing. Hau developed a tree diagram (Figure 2.5) to break down winning into its two main components: scoring more goals and losing fewer goals. The specific tactics and skills needed for each component were identified so the team could concentrate on them in practice. Hau decided to focus on skills that could be developed by everyone fairly quickly. On defense these were tightness of defensive positioning and listening to Murphy (i.e., learning from mistakes that resulted in goals). On offense these were counterattacking, shooting skills, crossing (passing the ball in front of the opponent's goal), and free kicks such as penalty kicks and corner kicks. Hau's coaching philosophy was to solidify the defense first and then concentrate on offense.

FIGURE 2.5 Tree Diagram of Success Factors

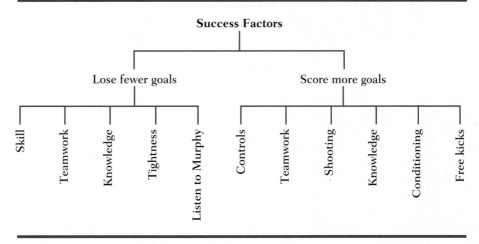

First Round of Data Collection

For each goal lost Hau kept track of what he and the coaching staff considered to be the root cause of the goal. Figure 2.6 shows the number of goals allowed in the first five games of the season along with the root cause of each goal. Breakaways,

FIGURE 2.6 Pareto Diagram of Goals Lost in Games 1–5

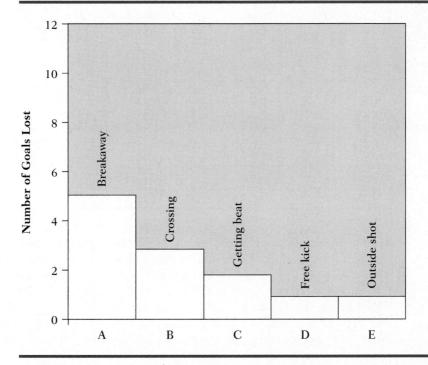

FIGURE 2.7   Cause-and-Effect Diagram of Breakaway Goals

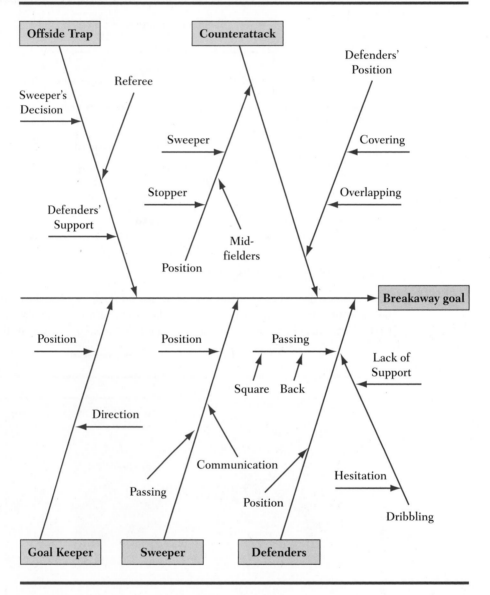

or allowing an opponent to get behind the defense with the ball, was the most common problem, followed by allowing the ball to be crossed into the goal area. The coaching staff decided to work on preventing breakaways, and they developed the cause-and-effect diagram in Figure 2.7 to identify root causes of breakaways. This diagram, also known as a fishbone diagram (for obvious reasons), begins with an effect (in this case, breakaway goals) and identifies the major causes of this effect, such as defenders' play, sweeper's play (the sweeper is the last defender), and so on. Each of these major causes is then broken down into more detail. Using this information, the coaching staff decided to work on several specific passing skills in practice to prevent further breakaway goals. The coaches thought defenders' passing skills could be significantly improved fairly quickly and that this would significantly reduce breakaways. The square pass, or pass from one defender to another in a parallel line to the goal, requires a great deal of skill and judgment to execute correctly without being intercepted by an opponent, resulting in a breakaway. Hau taught the defenders how to position themselves so that a square pass was not needed. The back pass to the goalie can also be dangerous because it sends the ball directly in front of the defenders' own goal. In addition to the possibility of interception, it can directly result in a goal if the goalie misplays it. Hau taught the defenders how to back pass to the side of the goal rather than directly toward the goal, which is a much safer play.

Second and Third Sets of Defensive Data

Figures 2.8 and 2.9 show the causes for lost goals in games 6–10 and 11–15, respectively. Note that the overall number of goals allowed is decreasing, demonstrating that defensive performance is improving during this time. In addition, the primary causes have changed. By focusing on eliminating breakaways, in games 6–10 breakaways are only the fourth leading cause of goals scored, and break-

FIGURE 2.8   Pareto Diagram of Goals Lost in Games 6–10

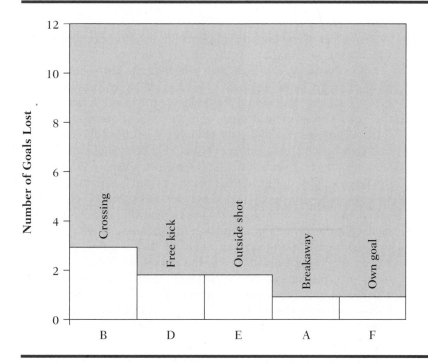

FIGURE 2.9   Pareto Diagram of Goals Lost in Games 11–15

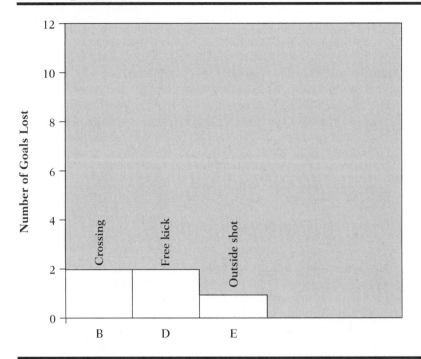

aways do not even show up in games 11–15. Allowing goals from balls crossed through the goal area is now the primary problem.

Hau's strategy to reduce goals allowed from crossed balls was to develop the girls' ability to "head" the ball, or hit the ball away from the goal area with their heads. Heading skills would also enhance the team's offensive performance from their own crosses. Unfortunately, practice drills on heading seemed to produce little improvement in the girls' ability to head the ball. The coaching staff realized that heading a kicked soccer ball is a fairly complex process that requires several individual skills: judging the height, speed, and angle of the ball; positioning oneself properly; timing the jump; and "snapping" the waist and neck to strike the ball with force. Hau developed a flowchart (Figure 2.10) to depict the overall process of heading, as well as the individual steps. He then developed individual drills to work on each specific step in the heading process. For example, to develop skill in anticipating where the ball is going, the girls were drilled by repeatedly having balls thrown in the air toward them from various positions. To learn to keep their eyes on the ball and time their jumps, they practiced heading a ball held in another player's hands. To perfect snapping the waist and neck without twisting the whole body, they practiced heading a thrown ball from a seated position. As they improved these individual skills, they began putting it all together to head a kicked ball.

FIGURE 2.10 Flowchart of Steps in Heading Process and Associated Drills

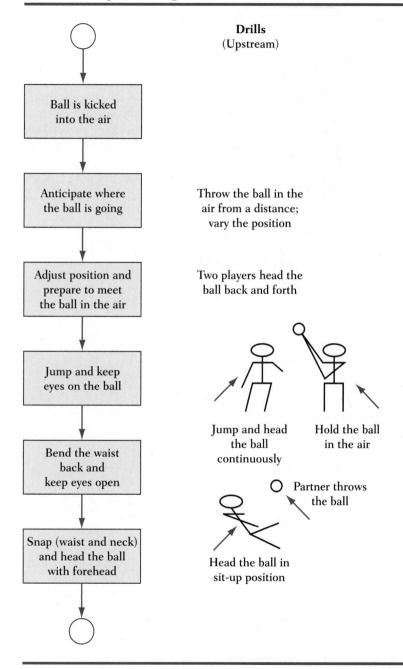

Drills
(Upstream)

Ball is kicked
into the air

Anticipate where
the ball is going

Throw the ball in the
air from a distance;
vary the position

Adjust position and
prepare to meet
the ball in the air

Two players head the
ball back and forth

Jump and keep
eyes on the ball

Jump and head
the ball
continuously

Hold the ball
in the air

Bend the waist
back and
keep eyes open

Partner throws
the ball

Snap (waist and neck)
and head the ball
with forehead

Head the ball in
sit-up position

Offensive Skills

The team began to solidify defensively, and Hau turned his attention to offensive performance. Because the team was working on heading skills and crossing provides a good opportunity to increase goal production, he decided to study crossing in more depth. Most of the players believed the greatest chance for a goal on a cross was to place the ball directly in front of the goal. This approach was not resulting in many goals, however. Hau conducted an experiment in practice to determine the best approach to placing crossed balls, and the outcome of this experiment is depicted in Figure 2.11. One player repeatedly crossed the ball from the right side, from the spot marked with an "X" in Figure 2.11. The offensive and

FIGURE 2.11 Location of Goals Scored on Crosses

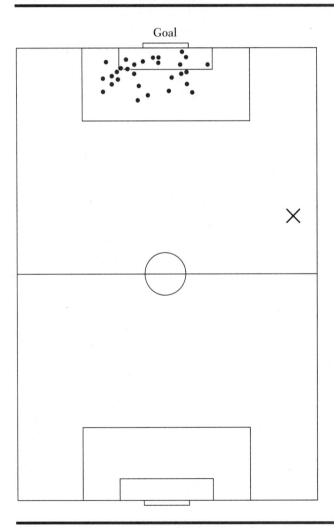

defensive players positioned themselves as if this were a free kick in a real game, and the individual dots represent where the ball came down on kicks that resulted in a goal.

First, it can be seen that more goals resulted from kicks coming down outside than inside the goalie box (the smaller rectangle directly in front of the goal). In other words, kicking the ball directly toward the goal was not the best strategy. Balls kicked directly toward the goal could be caught by the goalie (who is allowed to use her hands) before they could be headed by an offensive player. A second more subtle observation made by Hau was that fewer goals were scored when the cross followed a direct line toward the goal, regardless of where it came down. The highest concentration of goals occurred when the ball came down at a 90° angle, as shown by the cluster of points roughly forming a straight line from the middle of the goalie box toward the lower left of the penalty area (the larger rectangle). This was somewhat surprising and required an in-depth knowledge of the game of soccer to interpret.

Hau concluded that the high concentration of goals at a 90° angle was due to two effects. First, the goalie would be out of position and might have trouble getting in position because of congestion around the goal. A cross kicked directly toward the goalie would leave her in perfect position to either catch the ball directly or to defend a headed ball. Second, on a cross from this angle, defensive players could not follow the ball and also keep an eye on offensive players they were supposed to guard. The basic defensive position in this situation is to stand behind the offensive player, between the player and the goal. On a ball kicked directly toward the goal, the ball and the offensive player are in the same line of sight. On a ball crossed at a 90° angle, the difficulty in seeing both the ball and an offensive player who is moving is maximized. The concentration of balls on the left rather than the right side is simply due to the wider angle to defend because the ball was crossed from the right side of the field.

Next Round on Offense

As the team improved its offensive performance on crosses, Hau decided to focus on penalty kicks next. Penalty kicks are free kicks from 12 yards out with only the goalie allowed to defend (Figure 2.12). Penalty kicks are awarded for certain infractions occurring inside the penalty area and are also used to resolve ties in tournaments. Hau conducted an experiment to determine which players were most effective at scoring on penalty kicks. (The coach chooses the player who will take these kicks in games.) He had each potential kicker take 10 penalty kicks, for a total of 100 kicks, and repeated this trial each week for the top scorers. The player scoring highest in practice would take the penalty kicks in any games that week. Realizing that players would fatigue, especially the goalie, he rotated goalies so each player faced each goalie the same number of times. This is referred to as a "factorial design," and it prevents confusion between ability of the shooter versus ability of the goalies. Figure 2.13 shows the data for three of the players over 10 weeks. Of the 10 players who originally took 10 kicks each, 52 goals were scored, for an overall success rate of 52%. The benchmark for excellence is considered to

FIGURE 2.12 Penalty Kicks

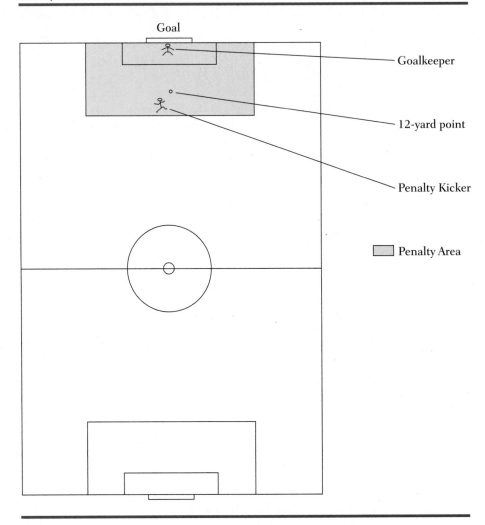

be about 90%, so this was poor performance overall. Hau decided to spend more time on penalty kicking in practice.

The performance of Player 5 was a big surprise to the coaches because she was not considered to be one of the "star" players. The skills required for penalty kicking are more specific than for performance overall, however, and the best overall players may not be the best penalty kickers. Testing all the players was important because most teams choose star players to take penalty kicks. Another key point is that all three players improved over the 10 weeks. This was due to the emphasis on penalty kicking technique in practice. Penalty kicking became critically important for this team during the state playoffs, as each of their games prior to the finals was decided by penalty kicks.

FIGURE 2.13   Performance in Penalty Kicking Drill

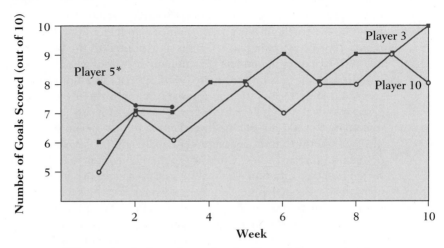

*There are only three data points for Player 5 because the player was injured after the third week.

Summary

Hau concluded that the statistical thinking approach had a significant impact on the performance of the team throughout the season. Specifically, he noted that a disciplined approach helped to identify the root causes of problems. Obtaining relevant data enabled decisions to be made rationally rather than on the basis of emotion, and knowledge of the game of soccer was critically important in correctly interpreting these data. The coaches and players found that having access to data generated a lot of fruitful discussion on tactics and avoided arguments. After reviewing so much actual data, everyone also had a much better appreciation of variation. The practical result of using this approach was that the girls' team placed second in the Wisconsin state tournament for the first time ever.

2.4   A Model for Statistical Thinking

The Commonality of Approach

The two case studies may appear at first glance to be radically different. They certainly involved different application areas and different tools. The first study involved an extensive analysis of a traditional business function—sales. The second involved a sports application that illustrated how statistical thinking can be applied to everyday situations. If we look a little closer, however, we can identify similarities in the approaches taken in both cases.

First, both applications illustrate that results occur through a *process* of doing work. To improve results, we must improve the process. For example, to improve the payoff from advertising, Busch needed to understand *how* advertising affected sales. Simply increasing the total amount of advertising could not have generated the same results. Similarly, the soccer coach realized that exhorting the team to win would not actually improve performance. Understanding the weaknesses of the team—that is, *how* the players' skills contributed to the outcome—provided the coaches with ways to improve the outcome by working to improve players' skills.

Both studies also reveal that *variation* exists in all processes. In both cases, understanding the variation in the data required the use of statistical tools. Because personal preferences for beer vary, asking only one person or looking in only one market would not enable investigators to draw meaningful conclusions. It is also noteworthy that the measurement analyzed in this case was variation between actual and forecast sales. Contrary to many business managers, the researchers recognized that variation would always be present in the actual sales figures. Similarly, the soccer coach needed several games' worth of data to properly diagnose the most common defensive flaws resulting in goals allowed.

Another important point of similarity is the *sequential nature* of the approaches. Each time data were gathered, something new was learned. Some questions were answered, but new ones were invariably raised. When the Busch investigators found some surprising results, they incorporated these findings in subsequent experiments to improve advertising effectiveness well beyond what anyone had originally thought possible. In particular, the surprising discovery that sales did not deteriorate for a year and a half after eliminating advertising in a region led to subsequent experimentation on optimum pulsing patterns and even greater improvements. In the soccer study, defensive adjustments were made to address breakaways. When this succeeded, other flaws became the most common.

Another key similarity is the *synergy between data and subject matter knowledge*. In both cases knowledge of the process led to wise choices of which data to collect. This in turn led to an increased understanding of the process, which guided further data collection, and so on. The Busch investigators originally had a somewhat vague hypothesis about the relationship between advertising and sales. The data obtained suggested a very different relationship, which caused them to reconsider their hypothesis. This led to further experimentation and further learning. Each time data were collected, the investigators' hypothesis about the relationship between advertising and sales was improved and refined. Data were critical to verify or improve subject matter theory, and subject matter theory was critically important to properly interpret the data. If the Busch investigators had not been familiar with typical advertising stimulus–response functions, they probably would not have been able to make any sense out of the V shape observed in the second experiment. Similarly, knowledge of the game of soccer was required to develop the list of causes of goals allowed, to decide what the actual cause of each goal was, and to choose specific actions to take to address each cause.

Both studies also recognized the *dynamic nature of the process* and consciously checked for stability. The third experiment in the Busch study deliberately replicated some of the same situations as the second experiment to ensure that market conditions had not changed significantly. This was necessary because each experiment was conducted over a 12-month period in a dynamic and competitive environment (sales and market share increased significantly). The researchers also recognized that changes other than their advertising interventions were influencing results. This is typical, and it is another reason statistical approaches are required to properly interpret data. The soccer coach consciously segregated data from games 1–5, 6–10, and 11–15 because he realized the teams' defensive performance was improving (changing) over time.

Finally, the *practical issues involved in collecting data or "sampling"* are also similar. Much of the theory underlying formal statistical techniques is based on the assumption of "random sampling." A random sample is one in which every possible unit or individual has an equal chance of being selected. In practical applications, however, this is rarely possible. Restrictions were placed on the areas to be used in the second Busch experiment because the CEO did not want any of the company's major markets to be included. Such restrictions limit generalization of the experimental outcomes. For instance, the results might not apply to major markets. *Knowledge of that business is critical in judging whether the nonrandom sample is likely to produce biased results.* For example, are the market dynamics of advertising different in major markets than in those studied? If so, how? The preliminary comparison of enhanced versus normal advertising also used a nonrandom sample. Other practical sampling issues faced were what to actually measure (recall, attitudes, actual sales, or deviation from forecast) and the inability to obtain data on the impact of pricing. In soccer, of course, one has to wait until a goal is scored to obtain data on causes of goals. Tips for practical sampling will be addressed in greater detail in Chapter 5.

Statistical Thinking Model

Figure 2.14 shows how statistical thinking can be applied to improve business processes. We will refer to this basic model throughout the text as we introduce the tools that can be applied to improve business processes. We begin by identifying, documenting, and understanding the business process itself. The discussion in the Busch study of the purpose of advertising, the different media used, and the inability of previous published studies to demonstrate cause-and-effect relationships to sales are illustrations of this step.

We almost always begin with some subject matter knowledge. Subject matter knowledge is everything we know about the process under study. This could be derived from experience or from academic study. This guides us in planning which data from the process would be most helpful to validate or refine our hypothesis. For example, the preliminary Busch study compared six areas with additional advertising to six with normal amounts of advertising; the soccer coach had to define the categories for causes of goals ahead of time. Subject matter knowledge helps us to determine where biases are likely to occur and to avoid or minimize them.

Once the data are obtained, statistical techniques are used to analyze the data and to account for the variation in the data. Although the analysis may confirm our hypothesis, additional questions almost always arise. The increased sales in the control areas in the preliminary Busch study is one such example. Sometimes the initial hypothesis is invalidated, such as the V shape observed between sales

FIGURE 2.14   Statistical Thinking Model

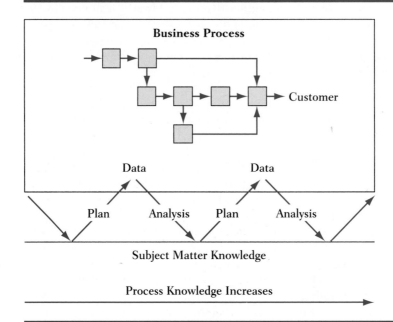

Source: Adapted from Box, Hunter, & Hunter, 1978.

and advertising in the second experiment. In either case *creativity is sparked by the data.* We are forced to rethink or revise our hypothesis in such a way that it could explain the observed data. For example, the potential of multiple beer segments, each with a mound-shaped advertising function, was a revision that explained the observed data rather than a rejection of the original hypothesis. This process leads to a desire to obtain additional data to validate or further refine the revised hypothesis, and the whole cycle is repeated. Most business applications are sequential studies involving a series of such steps. Fortunately, our knowledge about the process, and therefore our ability to improve it, increases with each step.

A complicating issue is the fact that business processes are not static over time but dynamic. When additional data are obtained at a later time, we must realize that we may be sampling from a somewhat different process and be cautious about combining these data with previous data. Statistical methods can be very helpful in determining whether the process has undergone significant change. In the Busch study some conditions from the second experiment were replicated in the third to determine whether market conditions relative to beer advertising had changed significantly. In the soccer study the team's overall defensive performance, and the most common causes for goals allowed, changed during the season.

Relationship to the Scientific Method and the PDCA Cycle

In its simplest form the scientific method begins with a stated *hypothesis* about some phenomenon, then an *experiment* is conducted to test the hypothesis, and *observation* of the results confirms or disproves the hypothesis. In application this method is also sequential. Observations from one experiment may cause us to revise our hypothesis, which may lead to another experiment to evaluate the revised hypothesis.

Statistical thinking uses the scientific method to develop subject matter knowledge and to gather data to evaluate and revise hypotheses. However, there are some important differences. First, statistical thinking recognizes that results are produced by a process and that the process must be understood and improved to improve the results. A second difference is the emphasis on variation in statistical thinking. The scientific method can be applied without any awareness of the concept of variation, which may lead to misinterpretation of the results. Key similarities are that both are sequential approaches that integrate data and subject matter knowledge.

The Plan-Do-Check-Act (PDCA) cycle is often referred to as the Deming cycle or the Shewhart cycle (Joiner, 1994). Walter Shewhart proposed this approach in the field of quality control in the 1920s, and W. Edwards Deming later popularized PDCA as a general management approach based on the scientific method (Deming, 1986). Using this approach, one *plans* the intended work, including desired results and methods. Next, one must *do,* or actually carry out, the plan. To ensure follow-through, one should also *check* to be sure that the plan has been carried out and that the anticipated results have been achieved. Finally, one must *act* on any discrepancies discovered in the check step. Unanticipated results may lead to altered plans, which leads into another cycle. The PDCA cycle (Figure 2.15) is generally presented in a circular pattern to emphasize its repetitive nature.

The plan step corresponds to planning the data we need based on our subject matter knowledge. The do step is conducting the experiment or survey to obtain data. The check step is analyzing the data statistically, and the act step is interpreting this analysis in light of our original hypothesis. The results will determine whether we need to repeat the cycle. Statistical thinking can be viewed as an adaptation of the PDCA cycle that focuses on the business process and an understanding of variation.

FIGURE 2.15   The PDCA Cycle

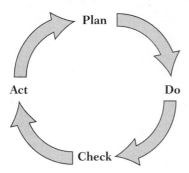

Plan

Act

Do

Check

2.5 Variation in Business Processes

It is important for business leaders to understand that variation occurs naturally in all processes. Sales, labor expenses, net profit, and even routine processes such as time to receive mail orders or transfer funds vary from month to month or even day to day. Sometimes this variation indicates a true change in the process, such as the stock market crash of 1929, and sometimes the underlying business process is stable despite variation in the output, such as Dow-Jones fluctuations of 5 or 10 points. Statistical thinking and methods help managers to correctly interpret variation. (For example, review how additional data were needed in the Busch and soccer case studies to reach sound conclusions.)

Figure 2.16 shows monthly closing values of the Dow-Jones Industrial Average from February 1990 to February 2000. A careful review of these data reveals an overall positive trend with some obvious change points wherein an important event caused noticeable movement in the Dow. Some background variation is also present and defies explanation. This background variation is sometimes referred to as "common-cause" variation because the causes of this variation tend to be common to all data points. These common causes may be normal fluctuations in exchange rates, pessimism or optimism based on recent financial publications or broadcasts, reactions to yesterday's Dow results, or normal business cycles. Common causes of variation tend to be numerous, but each has a relatively minor impact. Processes that only have common causes of variation are said to be stable or predictable. Causes of noticeable change points are often referred to as "special causes." These changes are usually preceded by a special event outside the normal common causes. An unexpected change to interest rates announced by the Federal Reserve Bank would be one example. When special causes can be readily identified, they are referred to as "assignable causes." Special causes tend to be

FIGURE 2.16 Dow-Jones Monthly Closing Prices, February 1990–2000

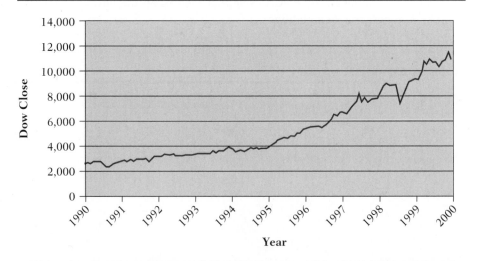

infrequent, but they may have a dramatic effect. Processes in which changes are attributed to special causes are said to be unstable or unpredictable. (We will discuss a third type of variation, structural variation, later in this chapter.) Let's review the differences between common- and special-cause variation:

Special-Cause Variation	*Common-Cause Variation*
Temporary and unpredictable	Always present
Few sources but each has a large effect	Numerous sources but each has a small effect
Often related to a specific event	Part of the normal behavior of the process
Process is unstable	Process is stable

These distinctions are very important because the proper approach for improving processes with common causes of variation (stable processes) is different from the approach used with special causes of variation (unstable processes).

If a process is unstable, eliminating the special causes will help to stabilize it. Because special causes are "assignable" to a specific event and are not a fundamental part of the process, they can often be addressed without spending a great deal of time, effort, or money. Special causes tend to have the most impact, so addressing them first typically will result in dramatic improvement. Special causes are often detrimental to the process, such as the impact of a computer crash on timeliness of financial transactions at a bank. In some cases, however, the special cause is positive, such as a sudden increase in productivity in manufacturing. Positive special causes must also be identified so they can be *institutionalized*. This ensures that the improvement will not disappear as suddenly as it came.

Once the process is stabilized, we can begin to study and understand it, allowing us to make fundamental improvements. Eliminating special causes is really fixing problems, or bringing the process back to where it should have been in the first place. True improvement comes by fundamentally changing the process. To do this, we must address the common causes.

If the process is stable, improvement generally requires careful study of the whole process to identify and prioritize the common causes and their relationships. In the Busch example the researchers assumed that the advertising process was stable, and they began studying the relationship between advertising (amount and type) and sales. If recent special causes had drastically reduced sales per advertising dollar, it would have made sense to deal with these first. In this case it was desirable to *increase the average* (increase sales revenue per advertising dollar). Very often it is desirable to *decrease the variation*. For example, customers do not want shipments to arrive either too early or too late. Generally, the objective is to reduce variation around the desirable average. In kicking penalty kicks, for example, proper technique and placement of the ball should result in a goal every time. Therefore, by reducing the variation in technique and placement we will increase the average, the percentage of goals scored on penalty kicks. This is not the exception, but rather the rule: *Reducing variation is key to improvement.* For example, Figure 2.17 depicts monthly yields on insurance policies, and Figure 2.18 depicts outsourced word processing (keying) costs. Of these two processes, clearly

FIGURE 2.17

Monthly Yields on Submitted Insurance Policies

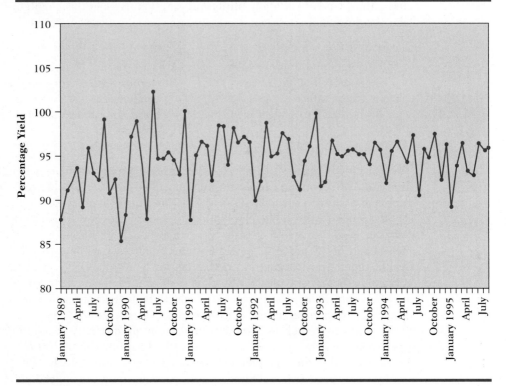

the insurance process is more consistent and predictable This predictability aids future planning. Note also that the variation has been reduced even more since April 1993. In contrast, outsourced word processing has been subject to special causes and cannot be easily predicted. Estimating costs for the next budget year would be risky.

Another reason the distinction between common- and special-cause variation is important is the tremendous amount of time wasted in many businesses attempting to "explain" common-cause variation. For example, Figure 2.19 shows data on another insurance process, the number of submitted applications. The process appears to be stable with no obvious special causes until July 1991. This does not imply that the results were good, only that they were stable. Managers who do not understand the concept of variation might demand an explanation of why applications were low in September or December of 1990. This could have a ripple effect within the organization, resulting in insurance agents spending a lot of time trying to come up with an assignable cause for low applications in these months. This is a fruitless exercise because there was no fundamental change to the process. There may be numerous causes for variation in insurance applications, which could include competition, general economic conditions, turnover in the company's agents, and so on, but there is no single reason to explain the lower application rates in September and December.

FIGURE 2.18
Outside Word Processing (Keying) Costs

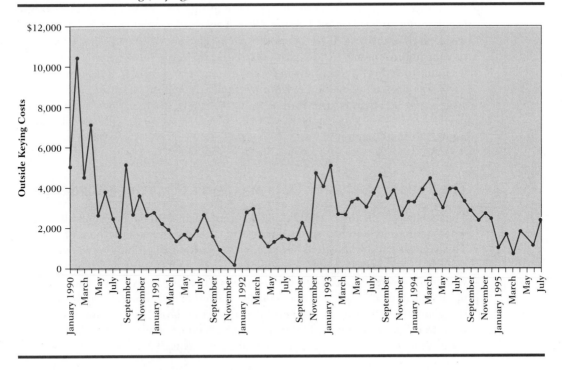

FIGURE 2.19   Submitted Insurance Applications

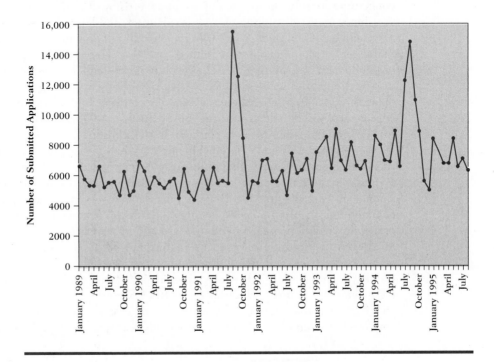

Conversely, note the dramatic increase in August 1991 and the more subtle increase beginning around January 1993. Clearly, a special cause occurred, and we need to identify it. If we overlook special causes, we are missing opportunities for improvement. In this particular case, the large increase in August 1991 was due to issuance of a new auto policy, and the subtle increase in January 1993 was due to acquisition of another smaller company. In most cases the answer is not so obvious, and formal statistical tools, such as control charts, are needed to determine evidence of special causes. These tools will be explained in more detail in Chapter 5.

2.6   The Synergy Between Data and Subject Matter Knowledge

The application of statistical thinking does not begin with data. It generally begins with some understanding of the process of interest and well-defined objectives. This subject matter knowledge is critically important in guiding us to collect the data that will answer specific questions. A great deal of time and money can be wasted in business collecting data that are not useful. For example, if a consumer product is targeted toward young females, collecting marketing research data from the general population is not helpful. Collecting data is costly, and businesses want to maximize the useful information obtained per dollar spent. This requires sound statistical survey and experimental design as well as knowledge of the process under study.

Perhaps even more important is the need to analyze and interpret the data in light of the subject matter theory. The revision of the Busch investigators' theory concerning stimulus–response curves for advertising after discovering the V relationship is an excellent example. As Box, Hunter, and Hunter (1978) state, "Data have no meaning in themselves; they are meaningful only in relation to a conceptual model of the phenomenon studied" (p. 291). This is a critical point to keep in mind when interpreting data. Before jumping to conclusions when observing data, one must think carefully about what hypothesis would explain the observed data. There is often more than one viable hypothesis, and subject matter knowledge must be used to determine the most plausible. Additional data can then confirm or invalidate the revised hypothesis. In the stimulus–response example, a naive interpretation of the V relationship might lead an investigator to conclude that this is the standard shape of stimulus–response curves in advertising.

Jumping to erroneous conclusions is a common practice in the media, but it is not good business practice. Data may occasionally be inaccurate, but it is more common that the *interpretation* is faulty. This occurs when data are naively interpreted or when data are deliberately presented in a misleading way. For example, in a recent presidential debate the incumbent bragged that "There are more people working in the United States today then ever before." The challenger responded by noting that "There are more people unemployed in the United States today than ever before." Who was right? In fact, both were. A person knowledgeable in the calculation of labor statistics would realize that the *number* of people employed or unemployed is a very misleading statistic. Overall, the population is increasing, resulting in both more people employed and unemployed. For this reason, those seriously interested in labor statistics generally look at the *percentage* of people employed or unemployed.

The converse is also true; that is, one should validate hypotheses with real data. Important business decisions are frequently made on the basis of a "gut feeling," intuition, or an unproven assumption. Corroborating data can often be obtained quickly and inexpensively. As the old saying goes, "In God we trust, all others bring data." In summary, we should have hard data to back up our theories and sound theory to back up and properly interpret our data.

2.7 The Dynamic Nature of Business Processes

Clearly there has been a lot of change in the Dow-Jones Industrial Average over the years (see Figure 2.16), even if one adjusts for inflation. This type of dynamic behavior is very common in business processes and complicates the use of statistics. For example, calculating an average of the Dow over this time period would be virtually meaningless. The financial processes that result in the Dow were certainly not operating at this "average" level for more than a few days. The average would also not help us predict where the Dow is headed. In contrast, the stable process depicted in Figure 2.17 could be summarized reasonably well by calculating an average, even though the variation has decreased. Unless the process is affected by special causes in the future, this average would provide some degree of prediction. It is generally true that summary statistics such as averages are meaningful for stable processes but are not useful for unstable processes. Unfortunately, many business processes are unstable, and summary statistics are misinterpreted.

Statistical thinking applications are most useful when data are gathered sequentially. Gathering data at one point in time gives us a "snapshot" but provides no information about dynamic behavior. The observed data may be out of date soon after they are collected. For example, the dramatic impact of Internet commerce has made a lot of historical business data meaningless. This can also be seen in modern presidential election polling, wherein public opinion tends to change significantly in the months leading up to the election. Another limitation of snapshot studies is the fundamental truth of the saying that "hindsight is 20–20." Once we have seen the results in a situation, it is easy to say what we should have done. Snapshot studies provide no opportunity to use hindsight.

Box, Hunter, and Hunter (1978) recommend that no more than 25% of the total project budget be spent in the initial round of an investigation. In this way we can gain valuable hindsight from the first round and use this information to guide subsequent rounds. This can clearly be seen in the Busch study. Unexpected results observed when advertising was eliminated in a particular region led the researchers to investigate this phenomenon further in the next study. Virtually all major breakthroughs in business and science occur from a sequential use of data gathering and analysis, as illustrated in Figure 2.14. This provides greater understanding of the process and increases our ability to improve it. In the soccer study, both the team's offensive and defensive performance improved significantly over the season. If data had been gathered at only one point in time, the conclusions would have been very misleading.

Gathering data over time, used in conjunction with sound subject matter theory, allows us to better understand how the process is performing today, and also where it appears to be going in the future. Determination of process stability is often a preliminary step in our investigations. As noted in Section 2.5, if the process is unstable, we generally attempt to identify the responsible causes and eliminate or institutionalize them. Once the process is stabilized, it is more amenable to study via statistical methods, which enables us to determine how to fundamentally change the process to improve it.

There are some "predictably unstable" processes; that is, some processes are technically unstable, but in a predictable way. The causes of this behavior are sometimes referred to as *structural variation*. For example, the number of customers arriving at a bank per hour tends to be unstable in that a "rush" typically occurs around lunchtime. This rush is very predictable, however. Similarly, the number of customers arriving at a bank per day tends to peak on Friday because many people are paid on Friday. Similar situations occur in most restaurants. Other examples include seasonal businesses such as swim wear, construction, or sporting goods. These processes are similar to stable processes in that their behavior can be predicted using statistical techniques that account for the consistent patterns.

Care must be taken when interpreting data from these processes. Knowing the average number of customers arriving at a bank per hour is not helpful to bank management in determining staffing requirements. It would be more helpful to know the average number arriving during the lunchtime rush and during off-peak hours so staffing could be appropriately balanced. This structural variation presents a problem, however, because adjusting staffing by the hour is difficult to manage. Reducing the structural variation in this case would simplify administration and enhance customer service. Some attempts to accomplish this include Saturday business hours to diminish the Friday rush and evening or early morning hours to diminish the lunch hour rush. In general, eliminating structural variation requires fundamental change to the process, much like common-cause variation. Figure 2.20 is a plot of military requisitions by quarter. In addition to some special causes, there is a consistent, predictable pattern of low requisitions in the second and fourth quarters. This pattern is likely caused by semiannual budgeting cycles.

2.8 Summary

1. Business results occur through dynamic processes; to improve results we must improve the processes.
2. Variation naturally occurs in these processes.
3. It is important to understand common and special causes of variation to choose the most appropriate approach to improve business processes.
4. Improvement generally requires a sequential approach.

FIGURE 2.20
Requisitions per Quarter

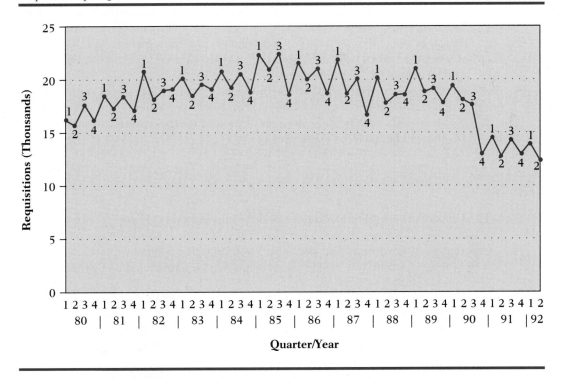

5. Data and subject matter knowledge are synergistic.
6. Statistical thinking is an overall approach to improvement based on these principles.

2.9 Project Update

To gain experience in applying statistical thinking to improve a real process, select a project that will enable you to apply the major elements of the statistical thinking model shown in Figure 2.14. The project you choose should have these qualities:

- Involve a process about which you know something so you can apply subject matter knowledge
- Allow collection of relevant data
- Provide an opportunity for improvement

The last point is particularly relevant in that the objective of applying statistical thinking is improvement.

Monitoring the price of a stock would not be a good project because there is little you could do to improve it. Trying to improve the return of an investment portfolio might be more appropriate because you could reallocate funds to improve the process. Work processes often provide good, relevant projects. Examples of work projects include inventory management, quality and productivity in manufacturing processes, or cycle time of accounting or billing processes. Everyday experiences or hobbies are also popular choices. Examples include golf, bowling, or other sports scores; perfecting recipes; losing weight; reducing commuting time to work or school; or caring for houseplants.

Prepare a written proposal for your project that includes background on the process, reason for selection, and objectives for improvement. At the end of each chapter we will provide hints or suggestions for how the material presented in the chapter can be applied to your project.

EXERCISES

1. Think of some attempts you have made to improve everyday processes in your life. How do these approaches compare to the statistical thinking approach? How might you have improved the results you achieved by incorporating statistical thinking concepts? Are important elements of your approach missing from the statistical thinking approach?

2. Answer the questions in Exercise 1 for improvement efforts you have experienced in an educational, work, or nonprofit environment, as opposed to your personal life. Consider, for example, any formal improvement efforts you may have experienced in school, in organized sports, on a job, or in a religious organization, such as a church, synagogue, or mosque.

3. Note the daily changes in the Dow-Jones Industrial Average (or other stock index) reported on the Internet or in the newspaper for at least one week, as well as the reasons given by financial analysts for the change. For which days do you feel this explanation is legitimate—that is, there has been a true special cause—and for which days do you feel the analysts are simply misinterpreting the common-cause variation in the Dow? Explain.

4. How could looking at the Dow-Jones Industrial Average for only one day, or the change from only one day to the next, lead to erroneous conclusions about the current state of the stock market?

5. Give an example from the media, or from your own experience, where data have been misinterpreted due to lack of subject matter knowledge.

6. Give an everyday example of a process relevant to you that can be measured, such as time to get to work or school, your weight, or your golf or bowling score. Identify both common and special causes of variation that could have an impact on this process measurement. Are there any structural sources of variation?

7. Give an example from your own experience where potential improvements were not achieved because a "one-shot" study was used rather than a sequential approach. How could a sequential approach have resulted in further improvements?

8. Explain how variation could be important to an Internet start-up company in the insurance business. What might be some key processes for this business? What are potential common causes of variation that would have an impact on these pro-

cesses? What special causes might be important? How might the company's business environment be dynamic and change over time? What data might the company want to obtain and analyze?

9. Answer the questions in Exercise 8 for a health maintenance organization (HMO) rather than an Internet start-up.

10. Respond to the scenario that follows. Assuming that the opinions expressed do represent those of many consumers, is this a statistical thinking issue? Why or why not? If so, how could application of statistical thinking help the beef industry?

Beef Industry Hits Hard Times—Experts Say Answer Rests in Satisfying Customers

(*USA Today*, March 5, 1996)

The $37 billion beef industry is loosing market share to chicken and pork: 34% share in 1996 down from 52% share in 1976. Annual beef consumption has decreased from 85 lbs/person in 1970 to 67 lbs/person in 1996, while annual poultry consumption has increased from 41 to 72 lbs/person during the same time period.

The problem is inconsistent product. "I can go into a store now and look through 20 steaks to find one that's acceptable. But a chicken breast is a chicken breast," says one consumer. The poultry industry has changed how its products are raised, prepared, packaged, and marketed. According to a 1991 study, consumers are not satisfied 1 in 5 times they cook or order beef. Complaints ranged from confusion over various grades of beef to difficulty cooking, to toughness or unpleasant beef.

And, in an era of increasing emphasis on healthy diets, consumers say they now avoid red meat because its fat and cholesterol content makes it a less healthy choice.

11. Respond to the scenario that follows. How might statistical thinking be used to help resolve the problems with this new federal mandate? Do you think the federal government could have used statistical thinking to avoid such problems in the first place?

Federal Mandate or Common Sense?

(From Galen Britz, 3M)

Nurse's aide Kari Taker was talking to one of her colleagues Phil Diperre at the Golden Acres Health Care Center.

"Why must we remove the seat belt from Helen Wheeler's wheelchair? You know what will happen when she tries to stand up. She will fall flat on her face."

Phil replied, "It will be okay. You know we have some new movement monitors that will alert us if she tries to stand up?"

"But Phil, you know how impulsive Helen is. She will try to stand up before anyone can get over to help her. Who came up with this dumb rule anyway?"

"Kari, I thought you knew that this is a federal mandate. It says and I quote: The resident has the right to be free from any physical or chemical restraints imposed for purposes of discipline or convenience, and not required to treat the resident's medical symptoms. I'm sure there have been abuses that led to writing this mandate. I read that there are studies that show the bad things that can happen to people who are overmedicated and/ or restrained."

"Yes, but I wonder if there are any studies that document what happens to people who are not properly medicated and/or are not restrained. You know as well as I do, Phil, that the definitions and guidelines are rather vague, and there is no specific list of medical symptoms. States are told that they will not receive federal money if they don't abide by the mandate."

"Yes, and now states are auditing the health care centers. Last week we had two auditors in our building, and you know what? Each auditor is interpreting the mandate differently. And, of course, we are threatened with loss of funding if our facility doesn't comply."

"Well, Phil, I guess we just have to go along with it. I hope Helen doesn't get hurt because of giving our residents their so-called rights."

Three days later: "Kari, whose monitor is going off?"

"Oh, no, it's Helen's! She's already fallen! I hope she is not badly hurt. What a price to pay for her rights!"

REFERENCES

Ackoff, R. L. (1978). *The art of problem solving.* New York: Wiley.

Box, G. E. P., Hunter, W. G., & Hunter, J. S. (1978). *Statistics for experimenters.* New York: Wiley Interscience.

Deming, W. E. (1986). *Out of the crises.* Boston: M.I.T. Center for Advanced Engineering Study.

Hau, I. (1990). *Quality improvement for a soccer team.* Unpublished manuscript.

Joiner, B. L. (1994). *Fourth generation management.* New York: McGraw-Hill.

3

Understanding Business Processes

The process produces results.

Peter F. Drucker

3.1 Overview

In Chapters 1 and 2 we introduced statistical thinking and discussed its use in improving business processes. The first fundamental of statistical thinking is that "all work occurs in a system of interconnected processes." The second fundamental is that "variation exists in all processes." Clearly, "process" is at the heart of statistical thinking and business improvement.

In this chapter we show how to view all the work within a business as a process and how to identify and critically analyze business processes—even before data are collected. These important skills are needed to improve business performance. The goal of process improvement projects is ultimately to improve business results, typically by better serving our customers (revenue generation) more efficiently (cost). Processes define how we do the work to serve our customers. In many cases little or no capital investment is required to make improvements, as was the case in the billing operation in Chapter 1.

The main learning objective in this chapter is to be able to identify a process—including the associated suppliers, process inputs, process steps, process outputs, and customers. Other learning objectives include being able to identify non-value-added process steps, to develop key process measurements, to utilize process benchmarking, to identify the interconnections of individual processes to form systems, and to identify measurement processes and their associated sources of variation.

Processes define how a product or service is created, and understanding the process helps employees better see their role in serving customers, both internal

FIGURE 3.1   Schematic of a Process: A Series of Activities That Convert Inputs into Outputs

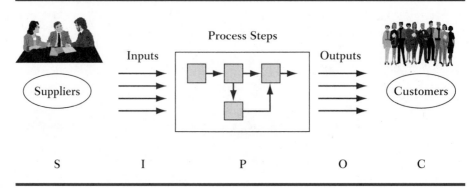

and external, and in reducing waste, rework, and other unnecessary costs. The process also provides the context for measurement. The generic SIPOC model for processes (see Figure 3.1) helps us identify critical measurements for improvement, resulting in better business results. Without a view of the entire process, people see only the work they do and perhaps the process steps just preceding them, for which they are the "customer." It is difficult for them to understand how their actions influence the process as a whole, and they may not, therefore, be fully supportive of process improvement efforts. Developing common understanding of the entire process via the SIPOC model and obtaining key process measurements are critical.

The focus of this chapter is on identifying, understanding, measuring, and analyzing business processes. Special attention will be paid to general business processes that involve flow of information and ideas. Typically manufacturing and engineering processes are easy to identify—you just follow the pipes! Business processes are more ephemeral and can be much more challenging. We begin by reviewing several examples of business processes. This is followed by presentation of the SIPOC model for processes with flowcharts to illustrate the process. The next two sections demonstrate how to use this model to analyze business processes. Following that we discuss relationships between multiple processes, which we refer to as "systems of processes." The measurement process—very important and common to virtually all improvement efforts—is then discussed. The chapter concludes with a summary of key points, the project update, and exercises.

3.2   Examples of Business Processes

In Chapter 1 we discussed a billing process for a publisher, and in Chapter 2 we looked at the process of advertising to increase sales along with several processes used to improve the outcomes for a soccer team. Here are some additional examples of business processes:

- Customer order delivery
- Developing budgets

FIGURE 3.2   Semiconductor Manufacturing Process

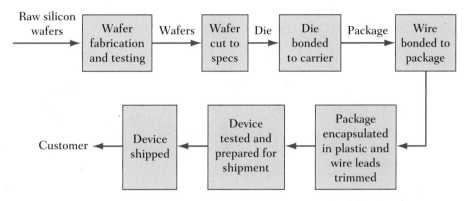

Many different semiconductor devices are produced through specified variations in the fabricating, cutting, die bonding, wire bonding, and encapsulating steps.

- Strategic planning and deployment
- New product development
- Manufacturing process design
- Employee recruiting and orientation
- Company mail delivery
- Equipment procurement
- Patent application
- Laboratory measurement

Processes are typically depicted in flowcharts, a graphical aid that helps us understand the process and create effective measurement, management, and improvement systems (Galloway, 1994; Rummler & Brache, 1995). Details on how to construct flowcharts will be provided in Chapter 5.

Figure 3.2 illustrates a manufacturing process, in this case semiconductor manufacture, a component used in computer chips. The manufacturing process has several elements: raw material inputs are obtained from one or more suppliers, several process steps modify the raw materials, and the output, in the form of a product, is shipped to customers. The overall flow of manufacturing processes is often easy to understand because we can observe the material flowing. As you walk through the factory, you can see where the raw materials come in, the sequenced equipment and machines used to make the product, and how the final product is packaged and shipped to customers.

Most business processes are not so clear, however. There is typically more human intervention, few machines and equipment, and no "line of sight" from the beginning of the process to the end. The work is often done in offices that may be on different floors in tall buildings spread out over the countryside or even around the world. For these reasons the flowchart is particularly useful in working with "soft" business processes. For example, a standards documentation team working in a Fortune 500 company commented: "Working with the flowsheet [flowchart]

FIGURE 3.3   Customer Order Process

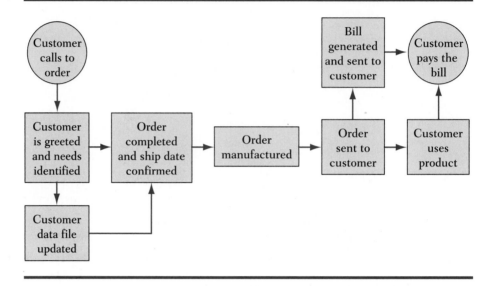

as a focal point made it easier to make changes in the process. We were not questioning each other's purpose or job; we were all looking at a common process and trying to figure out how to make it better. The flowcharting process helped to de-personalize and de-departmentalize the process" (DuPont Engineering Standards Documentation Team, 1990). Process flowcharts are also sometimes referred to as "process maps."

Figure 3.3 is a schematic of a customer order and delivery process, beginning with the "customer call" and ending with the "customer pays the bill." Note that a manufacturing process is embedded in the middle of this customer order and delivery process, illustrating that high-quality manufacturing does not guarantee satisfied customers. Other process steps may function poorly or perhaps completely fail, producing a late or defective product and an unhappy customer.

The order taking process shown in Figure 3.4 is typical of those used in call centers of telemarketing companies and other types of service operations. This particular process was created as a company standard, and all customer service representatives (CSRs) were trained in how to use the process. The process was pilot tested to verify its effectiveness in improving customer satisfaction and improving call effectiveness prior to rolling it out across the company. Measurements were monitored on call quality, customer satisfaction, call-handling time, repeat calls, and use of the process by the CSRs to maintain the performance of the process. Note that the order taking process is the beginning step of the overall customer order process shown in Figure 3.3.

Figure 3.5 shows a billing process, which is also a key aspect of the customer order process shown in Figure 3.3. Key measurements for this process are cycle time, process defects per unit (DPU), and customer defects per unit (DPU). It is easy to see the importance of cycle time for the billing process—time is money.

FIGURE 3.4   Order Taking Process

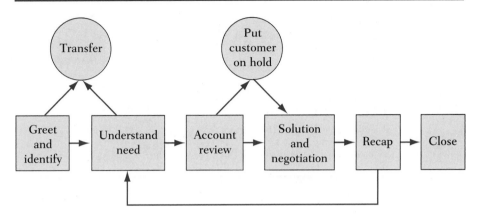

Key Measures
- Call quality
- Call handling time
- Repeat calls
- Customer satisfaction

FIGURE 3.5   Billing Process

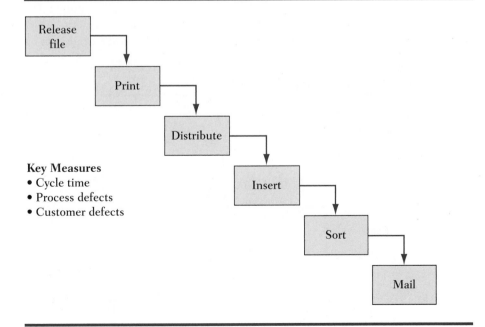

Key Measures
- Cycle time
- Process defects
- Customer defects

FIGURE 3.6   New Credit Card Account Process

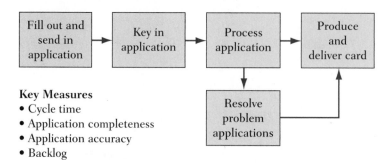

Key Measures
• Cycle time
• Application completeness
• Application accuracy
• Backlog

The faster the bills are sent out, the sooner payment will be received, enabling the company to earn interest on the money. Both the call process and the billing process are subprocesses of the customer order process shown in Figure 3.3. Subprocesses are discussed in greater detail in Section 3.4.

Figure 3.6 shows a map of a new credit card account process. A key observation here is the "Resolve problem applications" box, a clear source of wasted effort and rework. "Problem" applications here include such things as incomplete or illegible applications. Process improvement can reduce this waste, putting dollars on the bottom line, and likely improving customer satisfaction. We will discuss this point in greater detail when we address process analysis in Section 3.5.

Administrative functions also have processes. For example, Figure 3.7 shows a personnel requisition (hiring) process. A key measurement in this process is the time required to complete each of the steps. In most instances it is in the interest

FIGURE 3.7   Personnel Requisition Process

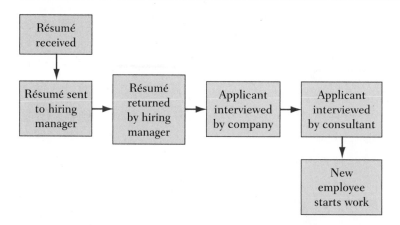

Source: Juran & Gryna, 1988, p. 21.

of all parties involved (candidate and company) to complete the hiring process as soon as possible. This process could be expanded to include a "decision" step—that is, whether to make an offer or not—because this is a key step in recruiting.

Process schematics have become a popular means of advertising and communicating how procedures work. For example, Figure 3.8 shows how mail from sponsors to children in El Salvador is handled. A flyer like this was distributed by World Vision, a nonprofit organization supporting needy children around the world. The process is depicted graphically and in two languages. Most of the sponsors reside in the United States and speak English, whereas the children in El Salvador who receive aid generally speak Spanish. This is a good example of how showing the entire process aids understanding.

3.3   The SIPOC Model for Processes

Business processes come in many sizes, shapes, and forms. By definition a process consists of a "series of activities done by an individual or group to produce a product or service for a customer." A *product* or *service* is any tangible output of the process, such as information, a report, admission to a hospital, a cleared check, or an approved loan at a bank. A *customer* is anyone who depends on the product or service produced by the process, not just an external, paying customer. For example, in an assembly line in a factory, the workers' primary customers are the people performing the next step. If the work coming to the people next in the process flow is already substandard, there may be little they can do to improve the product. Similarly, the primary customers of a computer programmer are the people who utilize the program. These customers will not be able to do their jobs properly if the program has "bugs" in it.

The SIPOC model is a very useful framework for discussing processes. The elements of the SIPOC model include the *suppliers*, who provide the *inputs* to the process; the *process steps* that convert the process inputs into process *outputs*; and finally, the *customer*, who receives or uses the process outputs. These terms are summarized in Table 3.1. Note that all of the processes discussed earlier can be

TABLE 3.1   SIPOC Elements

Element	Definition
Supplier	The individual or group that provides the inputs to the process.
Inputs	The materials, information, or other items that flow from the supplier, through the process, into outputs, and eventually to the customer.
Process steps	The specific activities that transform the process inputs into outputs.
Outputs	The product or service that is produced by the process.
Customer	The individual or group that utilizes the process outputs.

FIGURE 3.8   World Vision Flyer Showing How Mail Between Children and Sponsors Is Handled

Your Mail
Su Correspondencia

Did you know that the letter you send me has to go through a process?
Sabía usted que la carta que usted me envía tiene que pasar por un proceso?

You write me   It reaches my country in 15 to 30 days.
Usted me escribe   Llega a mi país de 15 a 30 días.

The letter is processed by the Post Office in 5 days.
Correos procesa la carta en aproximadamente 5 días.

Three or five days later, it is received at   World Vision El Salvador
De 3 a 5 días más tarde se recibe en   Visión Mundial El Salvador

and it goes through a process.
y lleva un proceso.

It is translated together with other 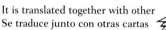 letters that are sent to other children.
Se traduce junto con otras cartas que son enviadas a otros niños.

External translators receive the letters for translation every week and submit the
letters already translated a week later.
Cada semana los traductores externos reciben las cartas para traducirlas y una
semana más tarde las entregan ya traducidas.

Twice a week, the field staff takes the mail to their Area which
is picked up once a week by the project manager.
Dos veces a la semana, el personal lleva la correspondencia a su Area donde el
líder del proyecto la llega a recoger.

 Two or three days later, the project manager hands me your letter.
Dos o tres días después, el líder del proyecto me entrega su carta.

Finally I receive your letter.
Finalmente recibo su carta.

When I answer your letter, the process is the other way around.
Cuando contesto su carta, el proceso es a la inversa.

NOTE: In El Salvador, the mail is quite inefficient. Although it is
a long process, once the mail reaches WV office it is secure.

NOTA: El servicio de Correos es algo ineficiente en El Salvador. A pesar de que
es un proceso largo, una vez el correo llega a la oficina de VM, está seguro.

viewed from this perspective. For example, the elements of the SIPOC model for the order taking process in Figure 3.4 are these:

Supplier	Person requesting the order
Input	Information defining the request
Process steps	Steps shown in Figure 3.4
Output	Completed order
Customer	Person requesting the order

This example illustrates a key characteristic of many service processes; the customer and the supplier may be the same organization or person. This situation can also occur in manufacturing processes. In some instances the supplier of the raw materials for the process is the recipient of the manufactured product. Such an operation is called a "tolling" operation because product passes through the process for a fee, the "toll."

The SIPOC model gives us a common framework to view and discuss any process. This is extremely helpful as team members work on perhaps a billing process, then an online ordering process, and then a manufacturing process on subsequent improvement efforts. The same methods used to analyze and improve the billing process can be applied to analyze and improve the manufacturing process.

Other process elements, such as people, equipment, information technology (i.e., computers), and facilities, can be added to the SIPOC model. Although they are not part of the process flow per se, these elements are often critical to the success of the process and should not be overlooked. A formal evaluation of these elements may lead to obvious improvements. For example, we may discover that our computer technology is antiquated, that our people are not properly trained or organized, that the equipment is substandard, or that the facility is not conducive to doing good work. Examples of poor facilities would be an office where teams who need to interact regularly do not have meeting rooms available to them or a training center with bad acoustics, a poor sound system, no white boards, and so on. Another rather ironic example comes from a consulting experience at a factory that makes light bulbs. The workers in this facility were asked in an employee survey what they felt the greatest barrier to doing better work was. Analysis of the survey showed that "poor lighting" was the problem most frequently mentioned by workers.

Understanding the process from end to end is critical to all the methods presented in this text. As noted earlier, in nonmanufacturing processes that typically involve considerable human intervention, it is difficult for employees to see the role they play in the process and the total impact of their work on that process. For example, one of the authors consulted with a paper mill that wished to improve their process for procuring and using recycled fiber. When developing a flowchart, they quickly discovered that no one in the company understood the process from end to end—that is, no single person could develop this flowchart alone. Each person only understood his or her step and the ones directly before and after this step. A team that included the corporate procurement department, the plant procurement department, sales and marketing, and the plant operations team had to cooperate to develop the overall process flowchart. This chart proved enlightening to the entire team because no one had seen or understood the entire process before.

We now address the first step in understanding and improving processes with the SIPOC model: identifying business processes.

3.4   Identifying Business Processes

In many cases, particularly where we cannot see the process, it is difficult to accurately determine the process elements. In fact, prior to training in statistical thinking, many people say that they do not have a process! It is only when you ask them to explain in detail how they do their work that they realize they do go through a sequence of steps and that they have inputs and outputs. Getting a team to agree on the process can also be difficult because everyone views the process from their own unique perspective.

It is often easiest to begin with the key outputs (products or services) provided. The reason this works so well is that the output is typically something tangible that everyone can relate to, such as an invoice, a product design, or a credit report. Most people are able to articulate what they produce. Next, the customers for each output can be identified by asking "Who wants or needs these outputs?"

We are now in a position to identify the process boundaries, or where the process begins and ends, as well as the key process steps along the way. Typically 5 to 10 process steps will do the job. We arrive at these steps by asking, "How do we produce these outputs?" It is important to agree on where the process of interest begins and ends. For example, in the billing process, we could choose to evaluate the flow of money coming back from customers, or we could choose to focus on getting the bills out. The job is completed by identifying the process inputs and suppliers of the inputs. For the inputs we ask, "What do we need in order to produce the outputs?" We always need something, whether it is information, raw materials, or customer input. Note that the inputs should not include parts of our process, the people or equipment that do the work. These are captured in the process and sometimes are referred to as "process resources." The suppliers are whoever provides the inputs to us.

If the team is having trouble agreeing on the sequence of process steps, it is often due to the fact that different people are viewing the steps at different levels of detail. Major processes can have hundreds, even thousands, of steps, depending on how finely we specify the process. Putting all the detailed process steps in a single chart may create more confusion than understanding. Much of this confusion can be avoided by first organizing processes into 5 to 10 major process steps, and then creating subprocesses that elaborate on these major process steps. For large processes the subprocesses can be broken down to even finer levels of detail as sub-subprocesses. Figure 3.9 shows how this might work at three levels of detail. The customer order process discussed earlier (see Figure 3.3) included two subprocesses—order taking (see Figure 3.4) and billing (see Figure 3.5). Note that each subprocess receives input from the preceding major process step and delivers its output to the next major process step.

The strategy typically used in identifying subprocesses is to display a 5 to 10 box model of the major steps in the overall process. Then a 5 to 10 box model of

FIGURE 3.9 Subprocess Schematic

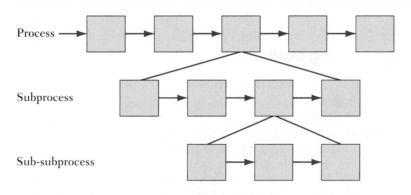

the subprocess for each major process step is displayed. In a similar fashion, sub-subprocesses are typically also displayed for critical steps identified in subprocesses. This enables us to have the best of both worlds—simplicity and clarity in the high-level process map and detail of important steps in the subprocess maps.

3.5   Analysis of Business Processes

Once business processes have been defined, we can begin to look for opportunities for improvement (OFIs). In many cases significant improvements to the process can be made even before collecting data. In the next sections we outline several methods for process analysis: identifying non-value-added work, reducing process complexity, uncovering the "hidden plant," making process measurements, and benchmarking. Each of these methods can identify opportunities for improvement.

Non-Value-Added Work

The work associated with any process can be divided into two categories, value added, and non-value added. Value-added work is required to produce the product or service, even when the process is functioning as desired. Cooking food in a restaurant is value-added work, as is giving a patient medication in a hospital or printing credit cards for customers. The term *value added* comes from manufacturing where it literally meant that monetary value is added to the product by going through this step. For example, a refrigerator is worth more money than the raw materials used to make it. However, storage, rework, inspection, and so on, do not add any value to a product. A customer is not willing to pay more for a product just because it sat in inventory for 2 months!

Non-value-added work is not necessary to produce the product or service, or at least it should not be if the process is functioning as desired. For example, a consumer credit business may make errors in a customer's credit card invoice. If the customer brings the error to the attention of the company, the company will have to correct the error in their records and issue a corrected invoice. Correcting the error and reissuing the invoice are non-value added because they are unnecessary work. If the invoice had been correct in the first place, this work would not have been needed. Non-value-added work often appears in the form of finding and correcting defects or errors in the process. Clearly this work represents an OFI because any reduction will not be missed by the customer and will go straight to the bottom line.

As much as 90% to 95% of the cost and cycle time of many processes is non-value added; that is, it is spent on activities the customer doesn't need and are not needed to run the business. Remember that "customer" and "product or service" have been defined broadly. A customer of the process may be the federal government, for example, which requires certain regulatory forms to be filled out. The work required to produce these forms is considered value added because the work is required to produce this "product" for this "customer."

Waste or *scrap* is one type of non-value-added work. This refers to work and materials that are completely lost, such as medication that has passed its expiration date while in inventory and must be discarded rather than sold. *Rework* refers to a defective product or service that can be corrected by additional work, which of course adds unnecessary cost and cycle time to the process. Restaurants experience rework when customers send food back to the kitchen because it is too cold or not completely cooked. If the order was recorded incorrectly or the meat overcooked, the order may have to be completely discarded, in which case it becomes scrap.

One way to analyze a process is to categorize each major step in the process as value added or non-value added. Figure 3.10 shows a graphical analysis of a polymer production process. The value-added steps are in the center of the figure, and the non-value-added steps are in the upper and lower parts of the figure. In this example the value-added work includes "materials," "react," "dry," "compound," and "deliver." All other activities are of no interest or value to the customer and are OFIs. This analysis is depicted in tabular form in Table 3.2.

TABLE 3.2 Polymer Production

Value-added work	Non-value-added work
Raw materials purchase	Scheduling
Chemical reaction	Storing
Drying	Blending
Compounding	Create buffer stocks
Delivery	Repair
	Recycle
	Rework
	Set aside
	Move around
	Inspect
	Scrap
	Accumulate batches

FIGURE 3.10

Value-Added and Non-Value-Added Steps in Polymer Production

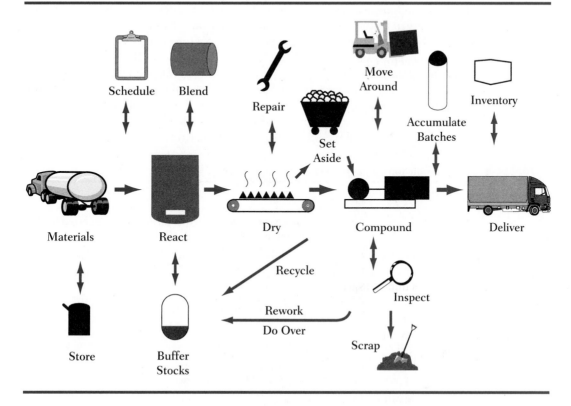

Table 3.3 lists the value-added and non-value-added activities associated with making a copy of a document at a photocopier (Scholtes, 1998). As with the polymer production process, there are lots of non-value-added activities. Each is an OFI whose identification was enabled by observing the process and evaluating the

TABLE 3.3 Assessment of Work Required to Copy a Document

Value-added work	*Non-value-added work*
Place original on machine	Wait to use copier
Select size	Clean dirty glass
Select page direction	Find and add toner
Select number of copies	Find and add paper
Push button	Check copy quality
Remove original	Stop and adjust copier to improve quality
Collect copies	Restart copier and remake defective copies
	Get help to find paper, toner, or to get copier to work

Source: Adapted from Scholtes, 1998, p. 131.

various activities involved. Unfortunately, much of the work done in business is non-value added. For example, most of the work done at customer service desks in retail shopping stores—returns of defective products, refunds for incorrect prices charged at the register, "rain checks" for out of stock items—is non-value added. The same could be said of most "1-800" customer service call centers, with the obvious exception of those taking orders. Identifying the non-value-added work and finding ways to eliminate or at least reduce the need for it provides major OFIs. This is not always obvious because traditional accounting systems do not segregate the costs of non-value-added work from the costs of value-added work. Tables 3.4 and 3.5 show examples of non-value-added activities in a variety of business functions and types of businesses, respectively.

TABLE 3.4   Examples of Non-Value-Added Activities by Function

Function	Non-value-added activities
Sales	Making repeat sales calls because the information requested by the customer was not available on the first visit.
Marketing	Changing a product image because the original image was rejected by consumers (e.g., the "Nova" automobile introduced in Mexico; in Spanish, *No va* means "it doesn't go")
Engineering	Designing out flaws in a product after they have been detected by customers who have lodged complaints.
Manufacturing	Disposing of waste or performing rework on defective products.
HR	Conducting two, three, or four interviews because not enough information was asked for previously to make a decision about a candidate.
Logistics	Shipping product from site A to site B and then to site C prior to eventually getting it to a customer.
IT Support	Solving employees' hardware or software problems two or three times because the problem was not solved correctly the first time.
Accounting	Performing manual account reconciliations to close the books.
Finance	Revising financial forecasts three or four times until senior management is willing to "accept" the numbers.
Procurement	Playing golf with the supplier's sales force. This is a very common "business practice," as is going to sporting events with the supplier's sales force.
Legal	Preparing extensive documents for a litigation only to find out that a decision has been made to settle out of court.
Environmental	Paying fines and preparing additional documentation due to violation of environmental standards.
Quality	Responding to customer complaints.
R&D	Beginning a research project and then abandoning it because funding was cut.

TABLE 3.5   Examples of Non-Value-Added Work by Business Type

Type of business	Non-value-added work
Health care	Second medical opinions obtained due to lack of confidence in the first opinion.
Financial services	Fixing errors in credit card bills or mortgage statements.
Food service	Disposing of spoiled food.
Transportation	Rescheduling passengers from a canceled flight.
Catalogue sales	Paying postage twice (to the customer and back) for goods returned because they were not what the customer ordered or were damaged in shipment.
Internet commerce	Restoring a "crashed" server.
Retail sales	Restocking returned merchandise.
Chemicals	"Blending" batches of finished chemicals because one or both did not meet physical property standards.
Microelectronics	Disposing of inventory that has gone from "state of the art" to "antiquated" while sitting in the distribution pipeline for a year.
Automobiles	All activities and expenses associated with recalls.
Telecommunications	All activities and expenses associated with reassigning area codes to existing phone numbers.

Value-added work may also provide opportunities for improvement because value-added work also takes time and costs money. Reduction of value-added work typically requires invention or innovation and a redesign of the process because this is required work. Reductions in value-added work are typically not the first place we choose to look for OFIs, but this may be required to keep up with the competition—or better yet, to create a competitive advantage. The microbrewing process shown in Figure 3.11 is an example of reduction of value-added work. The beer is produced and consumed on the premises with no handling by a bottler or a distributor. Bottling and distributing are value-added activities, because these are typically required to get the product to the customer. By redesigning the work, this business has combined the brewing operation with the retail sales operation, which "designs out" the distribution activities.

Process Complexity

Non-value-added work increases process complexity, which results in increased costs and increased demands on management's time. The more complex a process is, the more likely it is that errors will be made. Training new workers will be more difficult, and more supervision will be required. We would like to make every process as simple as possible while still getting the work done effectively. This is often referred to as the KISS principle, which stands for "Keep it simple, stupid!"

FIGURE 3.11
An Example of Reduction of Value-Added Work

IRON HILL BREWERY & RESTAURANT

BEER TASTES BEST WHEN IT'S MADE FRESH, NOT HAVING SPENT WEEKS IN A WAREHOUSE OR ON A SHELF. THAT'S WHY OUR BEERS ARE MADE RIGHT HERE, ON A DAILY BASIS. WE USE ONLY THE FINEST INGREDIENTS: DOMESTIC AND IMPORTED MALTED BARLEY AND HOPS, YEAST AND WATER. AND BECAUSE WE USE NO PRESERVATIVES, YOU ARE GUARANTEED A FRESH AND NATURAL GLASS OF BEER EVERY TIME.

1. Milling: Grains are selected depending on the beer style. The grains are milled into the grist hopper (A) to allow us to extract essential ingredients.
2. Mashing: The grains are moved from the mill room through a feed auger (B) to a mash tun (C). Hot water is added to form the mash. This process converts the grain starch to sugars.
3. Lautering: Hot water is passed over the grains to remove all sugars. The solution, or wort, is extracted from the mash tun and sent to the kettle (D).
4. Boiling: The wort is brought to a full rolling boil and hops are added for bitterness and aroma— giving each beer its unique profile.
5. Cooling: The hot wort is passed through a heat exchanger (E) where it is cooled to a temperature appropriate for the yeast to ferment.

6. Fermentation: The wort is transferred to a fermentation tank (F) where yeast is added and it is allowed to ferment, converting the sugars to alcohol and CO_2. Depending on the type of beer, this process takes 14 to 30 days.
7. Filtering: After proper conditioning the beer is sent through a filter (G) to remove all traces of the yeast before it is transferred to the serving tanks (H).
8. Serving: Finally, the beer carbonation level is adjusted and it is ready to be sent directly to our taps for consumption. From start to finish our beers travel less than 65 feet. There's nothing fresher!

Source: Illustration © 1996 Elizabeth Traynor, text © 1996 Iron Hill Brewery & Restaurant. Reprinted with permission.

Identifying unnecessary process complexity often uncovers OFIs and may lead to the detection of non-value-added work. For example, a professional journal one of us worked with was having difficulty reducing the cycle time for reviewing submitted articles. Upon analysis of the work flow, it was discovered that hard copies of articles were being mailed (via standard mail) from the editor in the United States to associate editors in Europe. The associate editors would then often mail hard copies back to reviewers in the United States. These reviewers would then mail reviews back to Europe, and the associate editors would compile these and mail them back to the editor in the United States. This process resulted in four separate mailings between the United States and Europe! Once this unnecessary complexity was discovered, the process was changed. Articles were mailed to U.S. reviewers by the U.S.-based editor, and reviewers' comments were collected electronically. This resulted in only two mailings between the United States and Europe and significant reductions in cost and cycle time.

Fuller (1985) presents a classic example of the effects of complexity on an electronics assembly process (Figures 3.12a and 3.12b). If we remove the unnecessary complexity caused by non-value-added work, the process would look like Figure 3.13. Most of the complexity in this process is due to missing parts, resulting in additional processes and supervisory functions to handle the missing parts and incomplete kits. Of course, this additional complexity is all non-value-added work. When this process is streamlined and operating well, one need only get a "kit of parts," "assemble it," and "move it to the stockroom." Pretty simple!

Every process flowchart has three distinct versions:

1. The way we think it is.
2. The way it should be.
3. The way it really is.

FIGURE 3.12a Process With Complexity

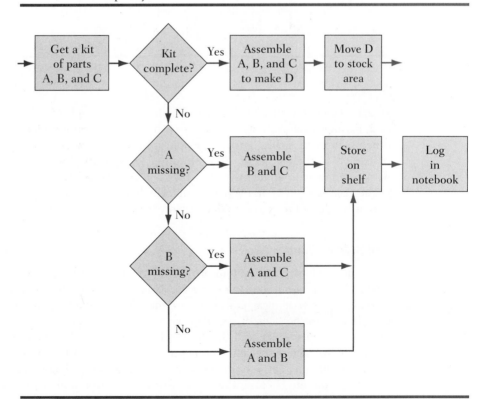

FIGURE 3.12b Supervisor Functions Needed to Run the Complex Process

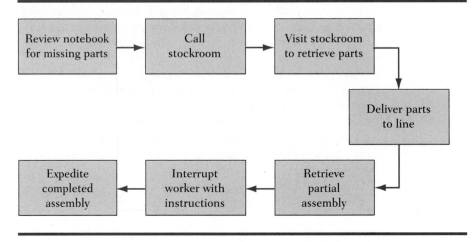

FIGURE 3.13 Process With No Complexity

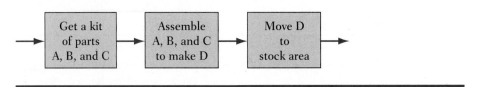

When asked to develop a flowchart of their process, people often depict version 1 or 2, perhaps because they do not wish others to see the inefficiency of their true process. In many cases, however, people are not even aware of all the complexity that exists in their process. This additional complexity—often non-value-added work—can take on an identity of its own and become a unique business function in the organization chart. When complexity becomes an integral part of the system, it is referred to as the "hidden plant," which we discuss next.

The Hidden Plant

The hidden plant is the system of non-value-added work and complexity in our business. In a manufacturing context this is the part of the plant where the re-working and scrapping operations are done. These operations are usually physically hidden in the back of the plant so as not to be visible to visiting corporate managers or customers, and this term reflects that fact. Manufacturing consultants jokingly suggest that one should always start a plant tour in the back of the factory so that one can see the piles of rework and scrap! The name *hidden plant* is also appropriate because the costs of this work are not allocated as non-value

FIGURE 3.14 Every Organization Has a Hidden Plant

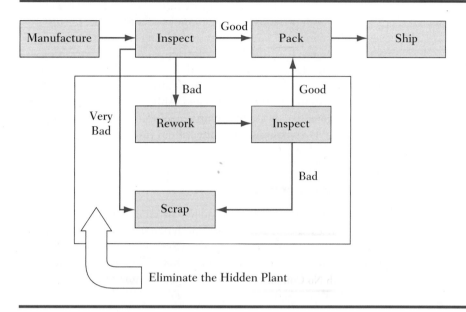

Eliminate the Hidden Plant

added in traditional accounting systems but are hidden under standard categories such as operations, processing, or manufacturing. Almost all processes, manufacturing and service alike, have hidden plants, as depicted in Figure 3.14.

In service processes the hidden plant may be particularly well hidden as there is no separate physical location where rework is done. Rework is often done in the same office, by the same person, who did the value-added work. For example, a secretary may reschedule a meeting four or five times due to changing plans of participants. This rescheduling is rework, just like fixing a defective product. The hidden plant may be so well hidden that it is excluded from our process flowcharts altogether. The rework loops in the hidden factory are not depicted in the process charts shown in earlier figures in this chapter. For example, the billing process depicted in Figure 3.5 does not show how we handle errors such as invalid addresses, incorrect computer files and reports, or customer complaints about errors in the bills themselves. Once identified, the hidden plant is a great place to look for OFIs. The goal should be to reduce or eliminate the need for a hidden plant.

The new credit card account process shown in Figure 3.6 illustrates one part of the hidden plant. The process step following the "Process application" step is "Resolve problem applications." Many problem applications are caused by missing or erroneous information on the credit card application. This results in rework and waste, the sole function of the hidden plant.

In consulting on one customer order process, we encountered a computer system whose sole purpose was to handle orders that had "fallen out" of the main process. This $2 million project was planned solely to expand the computer system to handle the large volume of orders that had fallen out. No effort was being expended on discovering why orders were falling out or on fixing the problem. This is analogous to purchasing more buckets to catch the water coming through a leaky roof rather than spending money to fix the roof. In this case the effort was eventually redirected and focused on finding the root cause of the problem. The root causes were identified and solutions were found, resulting in the defect rate dropping from 43% to less than 5% in a few months. This improvement produced $11 million in savings and a cost avoidance of $2 million in capital for the new computer system that was not needed. Although the fallout rate had been considerably reduced, 5% was clearly unacceptable. Additional improvement projects were initiated to further reduce the fallout rate, illustrating the iterative nature of statistical thinking.

Process Measurements

Process measures are key to the successful management and improvement of processes. Our experience, and the experience of many others, has been that lack of good data is typically the greatest barrier to improvement. Of course, we do not want just any data; we want the data that will enable us to study and improve the process. Process measurements tracked over time enable us to analyze the process in the following ways:

1. Assess current performance levels.
2. Determine if the process has shifted by comparing current performance to past performance.
3. Determine if the process should be adjusted (minor changes).
4. Determine if the process must be improved (major changes).
5. Predict future performance of the process.

These five key methods of analyzing the process with measurements are shown schematically in Figure 3.15. The billing process case study discussed in Chap-

FIGURE 3.15   Uses of Process Tracking

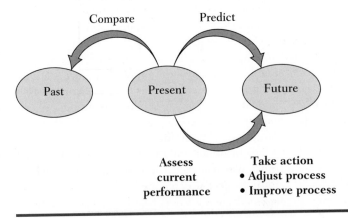

ter 1 illustrates the use of this model. Here is the information relating to the five major uses of measurements to analyze that process:

1. We are currently getting the bills out in an average of 8 days.
2. It used to take an average of 17 days; hence, we've reduced our cycle time by 9 days or 53% of the original 17-day cycle time.
3. We are continually adjusting the process to be consistent with standard operating procedures.
4. We will have to make significant improvements to our process if we are to get faster than the current 8-day cycle time.
5. Some special studies of our current performance predict that we can get our cycle time down to 5 days by removing the bottleneck at one of the steps in the process.

The SIPOC model helps us to determine specifically where to make critical measurements, as illustrated in Figure 3.16. These measures cover three key stakeholders of the process: customers, employees, and suppliers. Stakeholders include anyone who has a stake in the process—that is, anyone who directly affects or is directly affected by the process. When financial measures and environmental measures are added, two other key process stakeholders are usually considered: shareholder (owner) and community.

The customer feedback and employee satisfaction measures are usually collected using surveys (see Section 5.3). Two key process output measures are some type of total cycle time measure (the time required to produce the product or service) and defect levels. Defects can consist of inaccurate data, such as incorrect social security numbers on a credit report, or unacceptable physical measurements

FIGURE 3.16 Process Measurement

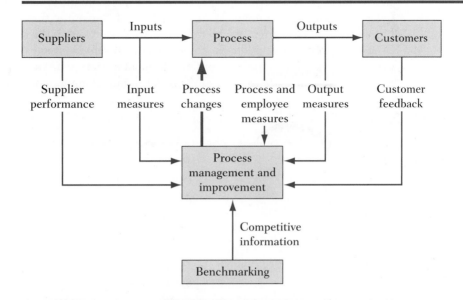

of a product, such as thickness of a contact lens. Defect levels and cycle time are often related because defects usually result in increased cycle time due to the hidden plant. Most of the measures for the order taking process (Figure 3.4), billing process (Figure 3.5), and new credit card application process (Figure 3.6) can be classified as either a defect measure or a cycle time measure. The key measure for the billing process improvement study discussed in Section 1.2 was cycle time.

We may also wish to measure cycle time and defect levels of individual process steps. This requires having an in-process measurement system as well as measuring outputs. Other important process measures include process downtime, capacity, scrap, and rework. An effective way to identify in-process measures is to compile a flowchart of the key steps in the process, identifying the key measures for each of the critical steps or control points. A set of transaction-based measures associated with a customer order taking process is shown in Table 3.6.

If process measurements are properly selected, they should be predictive of one another. For example, input measures should predict how well the process will function, and in-process measures should predict the output measures. This is consistent with our overall theme: To improve outputs, we must improve the process. The output measures should then predict customer satisfaction. If there is a breakdown in this sequence of prediction, then we have not identified the appropriate measures in some area.

A study reported by Fernandez (1998) showed that a key customer measure for telecommunication services was "honor commitments"—that is, did the company perform as they said they would? Fernandez found that the process output variable "appointments met (%)" was a good predictor of the customer's assessment of whether the company met the commitments to which they had agreed. The next step in the study was to identify the in-process measures, which are typically things the company could control, that were predictive of appointments met. Four important variables were identified: service order errors resolved (%), number of orders not assigned on time, number of orders that "had fallen out of the system" and required human intervention, and number of wrong addresses in the database (Table 3.7). Is any of these a hidden factory? This measurement system could then be used to manage and improve the order fulfillment process, with the goal of improving customer satisfaction as measured by the "honor commitments" variable.

TABLE 3.6 Transaction-Based Measurement System for Customer Request: "Please change my address to ____."

Process characteristic	Measure
Procedural accuracy	Calls answered in 30 sec. (%)
Quality of interaction	Professionalism rating Call quality rating
Address change cycle time	Completed in x days (%)
Computer system availability	Time transmission (% of standard)
Customer satisfaction	Survey: % excellent ratings

Source: AT&T Universal Card.

TABLE 3.7   Process Measurement Linkages

In-process measures	Process output measures	Customer measures
Service order errors resolved (%)	Appointments met (%)	Honor commitments
Orders not assigned on time		
Orders requiring human intervention		
Wrong addresses in database		

There are, of course, other important measures of customer satisfaction for this process. Similar linked sets of measurements were created to manage and improve these other aspects of customer satisfaction. (These systems of linked measurements were developed using the regression analysis methods discussed in Chapter 6.)

A key aspect of process improvement studies is to determine the effects of improving process performance measures on the bottom line. Some of the key relationships are summarized in Figure 3.17. The trick is to use these relationships to predict and then document bottom-line results to motivate and justify process improvement efforts.

FIGURE 3.17   Effects of Process Improvement

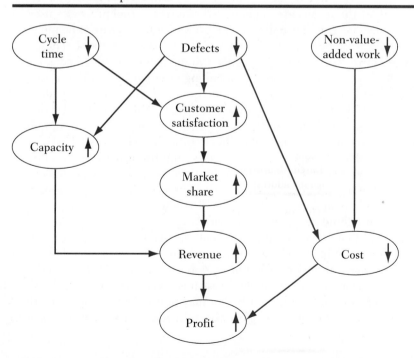

Benchmarking

Another method of critically analyzing processes is called benchmarking. Benchmarking is the act of determining how your organization's processes are performing relative to those processes in other companies (Camp, 1989). The external benchmark data—for example, DuPont's safety performance or L. L. Bean's product distribution process—are used to select process performance goals and methods, both long- and short-term. Benchmark data are typically most useful once we have some understanding of the process being benchmarked and the critical variables. We need to understand what and who to benchmark.

Benchmarking studies can be loosely grouped into two categories: output or results benchmarking and process benchmarking. Output benchmarking focuses on the level of performance obtained. Comparing one's own safety performance to DuPont's safety performance is one example of output benchmarking. People working in a process are often so close to their work that they do not realize it is possible to do better. They may feel that they are doing everything possible to improve safety and that no further improvements are possible. If DuPont's safety performance is significantly better than ours, however, it opens our eyes that further improvement is indeed possible. If DuPont can do it, why can't we? This can result in a breakthrough in people's thinking and a willingness to try to improve.

Output benchmarking is fairly easy to do because the output data needed are often available publicly. Output benchmarking has a serious limitation, however. It only tells you what level can be achieved; it does not tell you how to achieve it. This is where process benchmarking comes in. Process benchmarking looks at how the results are obtained—that is, the actual processes used to achieve these results. This is usually much more valuable in terms of helping us improve, but it is also much more difficult to obtain. True process benchmarking requires the active cooperation of the organization being benchmarked.

An advantage of benchmarking of either type is that it forces us to look beyond our own organization, or even industry, for better ideas. For example, a manufacturer of firearms was having trouble processing bullet casings without damaging them. To get ideas for improvement, they eventually benchmarked a cosmetics company, which made lipstick casings. The lipstick casings were very similar in shape to bullet casings, and the processes for both were similar. By copying certain aspects of the lipstick casing process, the bullet manufacturer was able to significantly improve their process. Similarly, a computer manufacturer could benchmark L. L. Bean's distribution processes, or a pharmaceutical company could benchmark Walmart's inventory control process.

Benchmarking gives us another method for analyzing our process. We can compare how well we are accomplishing this work with how others are doing it, either within our own organization or in a totally separate industry. We can also compare our output and process measurements to those obtained by other organizations to identify OFIs. We may not decide to copy the other organization's process, but we may gain critical insights that can be used to improve our process.

3.6   Systems of Processes

Up to this point we have focused on individual processes. We will now turn to systems of processes and why it is so important in business to view each process from the system perspective. A system is a set of processes that work together to accomplish an objective in its entirety. By "in its entirety" we mean that the system includes all related and necessary processes. For example, we can have a process for paying bills, but this process will not function if we do not have people or equipment to do it. Therefore, the hiring and equipment procurement processes are part of the bill paying system. It is important to view processes from the perspective of the overall system so that our process improvement efforts improve that system rather than simply push the problem from one department or function to another.

Intuitively, we might assume that any improvement in a process will automatically result in improvement to the overall system. Unfortunately, this is not the case. For example, a procurement manager may decide to reduce costs by buying lower grade materials. This will no doubt improve the procurement department's costs, but it may result in even greater losses in operations. Similarly, if the human resources department is reducing cycle time for hiring by doing less evaluation of candidates, the whole organization may suffer because inappropriate employees are hired. This is called suboptimization.

Suboptimization occurs when the performance of one or more processes is maximized but the overall performance of the system is not maximized. Here are some examples of suboptimization, which really just push problems or costs from one process or department to another:

- Manufacturing putting finished goods in railcars outside the plant for extended periods of time so they can list them as "shipped" and take them off inventory. Of course, someone else has to accept the costs of renting the railcars and holding the capital prior to it being shipped to a paying customer.
- Salespeople promising delivery dates that the organization cannot achieve in order to make a sale. The sales force gets credit for a sale, but logistics and customer service will have to "clean up the mess" when the customer does not receive what was promised. In all likelihood, a hidden plant will be developed to deal with unhappy customers and to expedite certain orders.
- Logistics depleting inventory levels at the end of a quarter to minimize taxes or to make the "official" numbers look good, resulting in out of stock situations and lost orders for the sales department.
- Salespeople from different territories in the same company "competing" with each other to land a nationwide account with a prospective client. As hard as it is to believe, we have seen different salespeople from the same company undercut each other's prices because each wants to land the account for their territory to get their commission or other incentives. The net result for the company as a whole is a less profitable contract.

Residents of New York City often complain about the difficulty of finding an apartment in the city. One reason for the shortage goes back several years.

Apartment rents began to increase dramatically because of an imbalance between supply and demand of apartments (see Kauffman, 1990). Rather than analyze the system to understand the root causes of the problem, the city government decided on a quick fix, which was to apply rent control, making it illegal for landlords to raise rents. At first glance, this may make intuitive sense. However, the rent control statutes caused developers to stop building apartments in New York City; they could achieve greater returns doing other things. Similarly, as apartments aged and needed more maintenance, many owners simply tore them down and sold the property or converted them to other businesses, such as parking garages, where they could make greater profits. The ultimate result of rent control was that the imbalance between supply and demand of apartments got much worse, making it even more difficult for people to find an apartment.

Now the lower rents are not always obtainable because a "black market" for apartments has become active. The black market arose because the legislated rent was so much lower than the market value of an apartment. People would obtain an apartment (often through illegal means, such as bribery) and then illegally sublet it at a rent far greater than the official "controlled" rent. The persons subletting the apartment usually did not complain about this to the authorities, because they would not otherwise be able to find an apartment. The situation got so bad that the city government partially relaxed the controls as an incentive to build, resulting in some apartments that were rent controlled and some that were not, adding complexity to the system. The issue would have been dealt with very differently if a systems view of the situation had been taken. In this case root causes of the imbalance between apartment supply and demand would have been identified and addressed, perhaps through tax incentives to build additional apartment buildings.

So how do we go about viewing each process from the systems perspective? First, we need to understand the role of this process within the larger system. We can do this by trying to visualize or map the core processes of the entire system. Core processes are the processes that are fundamental to achieving the objectives of the organization. For example, a large consumer credit organization may have an employee cafeteria in its headquarters, but providing food is not fundamental to making money in consumer credit, so it is not a core process. Obtaining accounts, servicing accounts, handling delinquent accounts, designing new credit products and services, and so on would probably be core processes for this business.

Organizations are recognizing the importance of focusing on business processes and are identifying, managing, and improving core processes. In some cases businesses have decided to outsource non-core processes so they can focus on the core processes. Hammer (1996) uses American football to illustrate core processes. Fans are the customers of football. The score and outcome of a football game (outputs) are determined in three ways: offense, defense, and special teams (kickoffs, punts, field goals). Hence, football has three core processes: offense, defense, and special teams. Now lets take a look at some business core processes.

The core processes of IBM-Europe (IBM-Europe, 1990) are shown in Figure 3.18. Another way to show core processes of a typical product-oriented business is in the form of a systems map that illustrates the key processes used to provide the product or service (Figure 3.19). The best systems map for an organization

FIGURE 3.18   IBM Core Processes

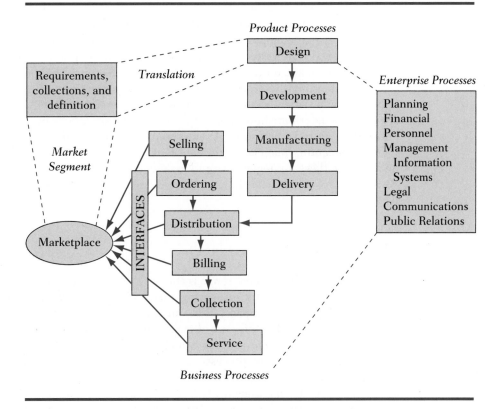

FIGURE 3.19   Core Business Processes

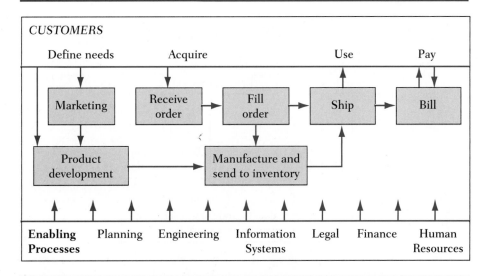

may depend on the specific application. Other formats are presented by Rummler and Brache (1995). Key supporting (enabling) processes are also shown on the systems map, but they do not directly "touch" the product or service. Some examples of key enabling processes are accounting, recruiting, personnel development, and legal. The systems map also illustrates the importance of linking the measurements made on the various processes (see Section 3.5).

Understanding the organization as a system is key to overall improvement. Suboptimization often occurs when the overall systems view is not taken. The goal should be to determine how to operate each of the core processes so that the output of the system is maximized as measured by product and service quality, customer satisfaction, financial performance, and so on. The use of the systems map helps define these strategies. Prior to making changes to one process, we should consciously consider what impact these changes will have on the system as a whole.

3.7   The Measurement Process

Measurement in its broadest sense is the act of assigning a number to an item. This could include a doctor measuring a baby's height and weight, counting the number of "hits" to a Web page, or calculating net income for a business. In this section we point out that the work of making measurements produces another key process, the measurement process, which must be managed and improved just like any other key business process.

Figure 3.20 illustrates the role of the measurement process in process management (Hoerl, Hooper, Jacobs, & Lucas, 1993). It also highlights the fact that all process measurements contain two types of variation: process variation (i.e., variation from the process we are studying) and variation created by the measure-

FIGURE 3.20   Interface Between Product or Service Process and Measurement Process

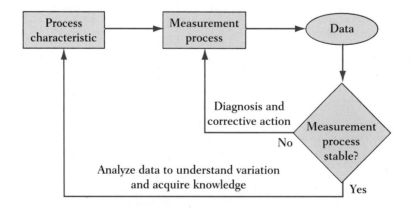

ment process. To understand process variation, we must understand the magnitude of the measurement variation. Understanding the measurement variation helps us better assess the process capability, improve the process, and determine whether the product or service meets the customer specifications. Analysis of process variation is discussed further in Chapters 4 and 5.

In a manufacturing environment the measurement process will typically include these steps:

- Selection of the item, product, or service to be measured.
- Transportation of the item to be measured to the measurement location.
- Preparation of the item for measurement or analysis.
- Measurement is made.
- The result of the measurement is stored in a database and reported to persons and organizations as appropriate.

Note that the process measurement in the above steps includes the components described by the SIPOC framework:

Supplier	Process operators
Inputs	Process samples to be measured or analyzed
Process steps	Activities to make and report the measurements
Outputs	The process measurements
Customer	Process operators

This measurement process, including sample collection, analysis, and reporting activities, is often automated. Some measurement processes involve sampling the product produced. Other systems measure 100% of the product, in which case sampling is not an issue. In the chemical industry many of the analyses are done in a laboratory using analytical chemistry methods of measurement (see Table 3.8). In manufacturing industries that produce pieces and parts, the measurements are often dimensions measured by gauges (see Table 3.9). Each measurement procedure should be well defined so that data collected are equivalent.

The measurements associated with nonmanufacturing processes (e.g., finance, hiring, and order entry) are usually less well developed than manufacturing measurement systems. In many cases, the people involved in these processes are not aware that they have measurement processes. The two key measurements for nonmanufacturing processes are cycle time and defects (accuracy). Just as with manufacturing processes, it is critical that the cycle time and defect measurement methods be well defined, documented, and used consistently by all involved persons. Operational definitions must be developed for all measurements detailing how the measurements are made and what constitutes a defect. (Operational definitions are definitions that will be interpreted the same way by different people in different situations.) For example, in an auto leasing business, different analysts may use different criteria for determining (measuring) the residual value of a leased car. In insurance, different agents may disagree as to whether a claim is actually covered under a client's policy. This is a measurement issue with discrete (yes/no) data. Clear, well-defined criteria (operational definitions) for calculating the residual value of a leased car or for determining

TABLE 3.8   Chemical Analysis Measurement Study

Operator	Sample	Run	Test 1	Test 2
1	1	1	156	154
		2	151	154
		3	154	160
	2	4	148	150
		5	154	157
		6	147	149
2	3	7	125	125
		8	94	95
		9	89	102
	4	10	118	124
		11	112	117
		12	98	110
3	5	13	184	184
		14	172	186
		15	181	191
	6	16	172	176
		17	181	184
		18	175	177

TABLE 3.9   Part Dimension Measurement Study

	Operator 1		Operator 2		Operator 3	
Part	Trial 1	Trial 2	Trial 1	Trial 2	Trial 1	Trial 2
1	415	360	358	357	370	407
2	313	324	323	314	324	364
3	335	243	269	284	334	253
4	403	381	354	367	391	417
5	284	264	266	254	284	304
6	64	139	15	5	213	205
7	353	367	359	362	357	412
8	430	368	367	366	416	378
9	321	355	295	324	365	394
10	698	803	651	646	813	746

whether a claim is covered would significantly reduce variation in these measurement systems.

Three key characteristics of a measurement system are accuracy, precision, and stability over time. *Accuracy* is the ability of the process to produce measurements that, on average, equal the true value of the item being measured. For example, if a clock has not been set ahead by 1 hour at the beginning of daylight savings time, it will consistently be off by 1 hour. This is an inaccuracy, or bias. The time measurement is consistently off by 1 hour versus equaling the correct time on average. *Precise measurements* exhibit low variation around an average value, which may or may not be correct (i.e., accurate). For example, if a professor grades the same term paper twice, will he or she give the same grade both times?

Stability of the measurement process refers to its ability to maintain accuracy and precision over time. In many cases, an accurate, precise measurement system will drift over time. For example, we can set our watch to the correct time, but over a month or so it may lose time (i.e., become inaccurate). One method to evaluate stability is to measure the same (or a similar) item repeatedly over time to see if the measurements drift. These data are often plotted on control charts, which will be discussed in Chapter 5.

It is critical that the measurement process be stable so that when we detect trends and shifts in the process measurements we can safely conclude that the observed changes are due to the process and not the measurement system. We do not want to change the process if the root cause of the problem is in the measurement system.

Various combinations of accuracy and precision are shown graphically in Figure 3.21, in which all four measures have a true value of 50 units. Measure A is the best, showing a small variation around an average value that is essentially 50 units; hence, Measure A is both accurate and precise. Measure B is also accurate but much less precise than Measure A. Measure C is precise but not accurate, averaging around 40 units, well below the true value of 50 units. Measure D is neither accurate nor precise, being biased to the high side and showing a lot of variation.

FIGURE 3.21   Four Measurements With a True Value of 50

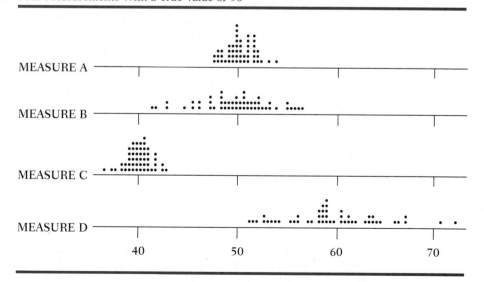

Lack of accuracy can be easier to deal with than lack of precision. To determine accuracy, we must have a standard item with a known value. For example, the National Institute of Standards and Technology (NIST) maintains standards that are exactly 1 inch, 1 pound, and so on. Historically, lack of defined standards inhibited accurate measurement systems and resulted in many measurements based on the human body, such as the "foot" or "yard" (defined as the distance from the king's nose to his thumb when the arm was fully extended). Once we have a standard, we can measure the standard to see if we obtain the "correct" value on average. In business settings we often define an "expert" opinion as the standard. For example, the Chief Legal Officer may provide a "final" answer as to whether an action by an employee is a dismissable offense according to corporate policy.

Measurements that are not accurate but that exhibit a constant bias can be used effectively to detect trends and process changes. Going back to the example of the watch that is off by 1 hour, we can still use this watch to monitor how long it takes to drive from home to work. In addition, correcting the problem of inaccuracy can often be accomplished by a simple adjustment. In a watch it would simply involve resetting the time. If someone setting residual values on leased cars is consistently high by 10%, we could simply ask them to reduce their estimates by 10%. In contrast, lack of precision often requires measurement improvement before meaningful process improvement work can be done.

Precision has two major components: repeatability (short-term test variation) and reproducibility (long-term variation). Short-term test variation is the variation we would observe if the same person measured the same item several times with the same instrument. For example, repeatability of a bathroom scale could be evaluated by standing on the scale and measuring your weight, then stepping off, and then stepping back on again. We would expect to see minimal variation in our weight, although there could be some due to the specific angle at which we were standing, where on the scale we placed our feet, and so on. Suppose, however, that we weighed ourselves on several different bathroom scales. We might expect to see more variation here because there may be differences between the scales. This is an example of reproducibility, in this case reproducibility between instruments.

A typical measurement study involves taking a sample from a process at two different times during the day. Each sample is analyzed in duplicate (two measurements) by two different operators. The resulting types of process and measurement variation (repeatability and reproducibility) can be evaluated as shown in Table 3.10. The determination of short- and long-term estimates of process and measure-

TABLE 3.10 Measurement Process Study Example

Source of variation	Time frame	Type of variation
Day-to-day	Long term	Process variation
Samples within days	Short term	Process variation
Between operators	Long term	Measurement reproducibility
Between repeat tests	Short term	Measurement repeatability

Sampling Plan: Two samples were drawn at random from each day's production. Each sample was analyzed in duplicate by 2 different operators. The sampling study was conducted over a 90-day period.

ment variation is key to the success of measurement studies. Short-term studies typically cover days and weeks, whereas long-term studies cover weeks and months. Further details on measurement process analysis can be found in Automotive Industry Action Group (1995), Snee (1983), and Wheeler and Lyday (1990).

3.8 Summary

The first fundamental of statistical thinking for business improvement is "All work occurs in a system of interconnected processes." We must understand our processes if we are to better serve our customers. We must also understand how our process relates to the larger system. Let's review the key aspects of understanding business processes:

1. All work can be viewed as a process.
2. Process thinking provides the focus and context for understanding, measuring, managing, and improving the process.
3. Improve the process and you improve the outputs or results.
4. We can often make significant improvements by analyzing the process itself, even before we obtain data.
5. Flowcharts of processes and subprocesses provide a picture of the process and enhance understanding.
6. The SIPOC model defines the key components of any process: **S**upplier, **I**nputs, **P**rocess steps, **O**utputs, and **C**ustomers.
7. The SIPOC model provides a framework for identifying key measurements.
8. Key process measures, particularly for nonmanufacturing processes, are cycle time and defects (accuracy).
9. Service and other nonmanufacturing processes typically have less well-developed measurement systems and involve more human intervention.
10. Measurements tracked over time enable us to assess the current performance of the process, determine if the process performance has changed, determine if the process needs to be adjusted (small change), determine if the process needs to be improved (major change), and predict future performance.
11. In many service processes the customer and supplier are the same person or group of people.
12. One useful strategy for improving processes focuses on finding and eliminating non-value-added work (i.e., finding the hidden plant).
13. The organization is a system of core processes that produce the product or service and enabling processes that support the core processes.
14. Use the systems view to avoid suboptimization.
15. Benchmarking provides external input on how similar processes are performing in other organizations and how the work is actually done.
16. Understand that collecting process measurements is a process that should be managed and improved like other processes.

Attention to these issues will provide the understanding and measurements needed to effectively manage and improve processes, thereby producing better business results.

3.9   Project Update

At the end of Chapter 2 you identified a project on which to apply statistical thinking during this course. You should now begin analyzing the process you selected, using the methods and tools from this chapter. Begin with a thorough SIPOC analysis. You may wish to develop flowcharts for some subprocesses after developing an overall, high-level chart. What larger system is this process a part of, and how does it fit into the system? How will you ensure that you do not suboptimize the selected process at the expense of the system as a whole?

Once you have produced your flowchart, look for opportunities for improvement. Is there non-value-added work in the process? Is there a hidden plant involving rework and scrap? Do you see signs of unnecessary complexity? What improvements can you make now without waiting until you have all the data you might eventually collect?

What important variables might you want to measure for this process? What evidence do you have that you can accurately measure these variables?

EXERCISES

1. **Touring a process.** Tour a manufacturing plant or a service organization and observe one or more key processes, the associated suppliers, inputs, process steps, outputs, customers, the measurement systems, and how the measurements are used to manage and improve the process. Report your findings to the group. Include these items:

    - Company visited
    - Process observed
    - SIPOC elements
    - Process measurements
    - Process management systems used

2. **Process maps.** Draw process maps of the processes you observed in Exercise 1, and comment on the insight you gained and the usefulness of the maps.

3. **Time value of money.** Calculate the impact on the bottom line of reducing the average of the monthly cycle time of a billing process from 17 days to 9 days. Assume that the process bills customers $12 million per month, all bills reach the customers within 7 working days, and all bills are paid within 30 days. State all the assumptions you have made to do this analysis.

4. **Core processes.** Identify the core processes of an airport. Draw a systems map showing the core processes and major supporting (enabling) processes. If you prefer, instead of the airport identify the core processes and create a systems map for another company or organization of your choice.

5. **SIPOC analysis.** Do a SIPOC analysis of each core process in Exercise 4.

6. **Choose your own process.** For a process of your choice, define the purpose of the process, SIPOC elements, and key measurements.

7. **Hospital processes.** Tour a hospital and identify the core processes and key process measures. Draw a systems map of the hospital showing core processes and major supporting processes. Do a SIPOC analysis of each core process.

8. **Tolling operations.** Identify a tolling operation and define the process suppliers, inputs, process steps, outputs, and customers.

9. **Supermarket processes.** Observe the process at a supermarket checkout line. Flowchart the key steps in the process. What are the key measures for this process? How would you define non-value-added work for this process? Does this process have a hidden plant? If so, where? Does the supermarket have a hidden plant?

10. **Personal processes.** Flowchart the process you use to "go to school" or "go to work." Where does the process begin and end? What are the key measures for this process? How would you define a defect for this process? How would you improve this process?

11. **You choose the process.** For a process of your choice, create a flowchart. Define the value-added and waste-producing process steps. Create a strategy for improving the performance of the process. Develop a measurement system for the process.

12. **Cycle time study.** Use the data in Table 3.11 for this exercise. The data presented are the average time to complete each step in the review process: actual time used and potential (the best time possible for this review process). The actual times are based on the review of 30 projects. The potential times are subjective engineering judgment estimates. With these data, answer the following questions:

   ■ What are the sources of value-added and non-value-added work in this process?
   ■ Where are the opportunities to improve the cycle time of this process both with respect to actual time used and the potential best times? What strategy would you use?
   ■ Step 10, resolving open issues, required 104 hours (potential) versus 106 hours (actual). Is there an OFI here? Why or why not? If so, how would you attack it?
   ■ What do you think are the most difficult critical issues to deal with when designing a sound cycle time study such as this one?

13. **Chemical analysis measurement process.** The data in Table 3.8 are from a study in which a series of runs were made by three operators to study the variation in a chemical analysis measurement system. The procedure consisted of taking a sample, treating the sample in a combustion-tube furnace, and then performing the chemical analysis. In the test three operators each took two samples and made three combustion runs on each sample and tested each run in duplicate. Using graphical techniques, determine the sources of variation in this measurement system. Is there variation in operators, runs, samples, or tests? Is this measurement system useable? If so, how and for what purposes? What steps would you recommend to improve the measurement system that produced these data?

14. **Parts dimension measurement process.** The data in Table 3.9 are measurements of a critical dimension of an extruded plastic part. Using graphical techniques,

TABLE 3.11
Basic Data Review for Construction Project Equipment Arrangement

Step	Description	Cycle time (hours) Actual	Potential	Difference
1	Read basic data package	4	4	—
2	Write, type, proof, sign, copy, and distribute cover letter	21.9	0.5	21.4
3	Queue	40	0	40
4	Lead engineer calls key people to schedule meeting	4	0.25	3.75
5	Write, type, proof, sign, copy, and distribute confirmation memo	25.4	2.1	23.3
6	Hold meeting; develop path forward and concerns	4	4	—
7	Project leader and specialist develop missing information	12	12	—
8	Determine plant-preferred vendors	12	12	—
9	Review notes from meeting	12	12	—
10	Resolve open issues	106	104	2
11	Write, type, proof, sign, copy, and distribute basic data acceptance letter	26.5	0.25	26.25
	Totals	267.8	151.1	116.7

evaluate the measurement system to identify key sources of measurement variation. What are they? Should this measurement process be used to monitor the process? If so, how and for what purposes? What steps would you recommend be taken to improve this measurement system?

15.   **Understanding process measurement variation.** To help you learn about measurement variation, try this experiential learning exercise. (We are indebted to Alan Goodman for bringing this exercise to our attention.)

You have started a new business providing height measurements of humans. Your customers expect accurate and precise measurements. You offer two methods of measurement: (1) yardstick or meterstick and (2) a tape measure. You need to test the two methods to evaluate their performance and provide the results to your customers.

Method 1 will be tested in this way: Ask a group of 20 or more people to measure the height of one person who is 6 feet, 2 inches or taller. The subject must stand in the middle of the room away from all walls. Each person will measure the height of the subject using the yardstick/meterstick and silently report the measurement to the data collector, who will tabulate the data and plot each measurement on a run or sequence chart. No deviation from the prescribed method is allowed.

In Method 2 you may use the same, or a different, group of 20 or more people and the same or a different subject. This time the group will use the tape measure in any way they desire. Again, each person silently reports the measurement of the subject to the data collector, who tabulates and plots each data point.

Now answer these questions about your data:

- Compare the accuracy and precision of the two methods using graphical and analytical methods. Which method was more accurate?
- Which method of measurement would you recommend? Why? Should different methods be used under different circumstances? Consider the role of different customer segments.
- Discuss the feelings the group(s) had when using the two methods. What were the differences between the two sets of feelings? Are these differences important?

16. **Processes on the Internet.** Search the Internet for processes and associated measurements. These may be in the form of flowcharts, verbal descriptions, or any other form of documentation. Consider searching for Six Sigma improvement projects that are good sources of process maps (see Chapters 1 and 4).

17. **Systems maps.** Construct a systems map of how global competition affects our society. Begin with the effects on business and then show how these effects influence you and your family and other aspects of our society such as education, health care, churches, and government.

18. **University as a system.** Draw a systems map for the university or college you currently attend or the highest level of school you have attended in the past. Show how the core processes of the educational institution fit together to educate students. Be careful not to omit any core processes: Education is more than going to classes and studying books.

REFERENCES

Automotive Industry Action Group. (1995). *Measurement systems analysis.* Suite 200, 26200 Lahser Road, Southfield, MI 48034.

Camp, R. C. (1989). *Benchmarking: The search for industry best practices that lead to superior performance.* Milwaukee, WI: Quality Press.

DuPont Engineering Standards Documentation Team. (1990, March). Personal communication to R. D. Snee.

Fernandez, M. M. (1998). *Measure alignment.* Paper presented to Metropolitan Section, The American Society for Quality, New York, November 19, 1998.

Fuller, F. T. (1985). Eliminating complexity from work: Improving productivity by enhancing quality. *National Productivity Review* (Autumn), 327–344.

Galloway, D. (1994). *Mapping work processes.* Milwaukee, WI: Quality Press.

Hammer, M. (1996). *Beyond reengineering.* New York, NY: Harper Business.

Hoerl, R. W., Hooper, J. H., Jacobs, P. J., & Lucas, J. M. (1993). Skills for industrial statisticians to survive and prosper in the emerging quality environment. *The American Statistician, 47,* 280–291.

IBM-Europe. (1990, June). Business process quality management. *Quality Today,* 10–11.

Juran, J. M., & Gryna, F. M. (1988). *Quality control handbook* (4th ed.). New York, NY: McGraw-Hill.

Kauffman, D. L. (1990). *Systems one: An introduction to systems thinking.* Minneapolis, MN: Future Systems, Inc.

Rummler, G. A., & Brache, A. P. (1995). *Improving performance; How to manage the white space on the organization chart* (2nd ed.). San Francisco, CA: Jossey-Bass.

Scholtes, P. R. (1998). *The leader's handbook.* New York, NY: McGraw-Hill.

Snee, R. D. (1983). Graphical analysis of process variation studies. *Journal of Quality Technology, 15,* 76–88.

Wheeler, D. J., & Lyday, R. W. (1990). *Evaluating the measuring process* (2nd ed.). Reading, MA: Addison-Wesley.

PART II

Improvement Strategies and Basic Tools

4

Process Improvement and Problem-Solving Strategies

To produce quality, you must have a system to improve it.

Thomas A. Edison

4.1 Overview

In this chapter we illustrate two detailed and specific approaches to applying the statistical thinking model presented in Chapter 2: the process improvement strategy and the problem-solving strategy. These are not the *only* possible approaches to applying statistical thinking, but they have proven effective in solving problems and making improvements in a wide variety of business processes. As the names imply, the *problem-solving strategy* is intended to address special-cause variation (i.e., fixing problems). The *process improvement strategy* is intended to address common-cause variation in processes that are stable but at an unsatisfactory level. These strategies use many of the same tools but in different ways. The Six Sigma process improvement approach (see Chapter 1 and Appendix D) is also discussed and compared to the process improvement strategy and the problem-solving strategy.

The most important learning objective in this chapter is to understand how the various tools fit together to form overall approaches to process improvement and problem solving. The reader should also be able to explain how these two strategies fit into the overall statistical thinking model. This understanding will be helpful in planning the rest of the course project.

We begin with two process improvement case studies. Using these as examples, we depict the process improvement strategy, indicating which tools are typically applied at each step. Two problem-solving case studies are then reviewed, and the problem-solving strategy is depicted in conjunction with the appropriate tools. Detailed instruction in the tools themselves will not be given until Chapter 5. By that time, you will have a good understanding of when and why each tool is applied.

4.2 CASE STUDY: Reducing Resin Output Variation

A production team at Ricoh's Numazu plant was in charge of production and inspection of raw materials used to make copy machine toner.* Their objective was to produce consistent resin quality and volume through daily operation of a sequential chemical process. The chemistry involved was extremely precise, resulting in tolerances on the order of one ten-thousandth of a gram. It was typical for the team to analyze and respond to various types of process data to achieve continuous improvement. A consistent, perplexing problem was that calculations of yield (ratio of actual production to theoretical output based on inputs) exceeded 1.0. Realizing that this was technically impossible, the team felt that the calculated values over 1.0 were due to excessive variation somewhere in the process. Their experience also suggested that overall process stability had a significant impact on resin quality. They therefore initiated a variation reduction project.

Figure 4.1 provides the macro-level flowchart developed by the team, which shows the major steps in PPC toner processing. Lots of raw materials go through two processing steps and then are divided in half. The two halves go through processing steps 3 and 4 separately. They are then automatically weighed and packed into large drums for sale.

After coming to a common understanding of the overall flow of the process, the team looked at stability by plotting the yield data over time; one example is shown in Figure 4.2. Although there was some detectable movement up and down, the overall yield was relatively stable, averaging between 99% and 101%, with several points above 100%. The time period in the middle of the graph that had the most points above 100% was investigated by the team, and they were able to verify that a drop in air pressure, since repaired, was the root cause. Because yield was simply output divided by a hypothetical number, the team decided to focus their remaining analysis on output.

The team's next step was to plot a histogram of recent output data (Figure 4.3). This plot displayed an unusual pattern, suggesting that two underlying distributions could be combined in the data. The two batches that resulted from the split following the second processing step (see Figure 4.1) were identified by the team as the likely source, and the histogram was stratified into individual histograms for the two batches. These data clearly revealed a consistent difference between the paired batches. In comparing the observed variation with the needs of the process, the team agreed on a target of 4300 kg, with a variation of plus or minus 5 kg, as their objective. These standards are noted on Figure 4.3.

---

*This case study is adapted from "The Quest for Higher Quality—The Deming Prize and Quality Control," by Ricoh, Ltd. See Imai (1986) for more details.

FIGURE 4.1 Resin Manufacturing Process

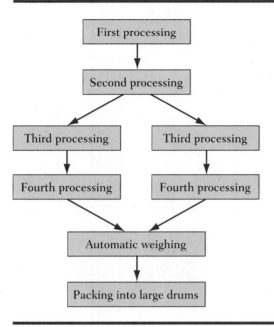

FIGURE 4.2 Yield Over Time

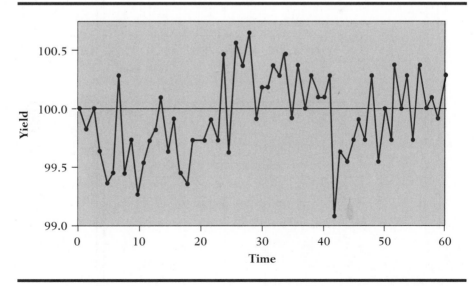

FIGURE 4.3 Resin Output Histograms, Combined and Stratified

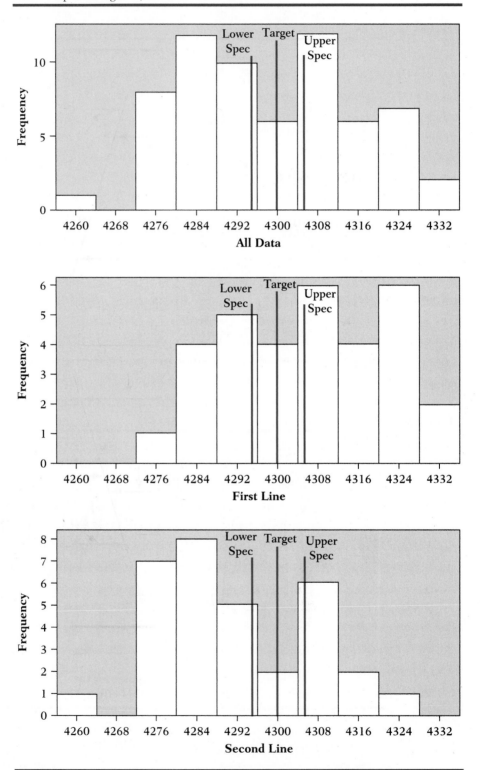

FIGURE 4.4
Cause-and-Effect Diagram for Variation in Resin Output

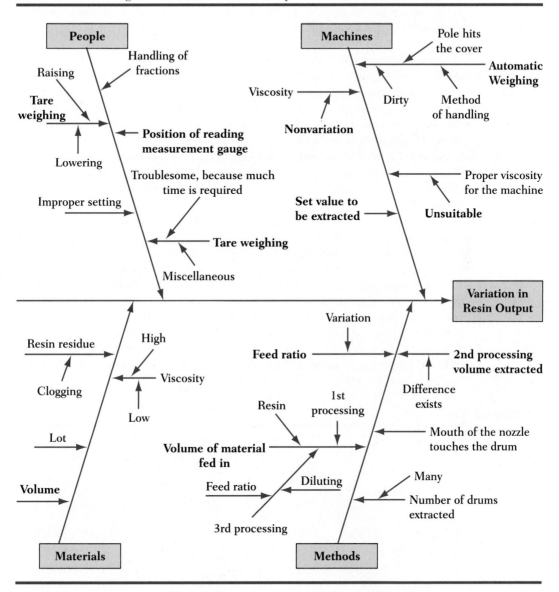

The team then brainstormed potential causes for the resin variation and documented their collective thinking in a cause-and-effect diagram (Figure 4.4). They highlighted areas that could explain the consistent difference between the paired batches. The highest priority issues selected by the team were the procedure for dividing the resin after the second processing step (most likely to be related to differences in the paired batches), the solvent feed ratio, and the in-process weighing, which was done manually by placing a container on a scale, whereas the final product was weighed automatically by an on-line scale.

Upon further investigation of the dividing step, it was noticed that some resin remained in the reaction tank, resulting in a lower output for the second batch. The dividing procedure was therefore changed to ensure an even split of the reaction tank. Subsequent data demonstrated no detectable difference between the two batches, but there was still an unacceptable level of variation in the process.

The team therefore focused on the next highest priority item on their cause-and-effect diagram, the feed ratio. Figure 4.5 shows a scatter plot of output versus feed ratio, which shows a relationship between the two. As the feed ratio went up, so did the measurement of output. The team did not believe actual output had increased, however, believing instead that only the measured value increased. Through investigation of the solvent feed, they determined that the feed ratio was affected by a measurement of specific gravity of the solvent. The specific gravity measurement was highly dependent on when the measurement was made, because the solvent took about 20 minutes to stabilize in the tank. The team changed the measurement procedure to ensure that the solvent had stabilized prior to measurement. Subsequent data demonstrated a reduced variation in feed ratio, and the dependence of measured output on feed ratio disappeared.

The output variation was still not at the desired level, however, and the team thought further improvement was possible, so they embarked on another round of variation countermeasures. The next issues they addressed were the weighing processes. For the in-process manual method, a person had to visually read a calibration line on the scale. The position read depended on the person's height, because different heights gave them different visual angles. To correct this problem, the

FIGURE 4.5   Output Versus Feed Ratio

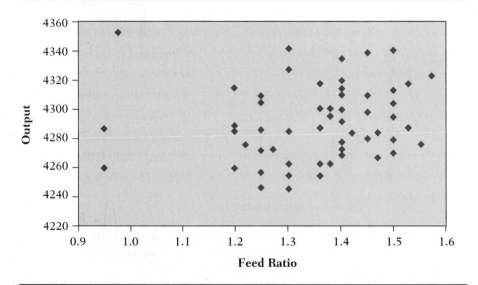

position of the calibration line was changed so that everyone had the same visual angle. This reduced in-process measurement variation.

In addition, the team noticed that on the final, automatic measurement, the pole of the scale and its connecting rod could physically touch the scale's protective cover, resulting in an inaccurate measurement. The protective cover was redesigned so that it would not come in contact with the moving parts of the scale. During this evaluation, very low output measures were recorded, which resulted in an emergency team meeting to diagnose the problem. Upon checking the scale again, it was noticed that contact points between the conveyor carrying the resin and the automatic scale were not in alignment, resulting in reduced sensitivity. This was corrected, and procedures were developed for checking the alignment on a regular basis.

As can be seen in Figure 4.6, which displays control charts of the average and range of batch outputs, the variation was significantly reduced, meeting their variation objective. (Control charts are basically run charts with statistically calculated limits that separate common- and special-cause variation. These will be discussed at length in Chapter 5.) Although the desired average of 4300 was not

FIGURE 4.6

Control Charts of Output Mean and Range by Time Period

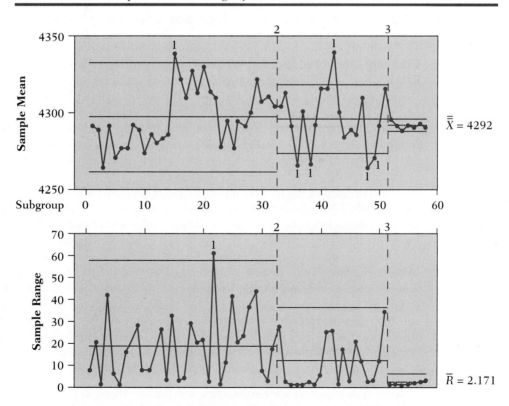

obtained, the resulting average of about 4292 was considered more than adequate, given the reduction in variation, which was the team's main objective. As an added benefit, the improvements in the process resulted in a reduction of variation in resin viscosity as well.

To standardize on the improvements and to prevent backsliding, the team implemented the following controls:

- Prepared a resin extraction procedures manual.
- Prepared a solvent insertion procedures manual.
- Revised the manual for synthesizing resin.
- Set a schedule for periodic assessment of the automatic weighing process.

Postscript   A few points are worth emphasizing from this case study:

1. Many improvement projects focus on reducing variation as opposed to changing the average. Excessive variation is often the key issue.
2. Calculating statistics on large data sets can be misleading. Very different subprocesses may be contained in the data, and these processes are not always comparable.
3. The measurement process is often a root cause of other problems.

4.3   CASE STUDY: Reducing Telephone Waiting Time at a Bank

The main office of a large bank averaged about 500 customer calls every day.* Customer surveys had shown that callers became irritated when the phone rang a long time prior to someone answering it. In contrast, a prompt answer gave customers a positive impression of the bank. The survey indicated that two rings or less was the customer expectation and that more than five rings was considered excessive. As a result of this survey, reduction of telephone waiting time was selected as a project. It was realized that this may be the first impression one has of the bank. In addition, the project coincided with a major initiative of the bank to be more friendly and responsive to customers.

The first step in the improvement process was to get a team organized, which included phone operators, and look at the way calls were currently handled. Figure 4.7 shows a frequent situation: A call from customer B came in, but the operator was still helping customer A find the appropriate receiving party for that call. This caused a delay in helping customer B. Two specific causes of delays were identified on the flowchart. The first problem uncovered was that the operator may not know where the customer should be connected. Additional time was then required to talk to the customer. Otherwise the call might be forwarded to the wrong person. The second problem was that the receiving party may not respond quickly, in which case the operator may have to try to locate them or find someone to take a message. The bank did not make widespread use of answering machines.

*This case study is adapted from "The Quest for Higher Quality—The Deming Prize and Quality Control," by Ricoh, Ltd. See Imai (1986) for more details.

FIGURE 4.7 Why Customers Had to Wait

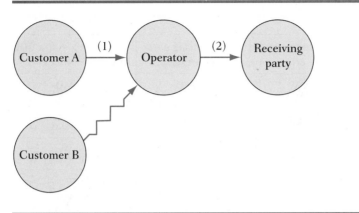

To document additional causes of delays, the team developed the cause-and-effect diagram shown in Figure 4.8, based on their experience. This is simply a listing, in greater and greater levels of detail, of the potential causes of the effect "makes customer wait." For 12 days each operator kept a formal checksheet to

FIGURE 4.8 What Makes Customers Wait

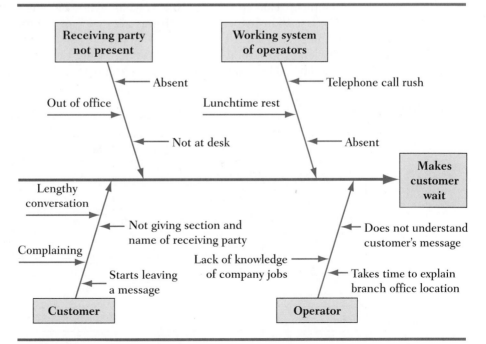

FIGURE 4.9
Checksheet Designed to Identify the Problems

Date \ Reason	No one present in the section receiving the call	Receiving party not present	Only one operator (partner out of the office)		Total
June 4	\\\\	卌 \	卌 卌 \		24
June 5	卌	卌 \\\	卌 卌 \\\\		32
June 6	卌 \	\\\	卌 卌 \\		28
June 15	卌	卌	卌 \\\		25

record the specific reasons for the delay of any calls that were answered after five rings. The results of this data collection are shown in the next three figures. Figure 4.9 is a checksheet. Operators added a check mark to the appropriate category each time a call took more than five rings to answer. Figure 4.10 summarizes these results, and Figure 4.11, a Pareto chart, is ordered by frequency. In this case, having only one operator was the most commonly occurring cause of delays.

The team agreed that when one operator was out of the office the operator on duty had too many calls to handle. An average of about 29 calls a day were not answered in less than five rings, which was about 6% of the total calls. With proper planning the team felt that virtually all the calls should be answerable in five rings or less, and several actions were taken to address the major causes of long waiting times. The first change was to make sure that there were always two

FIGURE 4.10   Reasons Callers Had to Wait

		Daily average	Total number
A	One operator (partner out of the office)	14.3	172
B	Receiving party not present	6.1	73
C	No one present in the section receiving the call	5.1	61
D	Section and name of receiving party not given	1.6	19
E	Inquiry about branch office locations	1.3	16
F	Other reasons	0.8	10
	Total	29.2	351

FIGURE 4.11 Reasons Callers Had to Wait

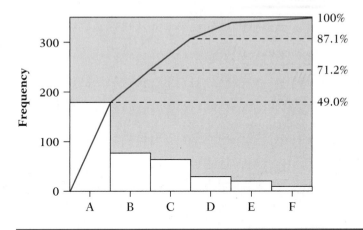

operators taking calls. To accomplish this, a clerical worker was brought in to cover for the regular operators when they were not available, such as over their lunch breaks. A second change was to institute a policy whereby all employees left a note on their desk as to where they had gone when they were not there. The objective was to simplify the task of trying to locate people who were not at their desks. A third action was to compile a notebook that listed each employee by their respective responsibilities. This enabled operators to find the appropriate person much more quickly.

A follow-up study, similar in design to the previous study, was undertaken to verify that these actions had produced significant improvements. The results are shown in Figures 4.12 and 4.13. Note that the overall reduction in long waiting

FIGURE 4.12 Effects of Improvement (Before and After)

	Reasons why callers had to wait	Total number Before After		Daily average Before After	
A	One operator (partner out of the office)	172	15	14.3	1.2
B	Receiving party not present	73	17	6.1	1.4
C	No one present in the section receiving the call	61	20	5.1	1.7
D	Section and name of receiving party not given	19	4	1.6	0.3
E	Inquiry about branch office locations	16	3	1.3	0.2
F	Other reasons	10	0	0.8	0
	Total	351	59	29.2	4.8

FIGURE 4.13   Effects of Improvement

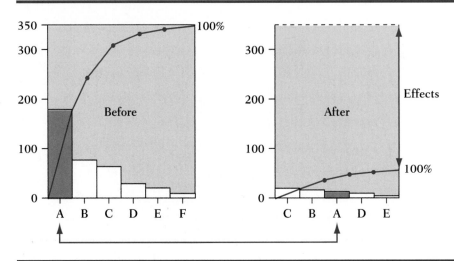

times was more than 80% and that the problem of having only one operator was reduced by more than 90%.

Postscript   A couple of points are worth emphasizing here:

1. A disciplined improvement approach is applicable virtually anywhere, even to processes that do not normally produce data.
2. Gathering data is an excellent way to gain team cohesion and focus everyone on the most important problem. Without data, everyone is an expert! Data also verify that improvement activities have actually produced the desired results.
3. It is important to standardize the improvements to prevent reoccurrence. For example, if no one is accountable for maintaining the notebook with everyone's responsibilities, it will soon become out of date and will not be helpful.

4.4   The Process Improvement Strategy

Although these two case studies are very different at first glance, they have some important similarities. Both clearly follow the overall statistical thinking model depicted in Chapter 2 and illustrate the key elements of this approach. For example, both viewed work as a process, gathered data in a sequential fashion, used subject matter knowledge to determine what data to collect and how to interpret it and revise original theories, and displayed understanding of variation and the dynamic nature of business processes. In addition, both utilized some of the same tools, such as flowcharts and cause-and-effect diagrams, in a consistent manner. This common approach is very similar to other published applications of statistical thinking.

We refer to this model as the process improvement strategy, and Figure 4.14 depicts the key steps in this approach, along with the tools typically used for each.

FIGURE 4.14   The Process Improvement Strategy

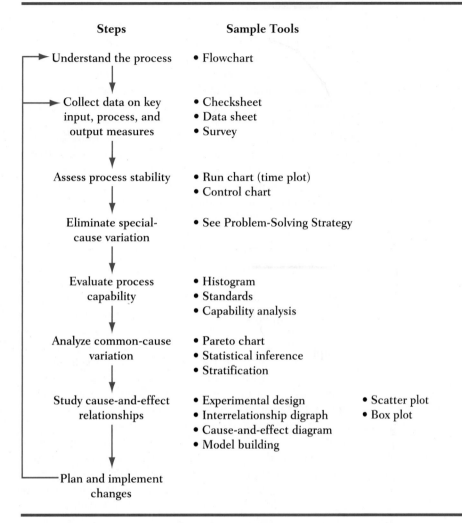

This list of tools is not exhaustive, nor is each tool used for every application. Note also that these steps are generic to any application and do not directly correspond to the specific steps in the case studies.

The first step is to document our understanding of the process involved using the techniques discussed in Chapter 3. This will generally involve a flowchart. Both case studies began with fairly high-level flowcharts. It is particularly important to flowchart nonphysical processes because one cannot physically see the process at work. We may also wish to benchmark how we perform this process by examining how competitors or recognized authorities perform a similar process.

The next step is to gather data on key outputs, inputs, and process variables. For measured or counted data (e.g., yield, output, or time to answer), this is typically recorded on paper or electronic data sheets. When the data are gathered from people, surveys or questionnaires are often used. Categorical data, where a variable falls in one of two or more categories, are typically recorded on a check-sheet such as that used in the telephone answering case. The cause of a call taking more than five rings to answer fell into one of six possible categories. Number of "waiting callers" would be considered an output measure, whereas cause of wait would be a process measure. Number of calls received could have been used as an input measure. As an aside, the team could have chosen as an objective to minimize the average number of rings before answer; then rings before answer would be the output measure.

At this point it is critically important to differentiate between stable or unstable processes. This is typically done with run charts, which plot data over time, or control charts, which are run charts with statistical limits placed on the chart. Stability generally implies a lack of special causes, in which case we need to study and understand the common-cause structure so we can fundamentally change the process. Instability generally implies the presence of special causes, in which case we need to identify and eliminate the root causes. Remember, stability does not imply that the process is performing satisfactorily, only that it is consistent over time.

Recall that the appropriate approach for addressing special causes is different from the approach for addressing common causes. (This was discussed at length in Section 2.5 and illustrated in Figure 1.4 in Chapter 1.) In short, special causes are generally few in number, have a large effect, and are related to a specific event. They are not a fundamental part of the system. Special causes are typically easier to identify and eliminate, and they produce the most dramatic short-term improvements. In other words, fixing special causes often represents "picking the low hanging fruit," at least in comparison to fundamentally changing the process. The elimination of special causes also results in a stable process, which makes study of the common-cause structure much easier.

Common causes are many in number, have a relatively small effect, and are a fundamental part of the process. In-depth study is typically required to understand them, and more drastic measures are required to address them. Resolving common-cause issues usually requires a fundamental change of the process. It is clearly important to distinguish between special and common causes because the process for improvement is different in each case.

In the resin variation study, the run chart (Figure 4.2) revealed that although there had been some movement of the process up and down, the recent data appeared stable—but at a theoretically impossible level (greater than 100% yield). Recall that a root cause for the unusually high data in the middle was identified and corrected. That is, the special cause was fixed. The team then looked at the rest of the data and found nothing unique in the weekly yields. Each was a typical output of the process. This common-cause variation led the team to study the whole process, with the intent of making a fundamental change.

In the case study involving telephone waiting time at the bank, an implicit assumption was made that the waiting times were stable. This led the team to rule out special causes in their analysis. Subject matter knowledge can sometimes be used in this fashion to skip a step in the strategy, but it is a good idea to gather data to verify these assumptions.

With stable processes, the next step is to evaluate process capability. This is typically done with histograms and standards. A histogram is a bar chart of the individual values. Standards are definitions of what we want the process to produce, and they are often stated as a target or the limits of variation (i.e., specifications, or "specs"). Standards may also be external, such as the performance level desired by the customer or the performance of a competitor. Plotting a histogram with standards noted on it allows us to see how close the actual average is to the target and whether the observed variation will fit within the desired limits. This helps diagnose whether we have an issue with the average, variation, or both. One can also calculate process capability indices such as "Cp" or "Cpk." These are ratios of acceptable variation divided by actual variation. A number of 1.0 or greater indicates that most, if not all, of the variation should fit within the standards or specification limits. (See Chapter 5 for more details on these process capability ratios.)

As demonstrated in the resin case, histograms can draw attention to other abnormalities in the data. Figure 4.3 showed an actual average slightly below the desired output of 4300 kg but with variation much greater than the desired ±5 kg. Note also that the height of the histogram decreases in the middle, near the average. Most histograms increase near the middle, producing the typical bell shape or so-called normal distribution. The appearance of two isolated peaks in this histogram indicates that we may have two separate processes combined and that the averages of the two processes are different.

Once we understand the actual performance versus the desired performance, the next logical step is to identify and analyze the major common causes. Tools such as stratification, statistical inference, and Pareto charts are helpful in analyzing common-cause variation. Stratification means breaking data sets down into logical subgroups to check for differences, and Pareto charts are bar charts of categorical data in decreasing order. Statistical inference consists of a number of statistical tools that infer characteristics of an overall process or population from sample data. This methodology is discussed in Chapters 8 and 9.

The resin team stratified their histogram because it appeared to have two peaks. Using subject matter knowledge, they identified the possible cause for this shape as combining two separate lots with different averages. By stratifying the data and plotting separate histograms for each lot, they confirmed this hypothesis and identified variation between lots as a key source of variation to be reduced. In the telephone case the Pareto chart of reasons for callers having to wait (see Figure 4.11) identified "one operator on duty (partner out of the office)" as the most frequently occurring cause for problem calls. The purpose of a Pareto chart is to objectively identify the most critical causes for a problem or issue when dealing with categorical data.

At this point, solutions may be obvious, or we may need to investigate further by identifying cause-and-effect relationships. Various tools can be useful here, including scatter plots, box plots, cause-and-effect diagrams, experimental design, interrelationship digraphs, or model building. *Scatter plots* are plots of two variables to detect relationships. *Box plots* are similar to scatter plots but summarize the distribution of the variable of interest (the "effect"). These and other plots will be reviewed in Chapter 5. *Experimental design* (Chapter 7) is a systematic approach, proactively manipulating and gathering data on several variables at once, revealing causal relationships. The *interrelationship digraph* (Chapter 5) is a knowledge-based tool used to indicate the causal direction of relationships (i.e., Does A cause B, or does B cause A?). *Model building* (Chapter 6) is a collection of various approaches to developing equations, or models, that explain and predict the behavior of a process.

The solutions to the major causes of delay in the telephone case were fairly straightforward, but the causes of resin output variation were more numerous and complicated. A cause-and-effect diagram was used to document the most probable causes and subcauses for variation in resin output (see Figure 4.4). The effect is noted on the far right—variation in resin output in this case—and major categories are shown as branches (people, machines, materials, and methods). More detailed causes in each of these categories are then added as subbranches, and so on. Because of the pattern of lines created, this diagram is sometimes referred to as a fishbone diagram. A cause-and-effect diagram was also used in the telephone case to identify the categories for the checksheet. This illustrates the fact that these tools can often be used for purposes other than as specific steps in the process improvement strategy. With experience in applying the tools to real problems, you will learn the most appropriate tool to use in a given situation. Fortunately, a variety of tools can provide the needed information for most real-world problems.

As one develops an understanding of causal relationships, it becomes easier to determine what changes would improve the process. In the telephone case, the critical changes were a modification to the lunch schedule and the addition of a relief operator from the clerical section. These changes ensured that two operators would always be present. Employees were also asked to leave messages when not attending their desks, and a personnel directory was compiled. These last two changes were implemented in an attempt to address secondary issues from the Pareto chart.

In the resin case several countermeasures were taken to reduce within-batch variation, including changing the volume extracted, defining a specific gravity measurement time, and revising work standards (procedures). The plots of feed ratio versus output are examples of scatter plots. In process improvement, several changes are typically needed to obtain the desired performance. Since the process is stable, no single special cause is producing the poor performance, hence there is no "silver bullet," as seen in these case studies. This reinforces the need to differentiate between special- and common-cause variation.

Once the process improvement strategy has been implemented, we need to verify the impact of changes by gathering additional data. To achieve the desired results, we may need to go through the process improvement strategy several times. Radical changes may warrant a revision to the original flowchart to document the new process. In both case studies additional data were gathered after implementing changes. The telephone waiting times had improved significantly, and the team was able to move on to other issues. Improvement was also made for resin output, but further improvements were necessary. This resulted in another loop through the strategy and another round of changes

The process improvement strategy in Figure 4.14 should be considered a more specific and detailed example of the statistical thinking model depicted in

Figure 2.14. Like the statistical thinking model, it begins with understanding the process, using a knowledge-based tool—flowcharts—rather than a data-based tool. The next step, collect data on key input, process, and output measures, uses this knowledge to determine what data should be collected. The third step, assess process stability, analyzes the data to diagnose the type of variation present—that is, to determine process stability. The conclusion reached adds to the subject matter knowledge about the process. The fifth step, evaluate process capability, requires additional data, or a different view of the existing data, and provides some tangible conclusion as to the ability of the stable process to achieve the desired standards. This step begins the next cycle of planning, analysis, and revision of subject matter knowledge depicted in Figure 2.14.

The sixth and seventh steps, analyze common-cause variation and study cause-and-effect relationships, often require several iterations through this same cycle. Each iteration is guided by the latest additions to the knowledge, based on the most recent data collected. The last step shown in Figure 4.14, plan and implement changes, is strictly based on knowledge and should be followed by a data-gathering step to ensure that the changes produced the desired results.

The process improvement strategy incorporates these key elements of the statistical thinking approach:

■ Improves results by improving the process.
■ Uses the synergy between subject matter knowledge and data.
■ Diagnoses and reduces variation.
■ Uses a sequential approach.

The process improvement strategy has been proven effective in improving stable processes in a wide variety of business applications. It will not always be the best approach to a given situation, however.

4.5 CASE STUDY: Resolving Customer Complaints of Baby Wipe Flushability

The Super-Wipe brand of baby wipes had been a very profitable business for Acme Baby Care Company.* The product provided acceptable performance in cleaning infants during diaper changing and was sold at a cost considerably lower than the leading brands. A key selling point, which Acme successfully marketed, was the flushability of the wipes. Unlike many competitive baby wipes, which had to be disposed of through trash removal or diaper services, Super-Wipe could be flushed down the toilet. This feature was achieved through use of a proprietary mix of natural and synthetic fibers.

This rosy picture turned into a disaster when the business team was informed that the Consumer Relations Department was being flooded with calls

---

*This case study is from the authors' consulting experience, and all the data presented are real. Because of the sensitive and proprietary nature of the problem, however, some details have been altered to protect confidentiality.

and letters complaining that Super-Wipe was not flushing properly. This was particularly perplexing because Acme had performed extensive septic system tests prior to originally introducing Super-Wipe, and it had passed with flying colors. A team of technical experts from research and development and manufacturing was assembled to quickly fix the problem. At the same time, consumer relations, marketing, and sales worked on a parallel path to maintain distributors' and consumers' faith in the product.

The team of experts had a diversity of areas of expertise, including specialties in synthetic fibers, natural fibers, processing chemicals, and manufacturing technology. Although they were under extreme managerial pressure to fix the problem before it resulted in a total marketing disaster, little progress was made initially as each expert pursued solutions in his or her individual technical area of expertise. A turning point occurred when the team jointly recognized that they really did not understand the problem well and would need to work cooperatively using a disciplined approach.

The manufacturing process was well understood by the team, and a flowchart (Figure 4.15) was developed to serve as a reference and to facilitate communica-

FIGURE 4.15   Super-Wipe Manufacturing Process

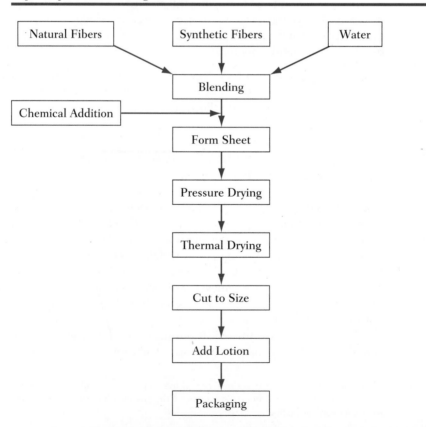

tion. In addition, the team obtained data on flushability complaints from consumer relations and plotted a run chart (Figure 4.16). This plot clearly showed that complaints had increased dramatically. Pinpointing the exact onset of the problem was difficult, however, because of the time lag between when the product was manufactured, when it reached retail stores, when it was purchased by consumers, when it was actually used, and when a complaint was received. Eventually the complaints were isolated to product from one manufacturing plant. Product produced at other plants did not receive high levels of complaints. This enabled the team to narrow their investigation somewhat.

But what was causing the increase in complaints? Most team members were very opinionated on this issue, and a structured brainstorming session was employed to capture everyone's ideas. No debate or discussion was allowed until all the ideas were captured. The team then discussed the various ideas about the root cause of the problem. These ideas included new synthetic fibers that had recently been added in the product, various chemicals being used, as well as some current process conditions. One idea was that the problem was related to introduction of low-volume toilets and was not associated with the product at all.

No clear consensus on the most probable cause was attained, and multivoting was used to prioritize the potential causes. This technique allows each participant to vote more than once on the issue. For example, each team member might give a 3 to their first choice, a 2 to their second choice, and a 1 to their third choice. The total points are then added up across all participants for each item being voted on, and a Pareto chart is made to prioritize the most likely causes. Obviously, subject matter knowledge must still be used to ensure that the most *probable* causes are identified rather than the most *popular* ones. This technique can be

FIGURE 4.16 Flushability Complaints by Quarter

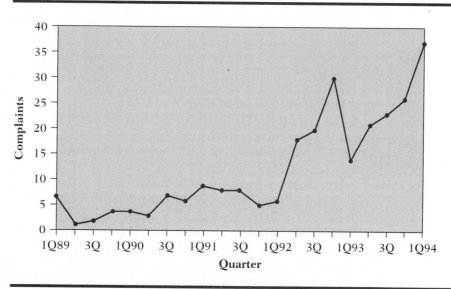

used to quickly identify the big hitters to investigate first. Use of multivoting in this case resulted in the following highest priority items:

- New synthetic fibers
- Use of the chemical X-Pro
- Chemical surfactants used in the process
- Trace chemicals in the plant water supply

All of these except the plant water supply were known to have changed around the time the complaints began to increase. X-Pro is a chemical that was added to the synthetic/natural fiber mixture when it was in a wet state to enhance the strength of the wipe and to aid mixing. Surfactants are chemical soaps added to the product to make it feel smoother and clean better. Another complicating factor was that there was no agreed-upon method of measuring flushability. The long-term septic system tests done prior to introducing the product were not considered practical to use on a routine basis. The team added a fifth high-priority action item to the top four potential root causes: Develop a simple, repeatable, flushability test.

The team split into five task teams, with some overlapping membership. Each pursued one of the five priority action items. The flushability measurement team had the highest urgency because progress on the other four would be held up by an inability to quantify flushability. This team visited the homes of some people who had complained to observe the problem firsthand. After several days of observing flushing, and even opening septic tanks, they concluded that Super-Wipe was definitely not flushing as designed. The problem could not be blamed on low volume toilets. In addition, it was observed that some fibers were floating indefinitely, even after the wipe had broken apart. This was very surprising because all the fibers used were denser than water and should eventually sink.

This discovery piqued everyone's interest and led the task team to consider whether the problem was flushability or simply floating fibers. The team began developing both a flushability test and a float test. The final flushability test measured how long the wipe took to disintegrate in water with mild agitation; the float test measured the percentage of fibers still floating after a fixed time immersed in water. Once these tests were developed and documented, baseline data on Super-Wipe and its flushable competitors were gathered and displayed in histograms (Figures 4.17 and 4.18). Clearly, Super-Wipe had both a flushability and a floating problem versus its competition.

At this point the other teams became more active. The synthetic fiber team designed an experiment to look at the effect of varying levels of three types of synthetic fiber. Four conditions, or cells, were manufactured, as shown in Table 4.1. The logic behind this design can be more easily appreciated by looking at the four cells geometrically, as in Figure 4.19. This is half of a "full" design, which would have included all eight vertices of the cube. Note that half of the conditions are with high levels of fiber 1, half with low levels. The same is true for fibers 2 and 3. With the same four data points, we can compare two observations at the high level with two at the low level of each variable. Although it appears that fibers do have an effect, Comparing the table data to the two histograms clearly shows that fiber is not the reason for the great disparity versus the competition.

FIGURE 4.17   Flushability Histogram

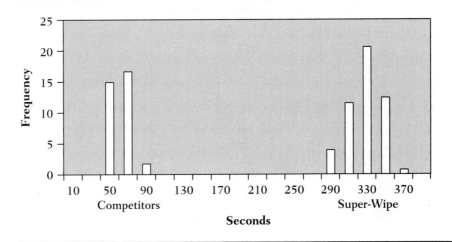

FIGURE 4.18   Float Test Histogram

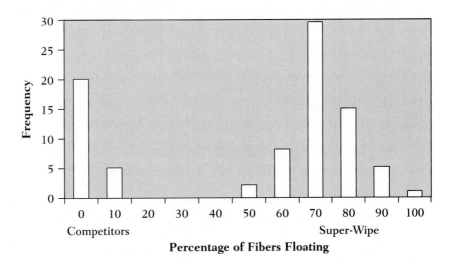

TABLE 4.1   Effects of Various Levels of Synthetic Fibers

	Design				Results	
Conditions	*Fiber 1*	*Fiber 2*	*Fiber 3*	*Filler*	*Average Flushability*	*Average Floating*
1	25%	0%	25%	50%	243 sec	65%
2	0%	25%	25%	50	372 sec	67%
3	25%	25%	0%	50%	262 sec	66%
4	0%	0%	0%	100%	235 sec	57%

FIGURE 4.19   Experimental Design for Studying the Effects of Three Fibers
on the Performance of Baby Wipes

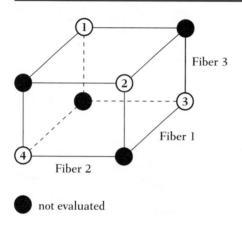

Fiber 3

Fiber 1

Fiber 2

● not evaluated

The X-Pro team had discussions with the vendor of this chemical, who
noted that X-Pro was not intended to be added while the fiber mixture was still
wet. It should be sprayed on in a later processing step. The vendor could not explain how this might cause flushability or floating problems, however. The team
planned to experiment with removal of X-Pro, but the plant canceled plans for
the experiment because they were concerned that removing X-Pro would hurt
productivity. Running laboratory experiments with and without X-Pro was not
considered useful because the issue was *where* in the process X-Pro was added,
not the presence of it. It was not considered feasible to replicate each of the process steps in the laboratory.

The water supply and surfactant teams both ran laboratory experiments with
these ingredients and found no impact on flushability or floating.

At this point suspicion focused on X-Pro by a process of elimination. While
waiting for a shutdown so that spraying equipment could be installed, one team
member began tinkering in the lab with various fibers and X-Pro. The data he generated are given in Figure 4.20. He felt that these data indicated both an X-Pro effect and a fiber effect, or perhaps an interaction between the two. Some team
members were willing to reconsider fiber as a potential root cause, but others were
skeptical of the data. There was a great deal of disagreement as to how to proceed,
which resulted in minimal progress.

The logjam was quickly shattered when the process had to be shut down
briefly due to a mechanical problem. When it was started back up, the flushability
and float test numbers showed significant improvement, at about the same level
as competition. Upon investigating this special cause, a team member at the site
discovered that the X-Pro line had accidentally been left in water-flush mode. The
process instrumentation indicated a flow of X-Pro, but water was actually being
pumped into the process. Faced with this evidence, the entire team was convinced
that adding X-Pro in the wet state was the culprit, and the process was shut down
to install the spraying equipment.

FIGURE 4.20   Additional Laboratory Data

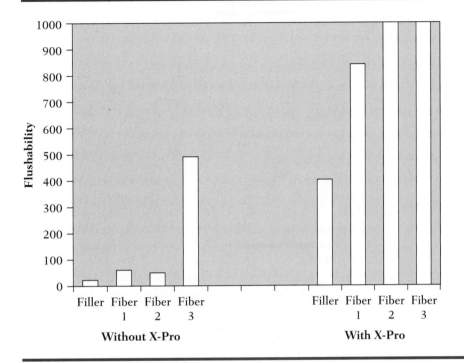

When the process was restarted, X-Pro was sprayed on according to the vendor's instructions and no flushability or productivity problems resulted. To protect against reoccurrence, the plant continued flushability testing on a periodic basis, but float tests were discontinued. Float tests are more time consuming, and floating only occurred with flushability problems, so testing for flushability was adequate. A technical explanation of the floating phenomenon was never determined.

Postscript   Several points illustrated in this case study are worth repeating:

1. Experts have a tendency to suggest solutions even before the problem has been documented and agreed upon. Careful documentation of the problem and use of a disciplined approach are critical to effective problem solving.
2. Difficulty in obtaining accurate, meaningful measurements is a common problem. Validation of the measurement process should be addressed early in the project.
3. Teamwork requires more than a collection of experts. Each member must embrace a common set of goals and work cooperatively to achieve them. (Appendix A contains more detail on effective teamwork.)
4. Putting permanent procedures in place to test for flushability was a critical step to rapidly detect reoccurrence of the problem. Without this step, the same problems tend to be solved over and over again. This is particularly true in organizations that value heroic "fire fighting" efforts more than planning and prevention.

4.6   CASE STUDY: The Realized Revenue Fiasco

The marketing manager for At-Care brand surgical supplies was in hot water.*
Her December net realized revenue had come in well below budget, throwing off
her fourth quarter results. The numbers were so bad that they caused the entire
Health Care Division of Atlas Industries to miss its budget for the year. Worst of
all, her boss, the Vice President of the Health Care Division, lost his annual bonus because of this. The marketing manager was given very clear direction that
this type of miss could never occur again.

The marketing manager's original reason for seeking help was to develop a
realized revenue forecasting model so that there would be no surprises in the future. We began with a discussion of how realized revenue is calculated, which led
to the flowchart in Figure 4.21. At-Care products were sold to large distributors
rather than directly to hospitals. To hide its pricing policies from the competition,
as well as to maximize profit, At-Care's list prices were rather high. Almost all sales
were originally booked at full list price, but distributors were offered a number of
discounts through a complicated rebate system. This approach enabled At-Care to
sell to major final customers at competitive prices (after rebate) while obtaining
high margins from smaller customers, or one-time buyers.

The rebate system was based on the final customer, typically hospitals, to
which the distributor sold. Some hospitals were part of nationwide networks that
negotiated volume discounts directly with Atlas. If distributors sold to these customers at discounted retail prices, Atlas would rebate the difference (retail minus
actual) to the distributor. The distributor would send copies of the sales contract
to the rebate center, and Atlas would send them a voucher for the appropriate
amount. This voucher could be deducted from the next Atlas invoice paid by the
distributor. Other allowable retail discounts included volume purchases by an individual hospital, special discounts to meet competition, or marketing promotions.

Although this arrangement was an inconvenience, it did have several advantages for Atlas. It allowed the sales force to focus on selling to distributors. It also
provided a great deal of market research information when the rebate center recorded exactly how much of which products each national hospital network purchased. In addition, as previously noted, it was very difficult for competitors to
determine Atlas's pricing for individual products. It was an administrative head-
ache to distributors, but it did enable them to benefit from Atlas's national contracts because these hospitals had to purchase from one of these large distributors
to get their discount.

When asked for data, the marketing manager supplied extensive data files
for the year, which contained detailed monthly information on sales, any price
discounts given directly to the distributor, and each specific type of rebate. This
information was broken down by customer, distributor, sales territory, sales region, and by several product groupings. In fact, there was so much data that it
was intimidating, and most people only looked at the bottom line, net realized
revenue. Due to the panic over missing fourth quarter earnings, the data had

---

*This case is an actual application from the authors' consulting experience. Because of the proprietary
nature of the pricing and rebate policies, some details have been altered and data have been disguised.

FIGURE 4.21
Calculation of Net Realized Revenue (NRR)

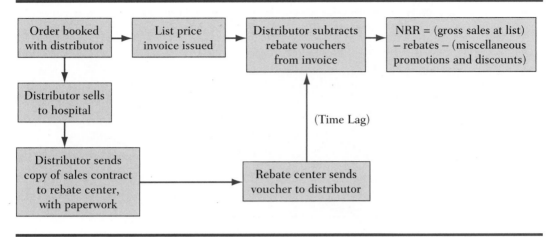

been researched sufficiently to suggest that unusually high rebates in December were the primary cause of low net realized revenue. Surprisingly, none of this data had been plotted.

A run chart of net realized revenue for the year is given in Figure 4.22. Clearly, December was a special-cause month. In discussing potential root causes, the marketing manager pointed to December rebates, which were higher than she could remember. A run chart of total rebates is given in Figure 4.23. December

FIGURE 4.22 Net Realized Revenue, One Year

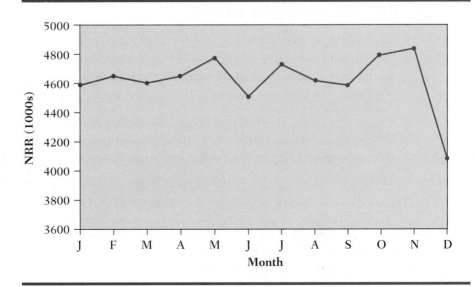

FIGURE 4.23   Total Rebates, One Year

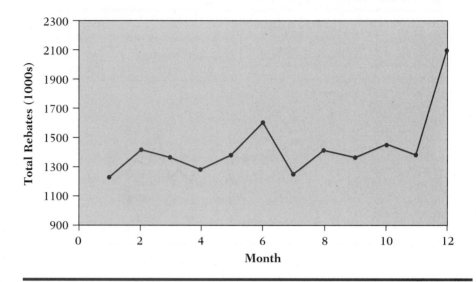

was also a special-cause month for rebates. This did not seem to be a seasonal effect, because distributors would naturally want their rebates as soon as possible. December rebates were high across virtually all customers, distributors, sales regions, and product lines. This fact ruled out many potential root causes. In addition, it was believed that rebating lagged the original sale by 1 to 3 months due to the flow of paperwork. It was suggested that a statistical forecasting model could be developed to model the lags and forecast rebates based on recent sales figures. Several time dependent, or "time series," models were attempted, but none provided any useful degree of prediction.

At this point we went back to the drawing board and took another look at the data. The more extensive historical files used in the attempt at time series modeling showed some interesting behavior. Figure 4.24 shows a run chart of rebates for the past 3 years. Note that December and June, to a lesser degree, were high each year. The marketing manager was shocked to see this plot, because she did not remember net realized revenue being low in December of the previous 2 years. Figure 4.25 reveals that it was not. In contemplating how this could be, she asked to see gross sales data (Figure 4.26). This revealed that unusually high December sales the previous 2 years had masked high rebates, resulting in an apparently typical month.

When questioned about the sales data, she noted that the past two Decembers the sales force had been asked to peak, or load. This is a common business practice whereby product for the following year is sold in December to inflate sales figures for the current year. Various incentives are given to customers to entice them to buy in December, such as significantly discounted pricing, carrying the inventory costs of the product until the customer actually needs it, and extending the billing terms so customers do not pay until they use the product.

FIGURE 4.24 Total Rebates, Three Years

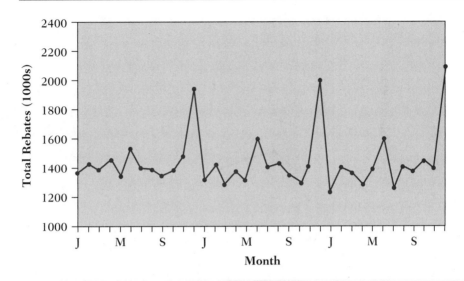

FIGURE 4.25 Net Realized Revenue, Three Years

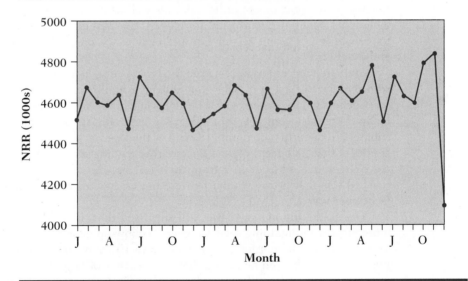

When asked about this, the marketing manager said they were doing well versus budget last December and did not need to peak to achieve their annual goals. Peaking incurs significant costs elsewhere in the system, and the manager was also concerned that if she exceeded budget by too large a margin, her budget would be set much higher next year.

FIGURE 4.26 Gross Sales, Three Years

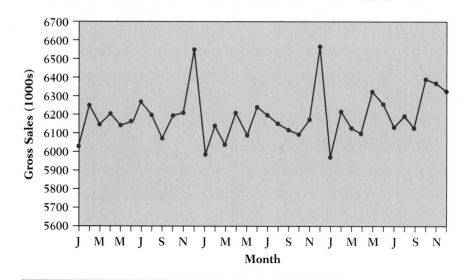

This explained why the high December rebates had not been detected previously, but it did not explain why they were occurring. Scatter plots of December rebates versus previous months' sales, as well as more sophisticated modeling techniques, had failed to provide any clue. At this point it was decided to travel to the rebate center, about 20 miles away, to talk to the people handling the rebates and vouchers.

At the rebate center we reviewed the process in greater detail with the people who actually worked in it. They had noticed a significant increase in paperwork for vouchers in late November. Reviewing some of the actual files, we noticed that many of the sales contracts sent in were several months old. It appeared that distributors frequently waited until the end of the year to fill out paperwork for sales contracts that had accumulated for months, sending them in just in time to receive the vouchers to deduct from their last bill for the year. The conclusion reached was that previous months were not as good as they appeared and that December was not as bad as it looked.

The marketing manager was still surprised that distributors would not be more prompt in obtaining their vouchers. Follow-up calls to key distributors revealed that they did not have enough employees to stay on top of the paperwork. Even though it delayed obtaining their rebate, they often put off completing the necessary paperwork until year end, or midyear if they were on a fiscal versus calendar year. Distributors on a fiscal (July–June) year explained the smaller increase in rebates in June.

The root cause of this problem was a predictable special cause—that is, one that could be expected to occur every December until the process was changed. In Chapter 2 we referred to this as structural variation. Structural variation cannot be fixed without changing the process itself, in this case, the rebate system. Although peaking could partially offset high rebates, it is an expensive business practice. It often hurts future sales at least as much as it helps December sales, and it requires significant giveaways to entice customers to buy in December.

The marketing manager decided to completely replace the information system being used. The new system would feature electronic billing and payment and would enable major distributors to send electronic copies of sales contracts with hospitals to the rebate center. The software would automatically determine any rebate, subtract it out of the next invoice to the distributor, and deduct this amount from the current month's net realized revenue. This solution is similar to what is commonly called reengineering. Reengineering is an improvement activity that attempts to wipe the slate clean and totally redesign a process from scratch, often using state of the art information technology. In following up with the marketing manager about a year and a half later, we learned that the new information system had resolved the problem of high June and December rebates and also had provided benefits in other areas.

> **Postscript** Several points are worth repeating relative to this case study:

1. A great deal of pertinent information was available, but it was not being used. In particular, none of the data had been plotted. The root cause of the problem was easy to diagnose when some simple plots were reviewed. These plots made sophisticated modeling, which was the marketing manager's original request, unnecessary. This confirmed the old saying (attributed to Yogi Berra) that "You can see a lot just by looking."
2. The problem went undiagnosed for years because of the lack of a process view. Until the December fiasco, people only looked at the key output—net realized revenue—and ignored indicators of the health of the process, such as sales, margins, and rebates. Although improved process outputs are our objective, we need to carefully manage the process to achieve these outputs.
3. Managing to budget is often inconsistent with the concept of continuous improvement. As in this case, the budget can be a restraint that prevents people from achieving even better results. Opportunities for improvement were overlooked as long as it appeared the budget would be achieved.
4. When looking at monthly data, it is helpful to plot over several years to diagnose potential seasonality. The obvious seasonal pattern in rebates was originally missed because only the current year's data were plotted.
5. To understand the finer details of a business process it is often very helpful to talk with the people who actually work in the process on a daily basis.

4.7 The Problem-Solving Strategy

All of the case studies in this chapter employed sequential approaches, viewed work as a process, made use of several tools, integrated data collection and analysis with subject matter knowledge, and recognized the importance of understanding and reducing variation. Upon closer examination, however, there is a critically important difference. The process improvement case studies developed deeper understanding of the *normal behavior* of the process so that it could be *fundamentally improved.* In contrast, the problem-solving case studies identified and diagnosed the root cause of *abnormal behavior* in the process so that it could be *returned to normal.*

Process improvement is required to take a stable process (i.e., one with common-cause variation) to a new level. Problem solving is required to eliminate the special causes in an unstable process. Because the special causes are, by definition, not an inherent part of the process, they can generally be eliminated without fundamentally changing the process. For this reason, it generally makes sense to eliminate the special causes first and then fundamentally improve the process. This sequence is depicted in Figure 4.14, where problem solving is applied to special-cause variation prior to the more in-depth process improvement steps.

The problem-solving strategy is depicted in Figure 4.27. This strategy is intended to follow understanding the process, gathering data on key process mea-

FIGURE 4.27   The Problem-Solving Strategy

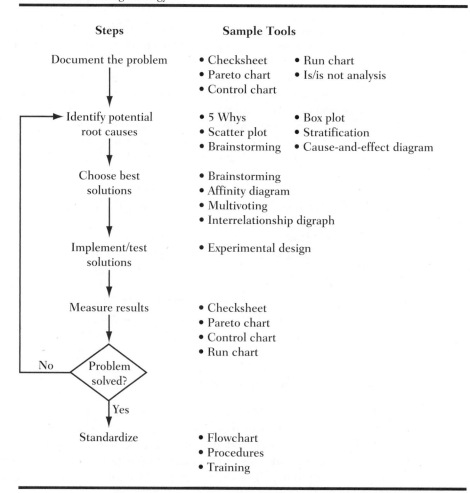

Steps	Sample Tools	
Document the problem	• Checksheet	• Run chart
	• Pareto chart	• Is/is not analysis
	• Control chart	
Identify potential root causes	• 5 Whys	• Box plot
	• Scatter plot	• Stratification
	• Brainstorming	• Cause-and-effect diagram
Choose best solutions	• Brainstorming	
	• Affinity diagram	
	• Multivoting	
	• Interrelationship digraph	
Implement/test solutions	• Experimental design	
Measure results	• Checksheet	
	• Pareto chart	
	• Control chart	
	• Run chart	
Problem solved? (No / Yes)		
Standardize	• Flowchart	
	• Procedures	
	• Training	

sures and outputs, and determining that the process is unstable. The emphasis in problem solving is on identifying root causes of abnormal behavior, therefore we focus on the abnormal behavior rather than the overall process, as we did in process improvement. This requires very thoughtful, thorough documentation of the problem and all related symptoms. The same run or control chart that detected the special cause can help us diagnose exactly when it occurred and whether it was evidenced by a gradual or sudden change. The run chart of complaints in the baby wipe case revealed the approximate time of a sudden increase in complaints and significantly narrowed the list of potential root causes. Checksheets and Pareto charts may be used to reveal changes in attribute data. In fact, a checksheet was used by the consumer relations telephone operators to obtain the complaint data used in the Pareto chart.

"Is/is not" analysis is used extensively in problem-solving strategies developed by Kepner and Tregoe (1979). This very practical technique suggests potential root causes by documenting where, when, what, and who the problem *is,* versus where, when, what, and who it could be, but *is not.* For example, the baby wipe flushability problem was at one Acme plant (where), but not at others, although it could have been. This information suggested that the root cause of the problem was related to a difference between this one plant and the others rather than a circuitwide cause such as common raw materials used by all. Such detailed documentation is extremely helpful in suggesting and testing various ideas that attempt to explain the abnormal behavior. Noting that December revenue was low across all sales regions and product lines saved time in finding the true root cause in the realized revenue case. For example, it was not necessary to look at differences from one sales region to another.

Once the problem is thoroughly documented, it is easier to identify potential root causes. A common mistake is to attempt to solve a problem that is not well defined or documented. Several tools can be helpful, including structured brainstorming. Some previously mentioned tools are also helpful in suggesting potential root causes. These include a mixture of knowledge-based tools, such as cause-and-effect diagrams, and data-based tools, such as stratification and scatter plots.

The "5 Whys" is a technique to go from symptoms to root cause. As a rule of thumb, we may have to ask the question "Why?" five times before we get to the root cause. Suppose we are investigating the cause of a hospital employee being injured at work. Why (1) was Diane injured? She slipped and broke her hip. Why (2) did she slip? She stepped in a puddle of water on the floor. Why (3) was a puddle of water on the floor? The water fountain was leaking. Why (4) was the water fountain leaking? We canceled our maintenance contract. Why (5) did we cancel our maintenance contract? The answer to this question is much more likely to reveal the *root* cause of the accident than is the fact that Diane slipped in a puddle of water. Stopping at question two might lead one to simply mop up the puddle, which would not prevent reoccurrence. Similarly, declining market shares, increased logistics costs, or declining price realization are symptoms rather that root causes of poor business results. In the revenue case, high rebates were a cause—but not the root cause—of low realized revenue.

Often a large number of plausible causes will be identified. The next step is to select the most likely potential solutions to these causes. Some solutions may address several potential causes. The key point at this step is to get the whole team aligned on a common path forward. This prevents several people from pursuing independent solutions in a haphazard manner, which often makes matters worse. Several soft tools are helpful here, including affinity diagrams, interrelationship digraphs, and multivoting.

Affinity diagrams are basically cause-and-effect diagrams in reverse. Rather than starting with known major categories and attempting to identify more detailed individual causes, we begin with a laundry list of unorganized, perhaps overlapping ideas and attempt to organize them into logical categories. This is particularly useful following a brainstorming session that produces a large number of ideas. It may not be feasible to follow up on all of these ideas individually.

An interrelationship digraph can then be applied to the major categories to prioritize them by determining which ideas are the primary drivers of the others, or the leverage points. Multivoting allows everyone equal input and often produces a clear consensus of priority items. This was the case in the baby wipe case.

The prioritized potential solutions may be relatively easy to implement, in which case they are generally implemented fully. In other cases, such as for baby wipes, full implementation would be quite expensive. For example, replacing all of the low cost fibers with more expensive ones could have cost hundreds of thousands of dollars. In these types of situations it is more sensible to test solutions on a trial basis. Experimental design techniques are the best known approach to learning about the effect of several factors with a minimal investment of time, money, and effort. Using this approach, the impact of three suspect fibers was quickly evaluated at a very low cost. Although none of these proved to be the root cause of flushability problems, this discovery enabled the team to move on to other potential solutions.

After permanent or trial changes have been implemented, it is important to quantify and document the results. Experimental design includes a number of graphical methods for interpreting results. For permanent changes we may use run or control charts for continuous data, such as money or cycle time, and checksheets and Pareto charts for attribute data, such as causes for complaints or injuries. These data will indicate whether the implemented solution actually solved the problem. A common mistake is to declare success after implementing a proposed solution before obtaining evidence that the problem has actually been solved. If the problem has not been solved, we need to reloop to select the next highest priority of potential causes and try something else. As in the baby wipe case, we may finally find the true cause by a process of elimination.

Even if the problem was solved by this solution, we are still not finished. It is very important to standardize the solution to prevent reoccurrence. This crucial step prevents having to solve the same problem over and over again, a common phenomenon. Documenting proper work procedures, often with flowcharts, and training employees in the use of these procedures are helpful in standardization. For example, if Acme had not instituted a procedure for routine measurement of flushability and for proper application of X-Pro, the same problem could have reoccurred in a couple of years and gone undetected for some time. Without formal standardization, the gradual turnover of people knowledgeable of the problem will frequently lead to reoccurrence.

The problem-solving strategy, like the process improvement strategy, should be viewed as a more specific example of the statistical thinking model (Figure 2.14). When we initiate the problem-solving strategy, we have already obtained data that indicate the presence of special causes. The Document the Problem step organizes the analysis of these data to enable easier diagnosis of the root cause. In the Identify Potential Root Causes and Choose Best Solutions steps we interpret these data carefully to refine and update any subject matter knowledge.

Accurately identifying root causes is primarily dependent on subject matter knowledge. In other words, we rely on our knowledge of the process to determine possible causes that could explain the specifics of the problem documented in the previous step. In some cases we may use data-based tools such as scatter plots or stratification to suggest potential causes. Use of these tools would comprise a separate cycle in the model. Once the list of potential causes has been identified, we typically need to prioritize it to select the best potential solutions. This is because the list of potential causes, even after testing them against the specifics of the problem, is often rather long. We usually do not have the resources to follow up on each one. If the prioritized solutions are tested on a trial basis rather than fully implemented, experimental design can provide detailed guidelines as to which data would be most informative.

The Measure Results step uses actual data to determine whether the problem has been solved or not. If it has not, then we must identify and test other potential solutions. The data collected here may provide further insight on root causes rather than just providing yes/no answers. This occurred in the baby wipe case when investigators noticed that fibers were floating indefinitely, to their great surprise. The Standardize step ensures that the original root cause will not reoccur.

The problem-solving strategy incorporates the key elements of the statistical thinking approach:

- Improves results by improving the process.
- Uses the synergy between subject matter knowledge and data.
- Diagnoses and reduces variation.
- Uses a sequential approach.

Each iteration between data and hypothesis increases and refines our subject matter knowledge. A fringe benefit of going through the problem-solving strategy is the knowledge gained about the process. This knowledge can prove valuable in the future when using the process improvement or problem-solving strategies or during routine management of the process. Table 4.2 summarizes the relationships between these two strategies and the statistical thinking model.

4.8 The Six Sigma Process Improvement Strategy

The Six Sigma approach introduced in Chapter 1 is also a process improvement strategy. It consists of four phases: **Measure**, **Analyze**, **Improve**, and **Control**. This is fondly referred to as the MAIC breakthrough methodology. General Electric has added a beginning step, **Define**, in which we clarify our objectives,

TABLE 4.2   Relationship of Process Improvement and Problem-Solving Strategies to the Statistical Thinking Model

Steps in statistical thinking model	Steps in process improvement strategy	Steps in problem-solving strategy
Determine need for data, based on subject matter knowledge	Understand process	Accomplished through process improvement
Obtain data	Collect data	Accomplished through process improvement
Analyze data	Assess process stability	Document problem
Interpret data in light of subject matter knowledge	Choose process improvement or problem-solving strategy	Identify root causes and prioritize solutions
Determine need for additional data	Evaluate process capability	Implement/test solutions
Reloop through model steps	Analyze common-cause variation	Measure results
Reloop through model steps	Study causal relationships	Identify additional potential root causes, if necessary
Reloop through model steps	Plan and implement changes	Identify additional potential root causes, if necessary
Reloop through model steps	Reloop through strategy, if necessary	Identify additional potential root causes, if necessary

constraints, and resources available to work on the problem. Here is a brief description of each phase:

Define. Select project and define the problem. Key process metrics are used to guide project selection. The resulting project and its objectives are summarized in a project charter.

Measure. Select the appropriate outputs to be improved, based on customer input. Ensure that they are quantifiable and that we can accurately measure them. Determine what is unacceptable performance (i.e., a defect). Gather preliminary data to evaluate current performance.

Analyze. Analyze the preliminary data to document current performance (baseline process capability) and to begin to identify root causes of defects and their impact. Obtain additional data as needed.

Improve. Determine how to intervene in the process to significantly reduce the defect levels. Several rounds of improvements may be required.

TABLE 4.3
Typical Six Sigma Black Belt Training Curriculum

Week 1—Measure	*Week 3—Improve*
Six Sigma overview	Design of experiments
Process improvement plans	■ Factorial experiments
The MAIC roadmap	■ Fractional factorials
Process mapping	■ Balanced block designs
QFD (quality function deployment)	■ EVOP
FMEA (failure mode and effects analysis)	■ Response surface designs
Organizational effectiveness	ANOVA
Basic statistics using Minitab	Regression (multiple)
Process capability	Facilitation tools
Measurement systems analysis	
	Week 4—Control
Week 2—Analyze	Control plans
Review of week 1 topics	Statistical process control
Statistical thinking	Advanced process control
Hypothesis testing (*F*, *t*, etc.)	Mistake-proofing
Correlation	Team development
Passive multivariable analysis	Wrap-up of tools
Regression (simple)	
Team assessment	

Notes:
1. Project reviews are done each day in weeks 2–4
2. Hands-on exercises are done on most days.
3. Allow three weeks of time to work on the project between sessions.
Source: Hahn, Hill, Hoerl, & Zinkgraf, 1999.

Control. Once the desired improvements have been made, put some type of system in place to ensure the improvements are sustained, even though additional Six Sigma resources may no longer be focused on the problem.

The tools used in the Measure, Analyze, Improve, and Control steps are summarized in Table 4.3. This table documents typical training given to individuals who lead improvement teams in Six Sigma and who are often referred to as "Black Belts." These tools are sequenced and linked in a unique way that enables the user to identify the appropriate tools to use in the appropriate phase of Six Sigma. The results of the use of each tool become the input to the application of one or more other tools. A more detailed discussion of Six Sigma is contained in Appendix D.

The Six Sigma approach has a lot in common with the process improvement and problem-solving strategies. Statistical thinking is fundamental to Six Sigma. It is a structured methodology with tools integrated with the different phases of improvement. Six Sigma differs from other approaches in that it focuses on the extensive use of fewer tools that are linked and sequenced in a very disciplined way. Six Sigma also uses failure mode and effects analysis (FMEA) and control plan tools not typically used in improvement and problem-solving approaches. FMEA enables us to do a risk assessment of the process and identify what happens when something in the process goes wrong, and helps identify problem fixes and appropriate reaction plans. The use of the control plan enables Six Sigma to "hold the gains" better than other improvement and problem-solving methods.

4.9 Summary

1. There are many specific paths to apply the statistical thinking model.
2. The process improvement and problem-solving strategies are two such approaches.
3. The process improvement strategy has been proven effective in fundamentally improving a wide diversity of stable processes (i.e., addressing common-cause variation).
4. The problem-solving strategy has been proven effective in identifying and fixing root causes of problems in a wide diversity of unstable processes (i.e., addressing special-cause variation).
5. Both strategies involve sequential use of various data and knowledge-based tools in a logical order.
6. These strategies are intended as general guides, not rigid, bureaucratic processes that must be religiously followed to the smallest detail.

4.10 Project Update

Now that you have flowcharted your process and understand the process improvement and problem-solving strategies, you are in a good position to plan the next steps in your project. The details of the individual tools will be explained in Chapter 5, but you should be able to plan the major steps now. The obvious next step is to obtain data on key input, process, and output variables to determine process stability. This requires identification of the key variables and collection of data *over time*. If all data are collected at one snapshot in time, it will not be possible to assess stability. Existing historical data may prove valuable and significantly reduce the time required to complete this step.

Thinking ahead to potential future steps, you may also wish to collect data on other variables you believe may be related to the key variables in question. These data may be helpful in determining potential causes or causal relationships through the use of scatter plots, stratification, or some type of formal statistical inference.

You may already have strong suspicions or even actual knowledge of the process stability. If so, you are in a position to plan to use the process improvement strategy, if stable, or the problem-solving strategy, if unstable.

EXERCISES

1. Discuss an actual situation you have faced where either the process improvement or the problem-solving strategy would have been helpful. Speculate on how the strategy might have been applied and what results might have occurred.

2. Make a list of everyday applications where either a process improvement or problem-solving approach should be used, but isn't. What results do you think might be obtainable? Some examples to stimulate your own thoughts would be simplifying the U.S. tax code, reducing waiting times at the local hospital or emergency room, and reducing difficulties trying to dial in to an Internet provider.

3. Find an Internet, newspaper, or magazine article that uses a statistical analysis. Contrast this analysis with either the process improvement or problem-solving strategy. What suggestions would you offer for improving the analysis that might lead to improved results?

4. Discuss an actual experience you have had where a problem-solving approach was used to address what turned out to be common-cause variation, or where a process improvement approach was used to deal with a special cause. What was the result? How might a better result have been achieved if the proper approach had been employed?

5. Critique one of the process improvement case studies in this chapter. What might you have done differently? Why?

6. Critique one of the problem-solving case studies in this chapter. What might you have done differently? Why?

7. The case studies in Chapter 2 do not rigorously follow either the process improvement or the problem-solving strategy. Which approach do you feel each case study is most similar to? Why?

8. Which aspects of statistical thinking are most critical to the success of the process improvement case studies in this chapter? Why?

9. Which aspects of statistical thinking are most critical to the success of the problem-solving case studies in this chapter? Why?

10. This chapter provides multistep strategies for problem solving and improving a stable existing process. Read the material on process design in Appendix H and develop your own strategy for designing a new product or process. Feel free to use ideas from the appendix, but be creative and develop your own strategy. What tools do you think might be appropriate for each step in this strategy? These could be tools mentioned in this chapter or any other tools of which you are aware.

11. How would you apply either the problem-solving or process improvement strategy to the following situations? Give as specific an answer as you can. (For example, if you suggest gathering data, what data would you gather?)

     ■ Improving the market share of a prescription medication
     ■ Attracting and sustaining more traffic to an Internet site offering personal loans
     ■ Winning more bids for municipal construction contracts (while ensuring profitability)
     ■ Responding to a downturn in business at a small management consulting firm
     ■ Responding to a rash of resignations at a research and development center
     ■ Reducing the time it takes the finance organization to close the quarterly books

12. Read Appendix D on the Six Sigma methodology. Do you think Six Sigma is more similar to the process improvement strategy or the problem-solving strategy? Why? Compare and contrast Six Sigma with the strategy you feel is most similar.

13. Respond to the following hypothetical scenario (from Britz et al., 1997).

Statistical Thinking in Health Care

Ben Davis had just completed an intensive course in Statistical Thinking for Continuous Improvement, which was offered to all employees of a large health maintenance organization. There was no time to celebrate, however, because he was already under a lot of pressure. Ben works as a pharmacist's assistant in the HMO's pharmacy, and his Manager, Juan de Pacotilla, was about to be fired. Juan's dismissal appeared to be imminent due to numerous complaints, and even a few lawsuits, over inaccurate prescriptions. Juan now was asking Ben for his assistance in trying to resolve the problem, preferably yesterday!

"Ben, I really need your help! If I can't show some major improvement, or at least a solid plan, by next month, I'm history."

"I'll be glad to help, Juan, but what can I do, I'm just a pharmacist's assistant."

"I don't care what your job title is, I think you're just the person who can get this done. I realize I've been too far removed from day-to-day operations in the pharmacy, but you work there every day. You're in a much better position to find out how to fix the problem. Just tell me what to do, and I'll do it."

"But what about the statistical consultant you hired to analyze the data on inaccurate prescriptions?"

"Ben, to be honest, I'm really disappointed with that guy. He has spent 2 weeks trying to come up with a new modeling approach to predict weekly inaccurate prescriptions. I tried to explain to him that I don't want to predict the mistakes, I want to eliminate them! I don't think I got through, however, because he said we need a month of additional data to verify the model, and then he can apply a new method he just read about in a journal to identify 'change points in the time series,' whatever that means. But get this, he will only identify the change points and send me a list, he says it's my job to figure out what they mean and how to respond. I don't know much about statistics—the only thing I remember from my course in college is that it was the worst course I ever took—but I'm becoming convinced that it really doesn't have much to offer in solving real problems. You've just gone through this statistical thinking course though, so maybe you can see something I can't. To me statistical thinking sounds like an oxymoron. I realize it's a long shot, but I was hoping you could use this as the project you need to officially complete the course."

"I see you're point, Juan. I felt the same way too. This course was interesting, though, because it didn't focus on crunching numbers. I have some ideas about how we can approach making improvements in prescription accuracy, and I think this would be a great project. We may not be able to solve it ourselves, however. As you know, there is a lot of finger-pointing going on; the pharmacists blame sloppy handwriting and incomplete instructions from doctors for the problem, doctors blame pharmacy assistants like me who actually do most of the computer entry of the prescriptions, claiming that we are incompetent, and the assistants tend to blame the pharmacists for assuming too much about our knowledge of medical terminology, brand names, known drug interactions, and so on."

"It sounds like there's no hope, Ben!"

"I wouldn't say that at all, Juan. It's just that there may be no quick fix we can do by ourselves in the pharmacy. Let me explain how I'm thinking about this and how I would propose attacking the problem using what I just learned in the statistical thinking course."

How do you think Ben should approach this problem using what he has just learned? Assume that he really did pick up a solid understanding of the concepts and tools of statistical thinking in the course.

REFERENCES

Britz, G. C., Emerling, D. W., Hare, L. B., Hoerl, R. W., & Shade, J. E. (1997, June). How to teach others to apply statistical thinking. *Quality Progress,* 67–80.

Hahn, G. J., Hill, W. J., Hoerl, R. W., & Zinkgraf, S. A. (1999, August). The impact of Six Sigma improvement—a glimpse into the future of statistics. *The American Statistician,* 208–215.

Imai, M. (1986). *Kaizen, the key to Japan's competitive success.* New York: Random House Business Division.

Kepner, C. H., & Tregoe, B. B. (1979). *Problem analysis and decision making.* Princeton, NJ: Princeton Research Press.

Introduction to
Microsoft Excel

Overview

The first four chapters of this text have focused on conceptual understanding of key statistical thinking concepts, and the strategies we use to apply them. There has been little focus on how to apply any specific tool or technique. Chapter 5, which presents the basic tools, will begin a shift in focus toward functional capability to apply the tools in addition to conceptual understanding. It is therefore an appropriate time to begin considering the use of statistical software to apply the tools. In contrast to many other texts, we have decided against providing step-by-step instructions in how to use statistical software packages. Once the software is updated, the step-by-step instructions are no longer valid and in many cases are actually a hindrance to appropriate use of the software.

Instead of step-by-step instructions, we will provide an introduction to Microsoft Excel at this point and introductions to Minitab and JMP prior to Chapter 6. Excel is not really a statistical software package, but it is commonly used in business and will perform many of the basic analyses presented in Chapter 5. For more rigorous statistical procedures, such as model building and design of experiments, a statistical software package is required. Minitab and JMP are two of the most popular and easy to use of these packages. For more detail on the use of Excel for statistical methods, see Berk and Carey (1995).

What Is Excel?

Many of you may be very familiar with Excel already. If so, you may wish to skip to the discussion on Data Analysis (p. 138). Excel is what is commonly referred to as a spreadsheet, a software application for the storage, analysis, and presentation of data. It differs from a database in that is it not designed for efficient storage of massive amounts of data. Storage, analysis, and presentation of data are basic to business, and Excel may very well be the most commonly applied software application in the business world. Because it is a Microsoft product, it is compatible with other Microsoft products used in a Windows environment, such as PowerPoint and Word. "Compatible" in this case means that one can create a graph in Excel and copy it directly into a PowerPoint or Word document.

Data Storage

Data can be entered into Excel either manually or by "pasting" (i.e., copying data from another compatible application). The data are stored in a series of columns as shown in Figure 1. To enter data, one moves the cursor into the "cell" (a given row of a given column) desired and types in the appropriate number. Text may also be entered in cells, but of course then the cell cannot be analyzed. By saving the spreadsheet, the data will be electronically stored exactly as they were when the spreadsheet was saved. Excel can handle thousands of data points, but it should not be confused with a relational database that efficiently handles massive data sets and keeps track of relationships more complicated than "rows and columns" (e.g., where row 1 of one variable corresponds to rows 2–5 of another variable).

FIGURE 1   Data Structure

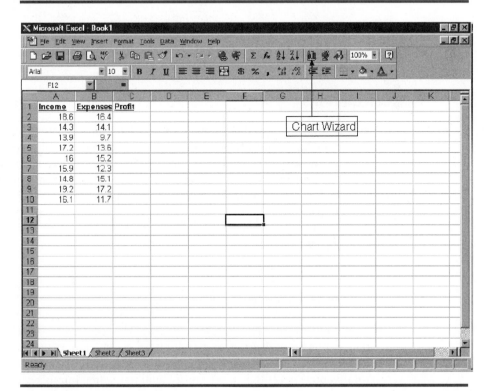

FIGURE 2   Calculating a New Column

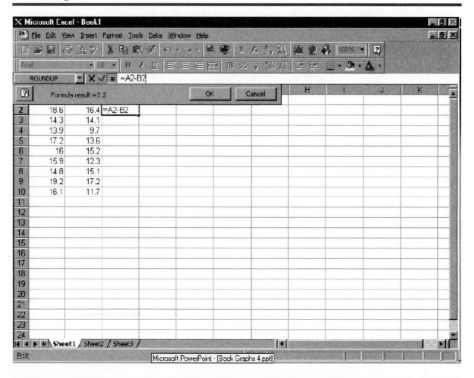

Calculations

In many cases, it is desirable to make calculations on data previously input. For example, if we have monthly income in one column and monthly expenses in another, we might want to calculate monthly profit (income minus expenses). This can be done easily in Excel by selecting a cell and entering an equation in the dialogue box at the top instead of typing in a number (Figure 2). There are several ways to set up an entire column to be calculated from other columns. Perhaps the simplest is to do this for one cell and then copy that cell into the other cells in that column. The steps required to do this are as follows:

- Select the cell of interest.
- Click on the "=" next to the dialogue box at the top. This tells Excel that this cell will be calculated from other cells.
- Enter the desired equation, using "A2" to indicate the second row of column A, without spaces. The standard symbols of +, −, /, and * are used for addition, subtraction, division, and multiplication. Much more complicated calculations can be done using preprogrammed functions in Excel. In this case we would enter "=A2−B2" to set profit equal to income minus expenses.
- Copy the cell you just entered using the copy icon in the toolbar.

- Select the rest of the column you wish to use to calculate profit and paste the copied cell into it.

The other cells in this column will not be calculated automatically.

Graphics in Excel

Most of the graphs discussed in Chapter 5, such as Pareto charts and scatter diagrams, can be created easily in Excel. These are typically done using the "Chart Wizard" in the Excel toolbar. This icon looks like a Pareto chart (see the note at the top of Figure 1). When you select the Chart Wizard, it will provide a sequence of menus asking you to select the type of graph you want, the data you wish to include in the graph, and a location in the spreadsheet to put the final graph. Excel allows the user to put text, or even graphs, into the cells of the spreadsheet. Figure 3 shows a sequence of menus the user might see for a run chart of income and expenses, as well as the final plot. Note that when a data range is requested

FIGURE 3
Creating Run Charts in Excel

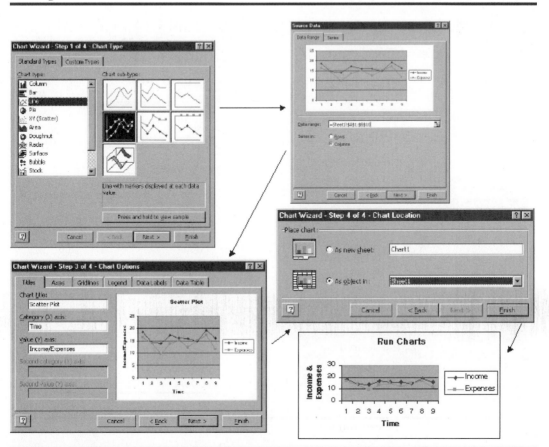

the user can select this range with the mouse and cursor rather than by listing the range in the dialogue box. The final plot can then be copied and pasted into another compatible application, such as PowerPoint. Excel has many advanced features that enable the user to significantly customize the graphs; see Berk and Carey (1995) for details.

Data Analysis

The feature of Excel that performs statistical methods, such as hypothesis tests and basic regression, is the Data Analysis option, which is located under the Tools menu (Figure 4). Note that this is an "add-in" optional feature of Excel. If you do not see the Data Analysis option when you select Tools, select Add-Ins and you should see an option to add-in the Data Analysis feature. Once you have selected

FIGURE 4
Using Data Analysis in Excel

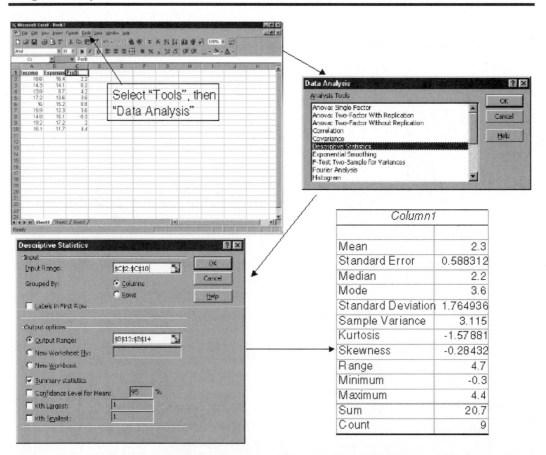

Data Analysis, a menu that includes several statistical and mathematical procedures becomes available. If you select one of these, you will be given a menu that asks for some of the same information as the Chart Wizard menus. Figure 4 shows a menu for Descriptive Statistics and the resulting information provided. Some of these statistics, such as the mean, standard deviation, and median, are very useful in practice. Others, such as kurtosis (a measure of "peakedness," the opposite of flatness, for a histogram), are not as common. Berk and Carey (1995) provide Excel commands to perform many other statistical methods for which Excel does not provide a "menu."

Summary

Excel is a very commonly used spreadsheet analysis tool. It is perhaps the best tool for performing basic graphs such as Pareto charts, histograms, run charts, and scatter plots. It can also be used for some statistical procedures, such as t and F tests (hypothesis tests to be discussed in Chapter 8), simple linear regressions involving only one x variable (see Chapter 6), random number generation, and Analysis of Variance (ANOVA, discussed in Chapter 8). For the more rigorous statistical methods, such as regression (Chapter 6) and design of experiments (Chapter 7), we recommend Minitab or another similar statistical analysis package.

REFERENCE

Berk, K. N., & Carey, P. (1995). *Data analysis with Microsoft Excel*. Pacific Grove, CA: Duxbury Press.

5

Process Improvement and Problem-Solving Tools

If you have an opportunity to make things better and you don't, you're wasting your time.

Roberto Clemente (Pittsburgh Pirates)

5.1 Introduction

Chapter 4 illustrated two specific approaches to applying statistical thinking: the process improvement strategy and the problem-solving strategy. In this chapter we present the individual tools used in these approaches. The main learning objective is to develop the functional capability to apply these tools to real applications of process improvement or problem solving. You should apply these tools to your own project to gain experience. To understand how each tool fits into the bigger picture, we discuss when and why each tool should be applied. These tools are frequently applied as part of the process improvement or problem-solving strategies, but they can also be applied in isolation—that is, separate from the use of one of these strategies.

Because this chapter provides details on a variety of individual tools, it does not "flow" as did Chapters 1–4. Now that you understand how the tools should be integrated, this chapter will simply present one tool after another.

The theoretical framework underlying these tools is discussed in Chapter 9, and intervening chapters discuss the following extensive tool sets: model building (Chapter 6), design of experiments (Chapter 7), and statistical inference (Chapter 8). Tools mentioned in Chapter 4 that are beyond the scope of this text and will not be discussed further include standards, procedures, and training. These are critically important tools that use statistical thinking, but they are really separate disciplines and cannot be adequately addressed here. Useful references for these fields are Imai (1986) for procedures and standards and Mager (1988) for training.

5.2 Relationship of the Tools to the Strategies

Individual tools are most effective when integrated in an overall approach applying statistical thinking. In this way the whole becomes greater than the sum of the parts. The process improvement and problem-solving strategies provide guidelines for using these tools that have proven effective in a variety of business applications. In some applications only one tool may be required, but these instances are infrequent.

Individual tools are organized under these three general categories: data collection tools, data analysis tools, and knowledge-based tools. Within each category the tools are presented in alphabetical order. This order of presentation does not represent a "sequence" in which to apply the tools. The process improvement and problem-solving strategies outlined in Chapter 4 provide logical sequences for applying the tools.

5.3 Data Collection Tools

In our haste to analyze data we may be tempted to grab whatever data are available and apply a formal statistical tool. Recall, however, that "garbage in, garbage out" applies to all disciplines. In other words, our analyses will only be as useful as the data examined. Give careful thought to proper data collection, and wherever possible, proactively collect data that answer the key questions about the process. It is a poor practice to rely on whatever data happen to be available.

Checksheet

Purpose Checksheets allow easy, consistent, and accurate collection of attribute (discrete) data that can be graphed subsequently in Pareto charts. Checksheets are typically used in the Gather Data step in the process improvement strategy.

Benefits Using actual data to guide decision making helps avoid overreliance on experience, gut feelings, or intuition and helps resolve disagreements objectively rather than on the basis of personality or hierarchy. Checksheets also save time and ensure consistency in how data are gathered.

Limitations Individuals using checksheets may try to "force" answers into preconceived categories. This may obscure some useful distinctions. Data must be collected manually for physical (nonelectronic) checksheets.

Examples Recall the case study in Chapter 4 involving reduction of telephone waiting time for bank customers. To gather data on the causes of long wait times, operators prepared a checksheet listing the potential causes. As they handled these calls, they recorded the actual causes. This enabled them to pinpoint the most important causes and dramatically reduce waiting times.

The checksheet shown in Figure 5.1 was used by operators who answered a 1-800 customer hotline for a consumer products company. One purpose of the hotline was to gather data on customer complaints and compliments for various products—in this example, toilet tissue. Detailed information was collected on each caller, and a new sheet was used on each call. These checksheets were on computer spreadsheets rather than on paper and were automatically summarized, so excessive paperwork was not required.

Without such a checksheet individual operators might record the same complaint using different words. For example, if a customer complained that the paper was rough, this could be listed as a softness issue, roughness, harshness, or scratchiness—all of which refer to the same fundamental problem. In many cases a checksheet simply lists potential causes, but in this case the checksheet incorporates additional information that may be helpful in looking for patterns in the data.

Procedure   The first step in preparing a checksheet is to carefully define the purpose of the data collection and identify the data to be collected. Next, individual items to include must be determined. This may require an educated guess, based on experience or subject matter knowledge. If some major categories are missed, new categories can be added as their importance becomes apparent. Next, terms must be carefully defined and clarified. For example, the odor category for toilet paper is intended to mean an unplanned detectable smell in the product, whereas the scent category is intended to mean the customer's opinion of the perfume scent that is deliberately added to the product.

The time frame of the study, the individuals who will collect the data, where data will be collected, and so on must be given careful thought. A physical or electronic checksheet must then be prepared and reviewed with those who will be using it. The checksheet will generally evolve over time, as more information is gained about the process being studied. Data gathered are periodically summarized (daily, weekly, or monthly), and the summary data naturally lead to data analysis tools, such as Pareto charts or run charts.

Variations   Some checksheets are more complicated than a list of potential causes. Figure 5.2 shows an example of a map that is used to note the details of automobile accidents for insurance companies. Another use of this type of checksheet is by sports analysts to convey information graphically to a television audience. One such example would be checksheets of a basketball court with check marks indicating where on the court a team has taken its shots. (A similar checksheet was used in the soccer case study in Chapter 2.)

Tips
- The input of those who will actually use the checksheet should be obtained as early as possible. These individuals typically know the process best and can ensure that the checksheet is easy to use in practice.
- It usually makes sense to include an "Other" category for abnormal or unforeseen responses. This avoids "force fitting" a response into existing categories.

FIGURE 5.1   Checksheet for Customer Complaints and Compliments on Toilet Tissue

Customer Name _____

Address _____

_____

_____

Area Code _____

Zip Code _____

Brand _____

UPC Code (if available) _____

Where Purchased _____

Date Purchased (if known) _____

Complaint _____   or   Compliment _____

Reason for Complaint or Compliment:
Color _____
Contamination (type, if known) _____
Count (sheets) _____
Dispensing _____
Flushability _____
Holes _____
Long Lasting _____
Odor _____
Packaging _____
Perforations (missing) _____
Scent _____
Softness _____
Strength _____
Thin _____
Value _____
Other (state cause) _____

FIGURE 5.2 Checksheet for Documenting Car Accidents

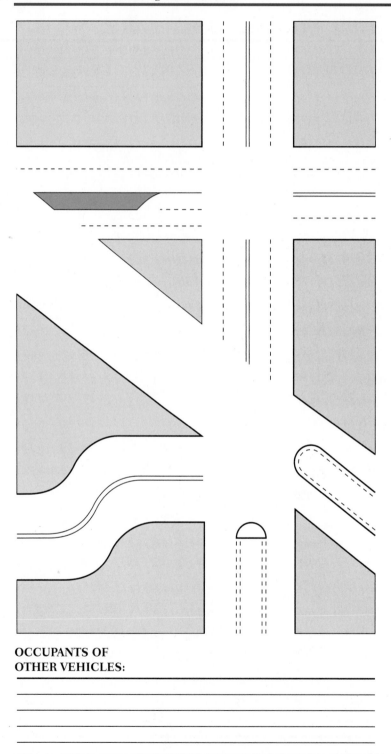

**OCCUPANTS OF
OTHER VEHICLES:**

If the "Other" category receives a relatively large number of checks, it probably should be broken up into several individual categories.

■ It is important to retain the order in which the data were collected. This is useful in determining whether the process was stable over the time period during which data were collected.

Data Sheet

Purpose Data sheets are used to record continuous (variable) data. Like checksheets, data sheets are also typically used in the Gather Data step in the process improvement strategy.

Benefits Makes data collection easier, faster, and more consistent.

Limitations Data sheets may not provide enough flexibility in gathering data that were not originally anticipated.

Examples In the resin variation case study in Chapter 4, operators manually wrote down data on data sheets, which simplified and standardized data collection. The "Reason for Complaint or Compliment" part of the checksheet in Figure 5.1 could have been set up as a data sheet by giving each cause of complaint a number and having operators record the corresponding number for each complaint.

Procedure Preparation and data-gathering procedures are similar to those outlined for checksheets: define purpose and identify data to be collected, use subject matter knowledge to identify categories, and add new categories as necessary.

Variations Computer spreadsheets such as Microsoft Excel are essentially generic data sheets with additional computational and manipulative capabilities. These spreadsheets can be adapted easily for use as data sheets by determining the appropriate headings and entering these data. Perhaps the best known (infamous) example of a data sheet is the IRS Federal Income Tax Form 1040. This data sheet asks the user to fill in specific data in boxes provided for that purpose and is available in both paper and electronic forms.

Tip It is important to retain the order in which the data were collected. This is useful in determining whether the process was stable over the time period during which data were collected.

Surveys

Purpose Surveys are used to gather "soft" data, such as opinions, from people as well as "hard" data, such as age and income. Surveys are typically used in the Gather Data step of the process improvement strategy. (Because surveys are such a broad and diverse topic, we have provided more detailed information on them in Appendix C.)

Benefits   Surveys provide an organized approach for gathering data from people. A prepared survey can help ensure everyone is asked exactly the same questions.

Limitations   It is very easy to bias the questions in the survey so that certain responses are more likely to occur. Nonresponse can also be a serious problem. Both of these issues are discussed in greater detail in the next section, Practical Sampling Tips.

Examples   The Chapter 4 case study involving telephone waiting time utilized a customer survey similar to that in Figure 5.3 to identify waiting time as an important concern. This survey is intended for use by internal customers of a corporate travel service and is primarily aimed at measuring customer satisfaction on a variety of factors. Measuring customer satisfaction is one common objective of surveys. Other uses of surveys include gathering information on political opinions, preferences for consumer products (i.e., market research), demographics, or "use" surveys to measure use of a product or service such as public television.

Procedure   The first step is to carefully consider and document the purpose of the survey. What do we wish to know that we do not know now? Next, determine the target population and the portion of that group that can actually be sampled. Choose a mode of data collection—mail, telephone, or in person—and develop a questionnaire and pretest it. Pretesting is evaluation in a trial setting to determine if there is any perceived bias, inconsistency, ambiguity, and so on in the survey. Questions should be clear, concise, and neutral (unbiased toward any particular answer). For example, a question that begins "Don't you agree that . . ." is clearly pointing toward a particular answer. Subtle changes in wording can have a significant impact on survey results. In general, use the minimum number of questions that will achieve the purpose of the survey. Long surveys tend to discourage people from responding or lead to quick answers without careful thought. Short-answer questions are generally more informative than questions that can be answered "Yes" or "No."

The next step is to select the group, or frame, from which the sample will be taken. The frame is the collection of people or items from which we will actually sample. For example, if we wish to contact registered voters in a state through a telephone survey, we will not be able to contact registered voters who have unlisted numbers or those who do not have telephones. Therefore, although the population of interest is current registered voters, the actual frame from which we can sample is registered voters with listed telephone numbers. Obviously, we would like the frame and the population to be as close as possible, but they will virtually never be exactly the same.

The next task is to decide on a sample size and select the sample. In general, surveys involve a sample size from 100 to 1000. The well-known national election

FIGURE 5.3 Satisfaction Survey for a Corporate Travel Agency

Please use capital letters, e.g.: SMITH

Last Name | Mid. Ini.

First Name

Business

Phone Number ___ – ___ – ____

Start Date of Travel: Month / Day / Year | Length of Trip: Days

In an effort to continually improve our service, we need to know about your experience with us. Please answer all questions by checking the appropriate box, then drop in the mail or fax. Thank you.

When you contacted your agent, how satisfied were you with:

	Very satisfied	Satisfied	Dissatisfied	Very Dissatisfied	Don't Know
1. Phone response time?	O	O	O	O	O
2. Courtesy/manner of agent?	O	O	O	O	O
3. Agent's response to your questions?	O	O	O	O	O
4. Agent's recommended options to reduce your travel costs?	O	O	O	O	O
5. The services provided by our strategic suppliers?	O	O	O	O	O

	Yes	No	Don't Know
6. Were your travel documents delivered as promised?	O	O	O

7. Were your travel documents . . .	Yes	No	Don't Know
Complete?	O	O	O
Accurate?	O	O	O

	Very satisfied	Satisfied	Dissatisfied	Very Dissatisfied
Overall, how satisfied were you with our service to you?	O	O	O	O

Are you answering as: O Traveler O Travel arranger

Comments

Thank you for answering these questions. Your comments will be used to better serve you.

polls typically use a sample of about 1000 to get precise information, and there is rarely a need to go beyond that size. The sample size is the number of *responses* to the survey. Because some people will be nonresponsive, the actual number of people included in the survey is usually higher than the sample size.

Ideally, participants should be selected randomly. That is, each individual or item should have an equal chance of being selected. People are generally incapable of truly selecting items randomly, so a computer-generated list using random number generation algorithms is used to provide a random sample. In a small-group setting, such as a small business, a complete sample (technically this is a "census") may be used. This sample includes 100% of the population of interest.

We are now ready to conduct the survey. Although this typically takes the most time and costs the most money, its success depends heavily on the planning steps just described. Once the data have been collected, they can be summarized in tables and graphs or subjected to some type of formal statistical analysis, depending on the original purpose of the survey.

Variations   Some surveys use a Likert scale, in which respondents are asked to answer statements using a scale such as, Strongly Agree, Agree Somewhat, Neutral Response, Disagree Somewhat, or Strongly Disagree. This provides a continuum of responses, and most people are able to interpret this scale well. Figure 5.3 uses Likert scales in questions 1–5 and the overall summary.

Tip   Real processes are dynamic. It is generally a good idea to save some of the budget to perform additional surveys to follow up on surprises that appear in the first survey or to determine if the situation has changed over time. For example, election survey results change almost daily leading up to an election.

Practical Sampling Tips

This section will focus on practical advice on how to sample, both in terms of how much data is required (data quantity) and how to get "good data" (data quality). (A formal discussion of formulas to determine appropriate sample size is presented in Chapter 8, and Appendix C has additional information on survey design.)

Data Quality   The quality of data is measured by the degree to which the data satisfy the needs of the analyst. For most rigorous statistical techniques, a random sample is assumed. As noted previously, a random sample requires that all items in the population have an equal chance of being selected. For example, a bank may wish to send a detailed questionnaire to its customers to estimate their receptivity to potential new financial services. To minimize costs, the bank may decide to send the questionnaire to only 100 of their customers. Taking the first 100 in alphabetical order or the 100 with the most money invested would not be random. These samples exclude smaller investors and those whose names begin with letters such as w, y, or z. Having a person select the names generally produces a biased sample. Individuals often choose people they know or names they like. It is almost impossible for people to be totally objective. The most viable approach is to have a computer randomly select 100 numbers from one to the total number of customers and use this sample. Pregenerated tables of random numbers can also be used for this purpose. This is conceptually equivalent to putting the customers' names on individual pieces of paper, placing these in a large hat, and selecting 100 out of the hat without looking. But "mechanical" methods of random sampling assume that the objects are thoroughly mixed and sampled without bias. This is often untrue, as has been documented in state or draft lotteries. For example, in 1970 the U.S. Armed Forces draft lottery was held by placing all 366 possible birthdays in separate capsules in a large bowl. The capsules were then removed one by one to determine the order for drafting individuals. Unfortunately, the capsules were not well mixed prior to sampling, and a disproportionate number of birthdays late in the year were selected. This led to numerous lawsuits and a switch to computer-based random number generation in the 1971 draft lottery (Fienberg, 1971).

Random sampling ensures that the sample data we analyze are conceptually representative of the total set of objects, or population of interest. Several practical concerns arise, however, with regard to this population. For example, the customer base of a bank is dynamic rather than static. By the time a list of customers is developed, new customers will have joined and others will have left. New customers not yet on the list will not be included in the sample, and the random sample of customers in existence at the time the list was developed may not be the group of interest today.

In addition, we rarely, if ever, obtain 100% response from questionnaires or surveys. Many people will simply not take the time to fill out and return them, even if reminded several times. This results in *nonresponse bias,* meaning that the results are biased because of some peoples' nonresponse. The group of interest was customers, but we received data only from those customers willing to fill out a questionnaire. It is a well-known fact in marketing research that individuals inclined to fill out questionnaires differ in many ways from those not so inclined. These differences, including educational and socioeconomic levels, may dramatically bias results. For example, most hotel or restaurant patrons do not take time to fill out the "Tell Us How We Did" surveys left on the table, and patrons who were extremely displeased are much more likely to fill them out, resulting in a negative bias.

Two techniques commonly used to reduce nonresponse bias are providing some incentive to respond (such as cash or free products or services) and follow-up. Follow-up contacts to nonresponders, encouraging them to respond, generally improves the response rate significantly. As a rule of thumb, a smaller sample with a high response rate (70% or more) is preferred over a large sample with a low response rate. The uncertainty associated with low sample size can usually be easily quantified statistically, whereas the uncertainty associated with nonresponse bias cannot.

Data from biased sources, such as restaurant customer satisfaction surveys, can be used for qualitative analysis, such as determining the main reasons for customer dissatisfaction, but we must be careful not to interpret it quantitatively as a random sample, by concluding that, for example, "More than 80% of our customers are dissatisfied." Although it is true that more than 80% of the customers who responded to the survey are dissatisfied, we do not know what percentage of all customers are dissatisfied. Other practical sampling problems include biased sampling (collecting data from our largest customers when in fact our population of interest is all customers), leading questions in surveys that encourage a particular answer ("Don't you agree that . . . "), dishonest responses to sensitive questions (questions about sexual practices sometimes asked of blood donors), and an inability to enumerate all items in the population.

An example of bias resulting from wording of survey questions was reported in *Readers' Digest* (1995). They noted that a majority of people polled agreed that "I would be disappointed if Congress cut funding for public television." However, a majority of the *same people* also agreed that "cuts in funding for public television are justified as part of an overall effort to reduce federal spending." Obviously, the specific wording of the question influenced whether people in this sample were for or against cutting public television funding.

An interesting example of biased sampling due to difficulty enumerating the entire population are famous erroneous forecasts of presidential elections. In the race between Alf Landon and Franklin Roosevelt in 1936, some polling agencies used telephone directories and automobile registration records to obtain names of individuals to poll. At that time phones and cars were primarily owned by those at higher socioeconomic levels, and these individuals tended to favor Landon. This biased sample led many pollsters to predict a Landon victory. Similar mistakes were made in forecasting a victory for Thomas Dewey over Harry Truman in 1948 (Parten, 1950).

Rather than sampling a target population about which we wish to learn more, we almost always sample from a *subset* of this population. The subset is often referred to as the *frame,* or actual sampled population. For example, in polling the Dewey-Truman election, the target population was voters. The frame, or population actually sampled from, was people with listed telephone numbers who indicated that they planned to vote. These two populations were different in several important ways, although they did overlap.

Two important considerations to keep in mind when sampling are to minimize the difference between the target population and the frame and to use subject matter knowledge to determine how important the difference might be. The ability to extrapolate the results of the sample collected from a particular frame to the entire population must be determined from subject matter knowledge rather than statistics. In the Landon-Roosevelt case, careful consideration of all voters versus those who had phones or cars would have revealed a socioeconomic difference. Is socioeconomic level important in voting preference? Obviously, it is. Data from biased sources may still be of value *qualitatively,* but we must avoid assuming that these data are representative of the entire population or process.

Business processes are dynamic, and the concept of a stable population generally has meaning only at a specified snapshot in time. We cannot randomly sample an ongoing process, because we cannot sample from future outputs of the process nor typically can we sample from historical outputs. Yet we are almost always interested in looking at trends over time before getting into too much detail at a particular snapshot in time. The practical result of this is that avoiding bias is more fundamental than strict randomness.

Of course, data quality also includes more obvious issues, such as having errors or missing values in the data and having the wrong data. An example of having the wrong data would be trying to improve market share with customer satisfaction data on current customers. To increase market share, we need data on potential and lost customers—those customers who could increase our share. In addition, data on customer needs are more helpful than data on customer satisfaction, because data on satisfaction do not tell us much about what products or services the customer might purchase if they were available. Customer satisfaction data, properly taken, does satisfy the need to know what our customers think of our current products and services, however.

Let's summarize the practical sampling tips relative to data quality:

- Clarify your original purpose for sampling to ensure you get appropriate data.
- Carefully consider the specific target population or process and the frame from which you can actually sample that most closely matches the target population.
- Use subject matter knowledge to determine how important the differences between the target population and the frame are.
- Minimize the potential biases, including nonresponse bias, that could have an impact on the result.
- To the degree possible, take proactive measures to ensure that errors or missing values in the data are minimized.

Data Quantity   A typical question asked of professional statisticians is "How much data do I need for my results to be statistically valid?" As seen in the discussion on data quality, a biased sample will produce misleading results no matter how large the sample is, so this question cannot be answered by citing numbers alone. But we have learned a number of things about data quantity that can help us decide on a sample size. Let's take a closer look at what we know.

In reality, *uncertainty* (inaccuracy) in our results decreases as sample size increases. The larger the sample size, the less uncertainty. For example, if we wish to estimate the average income in the United States for a specific year, we could ask one "randomly chosen" person. Even if we got an honest answer, there would be a great deal of uncertainty in using this one data point as representative of the average U.S. income. If we were to get access to IRS records, we could sample 100 records randomly and get a more accurate estimate (ignoring potential biases in reported income!) of the frame of people who file returns. Unfortunately, there is no magic cutoff where results go from uncertain to certain. Typically, the uncertainty decreases gradually, proportional to the square root of the sample size. For example, the uncertainty in our estimate of incomes from 100 records would be about one-tenth (square root of 100) of the uncertainty of our estimate from 1 record (not one one-hundredth, as might be supposed.) In this example uncertainty is being measured by a unit called the standard deviation, which quantifies how far off our estimate might be. This will be discussed later in this chapter and in even greater detail in Chapter 9.

Because doubling our sample size will not double our accuracy—but often will double our cost!—it typically makes economic sense to use a fairly small sample (less than 100). Chapter 8 will provide formulas for deriving the "exact" sample size required to achieve a specified level of accuracy with a specified probability. As we shall see, however, these formulas require knowledge of specific attributes of the process of interest, such as the degree of variation in customer opinions or the proportion of voters in favor of Candidate X. This creates an obvious problem: If we knew these already, we would not need to sample! Best guesses or "worst case" values are often used, but then the sample size formula is no longer exact.

An alternative is to use the "rule of thumb" for reasonable (but not exact) sample sizes. These rules are based on experience and statistical theory but do not claim to provide specific accuracy or probabilities. Often, they are used initially to get good estimates of the attributes required by the exact formulas. Most applications of statistical thinking involve one of two types of quantitative data: continuous data or discrete data. Continuous data are measured on a continuous scale, such as time, length, or weight. The rules of thumb for continuous data are:

- About 5 to 10 data points are required to get a reasonable estimate of the average (e.g., average time to pay invoices or average discount from list prices). This assumes that we have a constant average, or a stable process.
- About 30 data points are required to get a reasonable estimate of the level of variation, typically measured in standard deviation units (e.g., variation in time to pay invoices). This also assumes a constant level of process variation over time.

Discrete (attribute) data either have an attribute or not, such as purchase order numbers that are either correct or not. The rule of thumb for discrete data is:

- About 100 data points are required to get a reasonable estimate of a proportion, or percent (e.g., proportion of invoices that are accurate or proportion of credit card holders that are delinquent.)

A couple of points are noteworthy here. First, it is generally easier to estimate the average level than to estimate the degree of variation. Second, discrete data require larger samples than continuous data because these data contain less information. We should therefore strive to obtain continuous data wherever possible. For example, if we were trying to reduce cycle time for sending out invoices, it would be better to measure actual cycle time of the invoices rather than the percentage that were shipped late. The percentage shipped late may be useful as a managerial measure, but is not as useful for improvement purposes as the actual cycle times. We will discuss types of data in more detail in Chapter 9.

5.4 Data Analysis Tools

Data analysis tools are intended to evaluate, organize, and interpret data we have collected. These are the tools traditionally thought of as "statistics," and they are highly dependent on accurate, unbiased data. When invalid conclusions are drawn, the fault is often found to be biased or inaccurate data rather than faulty application of data analysis tools. We should therefore carefully consider the source of the data and the measurement process itself (Chapter 3) prior to formally applying the data analysis tools. With that caveat in mind, let's take a closer look at some of these tools.

Box Plots

Purpose   Box plots are graphs depicting the relationship between a discrete variable (such as region of the country) and a continuous variable (such as profitability). They are specifically designed to look for differences in location (average) and variation between the categories of the discrete variable. For example, we might want to know if the average and variation in profitability are the same between the five regions of the country. Box plots show percentiles of the continuous variable at each level of the discrete variable: They show the median (50th percentile), 25th percentile, and 75th percentile (not to mention extremes of the distribution) of profitability for each region of the country. Box plots are typically used during the Study Cause-and-Effect Relationships phase of the process improvement strategy and during the Identify Potential Root Causes phase of the problem-solving strategy.

Benefits   Box plots enable us to see the impact of discrete variables, which are often harder to plot. In addition, box plots can handle very large data sets because they plot percentiles of the continuous variable rather than the raw data themselves. If we have massive data sets with perhaps millions of data points (now readily available over the Internet), standard plots lose their ease of interpretation—they look like one large blot of ink—whereas box plots do not.

Limitations   Box plots do not plot individual data, so some subtleties in the data could be missed. The box plot may also be harder for those not trained in statistical thinking to understand because it plots percentiles.

Examples   Figure 5.4 shows a box plot of the hourly consulting fees a large conglomerate paid to public accounting firms in the past year. Senior management felt that costs for public accounting services were getting out of hand, and they wanted to find out where all the money was going. Several analyses were done to address this concern, one of which produced this box plot. The data were obtained by sampling the invoices received from the several public accounting firms the conglomerate did business with, and each data point represents an hourly rate charged by an accounting firm. Note that the discrete variable in this case is level of person doing the accounting work. The categories ranged from partner in the firm to staff, which generally identified an entry-level accountant.

FIGURE 5.4 Box Plot of Hourly Rates Paid to Public Accounting Firms

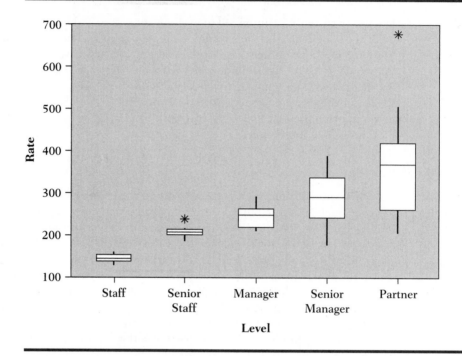

The purpose of constructing this graph was to see how fees varied by level of consultant. It was expected that the charge-out rates for entry-level accountants would be lower than for managers or partners, but the size of the difference was not known. Figure 5.4 shows that not only was the average rate for the higher levels greater (the location of the box relates to average level) but also that there was much more variation (variation is related to length of the box) in these levels. This was interpreted to mean that there were generally accepted standard rates for entry-level accountants but that there was less standardization for the higher levels. The investigation therefore began to focus on the higher levels: Why were these levels needed? and Why did their charge-out rates vary so much? Eventually, this led to standardization within the company for when higher levels would be used and standard rates for these levels. Recall that standardization is a means of reducing variation. In this case the company lowered the average by never paying more than "company standard" rates. The company might still pay less than this rate, however, hence the reduction in average level as well as in variation.

Procedure   Box plots will almost always be computer generated because of the need for percentile and other calculations. Minitab, JMP, and other statistical software packages can create box plots using the data for the continuous variable, such as hourly rate, and the discrete variable, level of consultant in this case. The software then calculates the median (50th percentile), the 25th percentile, and

FIGURE 5.5 Box Plot Construction

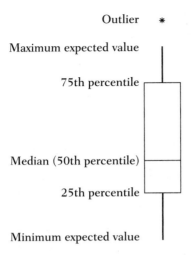

the 75th percentile of the continuous data. These are also the 1st, 2nd, and 3rd quartiles. This is done separately for each category of the discrete variable. The box plot then consists of a "box" whose top is the 75th percentile and whose bottom is the 25th percentile, as shown in Figure 5.5. The median is then drawn as a line within the box. Note that the median line need not be exactly in the middle of the box. The lines draw above and below the box are called "whiskers" and represent the extremes of the data.

The whiskers are determined by first calculating the maximum and minimum values we would expect to see based on the 25th and 75th percentiles. These are called the upper and lower adjacent values; the specific formulas typically used are:

Upper Adjacent = 75th + 1.5 × (75th − 25th)

Lower Adjacent = 25th − 1.5 × (75th − 25th)

These adjacent values represent the largest and smallest values we would expect to see in the data, if there were no outliers (outliers are extreme data points that are inconsistent with the rest of the data). The difference between the 75th and 25th percentiles (1st and 3rd quartiles) is referred to as the interquartile range and is a measure of variation in the data. These are theoretical values; they do not represent actual values seen in the data set. We then determine the maximum actual data point equal to or less than the upper adjacent value and draw a whisker from the top of the box to this value. The lower whisker is determined similarly using the lower adjacent value. For example, if the upper adjacent value is 100, we would identify the largest data point less than or equal to 100 and draw the whisker up to it. In this way the whiskers always represent actual data points and are the largest and smallest values within the range of what we would expect to see theoretically. Any data points beyond these whiskers (such as that in the upper right corner of Figure 5.4) represent outliers, which are perhaps due to special causes.

Variations   Numerous variations to the standard box plot exist. One can choose percentiles other than 25th and 75th for the box itself, and most software allows the user to draw whiskers to adjacent values as opposed to the maximum and minimum actual data points within the range of the adjacent values. The average can also be used instead of the median.

Tip   Box plots work best with fairly large data sets (100 or more values for each discrete category). With small data sets the shape of the box plot, particularly the length of the whiskers, can vary dramatically. This is because the whiskers are drawn out to extreme data points. If one such point is omitted, the resulting whisker could be much shorter. With large data sets the shape is much less dependent on any one data point.

Capability Analysis

Purpose   Formal capability analysis, which generally means calculation of capability ratios, is performed to quantify how well the current process meets or could meet its requirements (specifications). By plotting the data of key variables versus their specifications, we can determine graphically whether we have a problem. Calculation of capability ratios supplies a number to document how well we are, or can be, doing.

Benefits   A benefit of capability ratios is that they make discussion of process capability more rational and less dependent on opinion. For example, when people look at a graph of a key output versus its specifications, they might reach different conclusions about how good or bad the situation is. A number helps to quantify the degree of "goodness."

Limitations   Statistical capability ratios became a very popular tool in the U.S. auto industry, and among their suppliers, in the 1970s. Unfortunately, this popularity led to a gross overreliance on capability ratios. It was often assumed that if one knew the capability ratio, then one knew everything important about the process. This is obviously not true, as no one number can tell us everything we need to know about a process. Business contracts often stipulated specific capability ratios that must be achieved. By sampling or subgrouping the data in unusual ways, however, one can often manipulate the ratio to "look good." This led to a lot of game playing with capability ratios. Data were often collected during a time of process instability, which renders the capability analysis confusing, if not meaningless, because it does not represent one consistent process. In addition, a bad capability ratio could be due to the average being off target, or the variation being too great, or even to a specification being unreasonably tight. The ratio itself does not diagnose the root cause of the problem. This is why plots of the data are generally to be preferred.

Examples An engineering consulting organization one of the authors worked with was trying to control unapplied labor costs—labor costs that cannot be charged out to a client. These costs include vacation, sick time, internal meetings, and lost time (time the consultants were available for work but had no billable work to do). In the consulting business one must maintain a steady flow of billable hours. This is the primary, and perhaps only, income. Because unapplied labor costs cannot be recovered from a customer, minimizing unapplied labor cost is critical to maintaining a positive cash flow.

The business collected data on unapplied labor costs (as a percentage of total labor costs) for 18 weeks. The histogram is given in Figure 5.6. We would typically like to have more data to perform a capability analysis, but in this case that was all the data they had collected (previous data had been lost). This organization was targeting an unapplied labor rate of 15% and trying to prevent any individual month from being above 25% or less than 5%. Although it might not make sense at first glance to have a lower limit here, the organization realized that very low unapplied rates (such as zero) in a given month meant that no one was taking vacation. The natural consequence would be that in future months many people would be taking vacation, resulting in wide variation in available engineers. In addition, they did not want to totally eliminate internal meetings designed for team building, training, discussions on business strategy, and so on.

The data collected fit within the specifications (5% to 25%) reasonably well, with the exception of one point. How could we quantify the capability of this process, assuming it is stable and continues to perform this way? We calculated

FIGURE 5.6 Unapplied Labor Rate

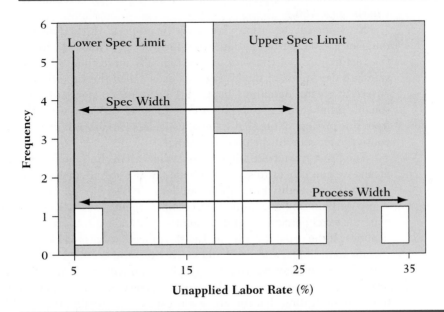

a C_p, one of the most commonly used capability indices that focuses solely on variation, and obtained a value of .84. In general, we want at least a value of 1.0, and preferably higher. Then we calculated an index called C_{pk}, which is similar to C_p but also focuses on the location of the average versus the target, and obtained a value of .62. The C_{pk} is typically lower than the C_p whenever the average is off target. Both of these values indicate an opportunity for improvement.

Procedure   Capability ratios are calculated by standard statistical software, such as Minitab or JMP. To understand the interpretation of these ratios, we will review the formulas, which are based on the average, target, specifications, and standard deviation. Recall that the standard deviation is a measure of variation (see Chapter 9 for more detailed discussion of its statistical properties). The formula for C_p is:

$$C_p = \frac{\text{Specification Width}}{\text{Process Width}}, \text{ where}$$

Specification Width = Upper Spec Limit − Lower Spec Limit

Process Width = 6s, where s is our short-term estimate of variation

The specification width and process width are illustrated on Figure 5.6. A short-term estimate of variation is one that is not artificially inflated due to special causes, such as process drift or upsets. Our objective is to estimate the common-cause variation—the variation we would experience if we eliminated all special causes and had a truly stable process. This is why these ratios are referred to as capability ratios in comparison with performance ratios. To estimate the short-term variation, we typically employ a moving range calculation. (See Control Charts, page 160.)

Six standard deviations are used for process width. This corresponds to a range of ±3 standard deviations from the average. Theoretically, we would expect the vast majority of the data points, typically more than 99%, to fall within ±3 standard deviations of the average. (This is called the Empirical Rule and is discussed in greater detail in Chapter 9.) Therefore, 6 standard deviations is a reasonable number to use to represent the range of variation we would expect to see from this process. Note that C_p only considers variation; it does not take into account the location of the process average.

The process width in Figure 5.6 is wider than the specification width, which means we can expect to see data outside of the specifications in the future. The smaller the capability ratio, the more data are likely to fall outside the specifications. With an understanding of probability theory (see Chapter 9), we can estimate the exact percentage of data points we would expect to see outside the specifications. If the capability ratio is 1.0, the process variation exactly "fits" within the specification limits, and we would rarely expect to see data outside the specifications. If the line representing the specification width is longer than the line representing the process width, this would correspond to a capability ratio greater than 1.0, indicating that the process is capable of meeting the specifications and that there is a very low probability of seeing data outside the specification limits.

In Figure 5.6, however, the process average is 17.67, and the short-term estimate of variation from the moving range calculation is 3.96. Therefore, C_p is:

$$C_p = \frac{(25-5)}{(6 \times 3.96)} = \frac{20}{23.76} = .84$$

A similar ratio that looks at the location of the average, in addition to the variation, is called C_{pk}. This is a logical extension of C_p, because one could have a C_p greater than 1.0 but be experiencing data outside the specifications because the average is off center. The process appears to be slightly off target in Figure 5.6, with the average being greater than the target of 15. The formula for C_{pk} is a little more complicated than for C_p:

C_{pk} = Minimum (C_{pkL}, C_{pkU}), where

C_{pkU} stands for upper specification C_{pk}, and equals

$$\frac{\text{upper specification} - \text{average}}{3s};$$

C_{pkL} stands for lower specification C_{pk}, and equals

$$\frac{\text{average} - \text{lower specification}}{3s}; \text{ and}$$

Minimum means we take the lower of C_{pkL} and C_{pkU}.

Note that if the process average is located exactly in the middle of the specifications, which will typically be the target value, then both C_{pkL} and C_{pkU} are nothing more than half the C_p specification range divided by half the C_p process range, which will just equal C_p. If the average is off center, however, then one of the C_{pkL} or C_{pkU} values will be greater than C_p, and the other will be smaller. Because C_{pk} takes the minimum of these two, however, C_{pk} will always be less than C_p when the process average is off center. In this case the process average is 17.67, which is greater than the target of 15, which should produce a C_{pk} less than the C_p value of .84. Using the C_{pk} formula, we obtain:

$$C_{pkU} = \frac{25 - 17.67}{3 \times 3.96} = \frac{7.33}{11.88} = .62$$

$$C_{pkL} = \frac{17.67 - 5}{3 \times 3.96} = \frac{12.67}{11.88} = 1.07$$

C_{pk} = Minimum (.62, 1.07) = .62

Note that the C_{pk} is lower than C_p because the average of this process is slightly off center. This results in a lower C_{pkU} and, hence, a lower C_{pk}.

Variations Because C_p and C_{pk} are both based on a short-term estimate of variation, they are really measuring capability, or what the process could do if we only had common-cause variation. If the current process has a lot of drift, or other special causes, much of this variation will be filtered out when we calculate the short-term estimate of variation. For this reason, the ratios can be unrealistically optimistic if they are interpreted as current performance. Calculation of these ratios can be made using a long-term estimate of variation, which combines common-cause and any special-cause variation that may be present. These ratios are called P_p and P_{pk}, where P stands for performance as opposed to capability. To utilize these ratios, we would use the C_p and C_{pk} formulas but replace the short-term variation estimate with a long-term estimate. This long-term estimate is typically the standard deviation of all the data grouped together.

Tip   Always plot the data prior to using capability ratios. This will avoid overlooking obvious issues such as extreme outliers or special causes that may inflate even short-term estimates of variation. If the process is drifting, there really is no consistent average and a C_{pk} has little practical meaning.

Control Charts

Purpose   The primary purpose of control charts is to differentiate common-cause variation from special-cause variation over time, which will determine whether a process improvement or problem-solving approach is warranted. This type of application typically occurs in the Assess Process Stability step. Control charts are also often applied to a stable process to maintain performance. Special causes can be detected rapidly using these charts, and problems can be quickly diagnosed and eliminated. In this type of application the charts can be thought of as signals for human intervention.

Benefits   Control charts prevent misinterpretation of data by distinguishing between special and common causes of variation. All too often businesspeople interpret any change in a financial figure as evidence of a change in the system, a special cause, and a great deal of time is wasted looking for the reason for the change. Control charts also enhance our understanding of the process by helping to generate hypotheses to explain patterns that we did not anticipate. The charts are therefore a tool for learning about the process.

Limitations   The control chart itself does not eliminate the special causes, it only detects them and provides clues to help us understand the process. We need to intervene with the process to identify and remove the root causes. A control chart is also of little value in improving a stable process because it will only continue to tell us the process is stable.

Examples   The resin case study in Chapter 4 used a control chart to detect special causes in resin yield over time and to demonstrate improvement. The data average was plotted to detect special causes affecting the average level of the process, and the data range was plotted to detect special causes affecting the process variation.

FIGURE 5.7 Individuals and Moving Range Chart of Unapplied Labor Rate

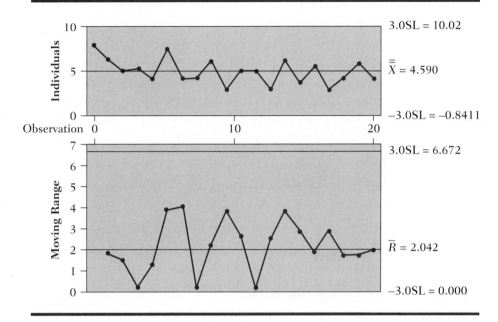

Here are some additional examples of control charts. Figure 5.7 charts the unapplied labor rate at a research institution, showing hours charged to "unapplied" (not billable to a contract) divided by total hours. Figure 5.8 charts pricing accuracy, showing the proportion of pricing quotes given to customers that

FIGURE 5.8 *P* Chart of Pricing Accuracy

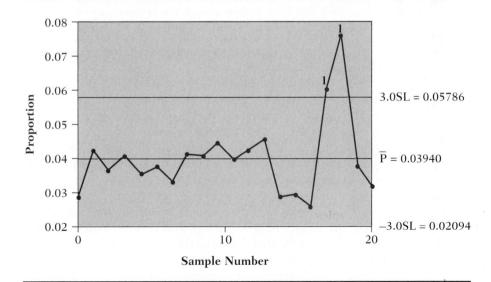

FIGURE 5.9   X-Bar and R Chart of Cycle Time for Credit Approval

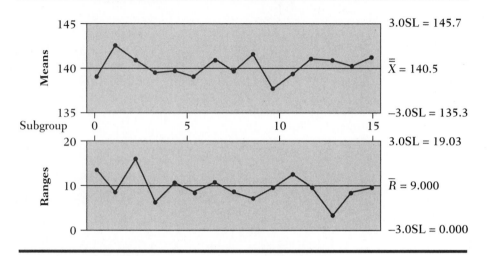

turned out to be incorrect. Figure 5.9 charts cycle time for credit approval for equipment leases. These data sets were supplied by H. A. Lasater.

Procedure   The basic steps for control charting are as follows:

1. Determine the purpose for the chart and the variable to be plotted.
2. Determine the most appropriate type of chart.
3. Select a sampling scheme for initial data collection.
4. Gather the initial data.
5. Plot the data on the chart.
6. Interpret the chart and reevaluate the original sampling plan.
7. Maintain the chart over time to improve or control the process.

We will now explain these steps in greater detail. Note that only Step 5 is typically performed using computer software.

Step 1: Purpose and variable to be plotted. There are several possible purposes for constructing a control chart. In the early stages of our improvement efforts a control chart can help diagnose the degree of stability in the process. The control chart also provides insights into the behavior of the process that cannot be seen from the numbers themselves. Alternatively, after completing major improvements a control chart can be used to maintain the gains by rapidly detecting and eliminating special causes occurring over time.

Selecting the specific variable to be plotted will depend on the purpose of the chart. This can be a measured value (such as length or purity), counts of defects or incidents, money, or time (such as time between system failures or cycle time). The variable to be plotted should be the variable we are interested in improving in terms of its average, degree of variation, or both.

Step 2: Determine the most appropriate type of chart. There are many types of control charts. Typical charts include the X-bar and R chart (average and range),

X-bar and S chart (standard deviation, another measure of variation), and individuals charts (plots of individual data points) for variable data measured on a continuous scale, such as cycle time. For discrete data that fall in only one of two categories, such as defective or not defective, typical charts are the p (proportion defective) chart and the np (number defective) chart. These are basically the same charts, except that one (p) is on a proportional scale and the other (np) is not. As we shall see, the p has the advantage that groups of data with different sample sizes can still be plotted on the same chart.

For discrete data, where we count the number of defects or instances of an attribute, such as the number of customer service calls received, a c (count) chart or u (unitized count) chart is typically used. The unitized count chart simply converts the data to a common scale when the opportunity for defects varies from data point to data point. For example, when plotting typographical errors in legal documents over time, suppose that some documents contain 2 pages and others contain 200 pages. A more logical plot than the count of errors would be errors per page, which is a unitized count.

Although the number of possible types of control charts may seem confusing, they all plot data (or statistics calculated from the data) over time and have statistical limits, which represent the range of common-cause variation to be expected. The individuals chart is a generic chart, and it can be used in virtually any situation, including with discrete data. The individuals chart is therefore a good default chart to use when you are uncertain about the most appropriate chart to use.

Figure 5.10 is a flowchart that will help you determine the appropriate chart for a particular application. This is based on a similar chart developed by Hacquebord (personal communication) and should be used as a rule of thumb rather than rigorous science. The questions are based on the underlying statistical distributions upon which the charts are based. Different distributions have different statistical properties and, hence, different formulas for the limits on the charts. For example, the continuous charts are loosely based on a normal (bell-shaped) distribution, although having this distribution is not an absolute requirement. The c and u charts are based on the Poisson distribution, and the p and np charts are based on the binomial distribution. (These distributions, and their theoretical implications, are explained in more detail in Chapter 9.)

Answering the first question, "Can we count discrete occurrences?" tells us whether we are dealing with continuous or discrete data. If we are not counting some discrete occurrence, it is not discrete data. For example, waste in a chemical factory is sometimes expressed as a percentage of total production, but it is still a continuous measurement (typically weight). The percentage of checks produced with incorrect amounts, in contrast, is a count of a discrete occurrence and is discrete data.

Answering the next question, "Can we count nonoccurrences?" determines whether the data are like "defectives" (i.e., something is either defective or not) or like "defects" (wherein one item can have numerous defects). In counting the number of credit cards produced with the name incorrect or misspelled, we can count both the number of incorrect cards and the number of cards with correct names. These would be defectives data. But if we are counting the number of customers arriving in a restaurant, we can count the number of customers arriving

FIGURE 5.10
Selecting a Control Chart

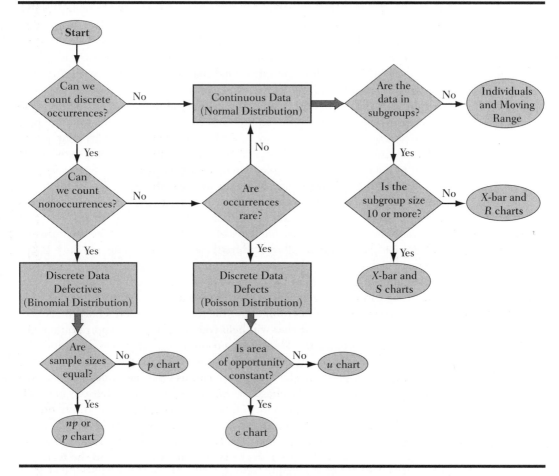

but cannot realistically count the number of people who are not coming to the restaurant. These data would be treated as defects data. The terms *defects* and *defectives* were originally coined when control charts were applied to quality control in manufacturing. The terminology stuck, even though we now use control charts for more general applications.

The third question, "Are occurrences rare?" may seem perplexing. The reason for including this question is that some data that are very similar to defects data do not satisfy a statistical requirement for use of the Poisson distribution (see Chapter 9). These data are better treated as if they were continuous data, even though they technically are not continuous. For example, in an accounts payable center we can certainly count the number of invoices processed daily, but we cannot realistically count the number not processed (theoretically there could be an infinite number processed). Invoices are being processed continuously while the office is open, however, and we should therefore treat the number of invoices processed as continuous data for statistical purposes, even though it is actually a count.

The term *rare* is intended to be interpreted at a particular location and point in time. For example, some people might say that airline crashes are not rare because they read about them in the newspaper from time to time. But the probability that a particular flight on a particular day will crash is quite low—less than one crash occurs per million flights. Similarly, if we are counting visual defects on automobile finishes, we might frame the question this way: "If we randomly selected a particular spot on the car ahead of time, how likely is it that a defect will occur precisely at this spot?" Having visual defects in the finish may not be rare, but having a visual defect in a specific location probably would be. Typically we define rare as a probability of .10 or less (i.e., less than a 10% chance).

For variables data we typically use X-bar and R charts for grouped data. For large subgroup sizes (10 or more) we typically use X-bar and S charts. The standard deviation is a much better measure of variation for larger subgroup sizes because it uses all the data, not just the largest and smallest value. For individual values (subgroup size 1), we use individuals and moving range charts. These data occur quite frequently with business processes, such as quarterly earnings, inventory levels, sales, or cash flow. It would not make sense to ask for a subgroup of five earnings figures this month when only one figure exists.

For defectives (binomial) data, we may have equal or unequal subgroup sizes. For example, if we inspect purchase orders for accuracy, we may not be able to inspect the same number every time due to circumstances beyond our control. If the sample sizes are the same, we can use the np (number defective) chart. The p (proportion) chart can be used with either equal or unequal subgroup sizes, however, because it is calculated on a proportional basis. Similarly, the c (count) chart is typically used for defects (Poisson distribution) data when the area of opportunity remains constant, but the u (unitized count) chart must be used if this varies. For example, if we are counting defects in automobile finishes for subcompact and large luxury cars, it might make more sense to plot defects per square foot or meter of finish. If subgroup sizes vary by no more than about 20% of the average subgroup size, this is close enough to consider the subgroup sizes equal.

Step 3: Select the sampling scheme for the initial data collection. To calculate the limits on the chart and interpret the behavior of the process, 20 to 25 subgroups of data are needed. If this is not feasible, we may choose to start calculating limits with 10 subgroups, but the limits should be updated until we have about 25. Once we have 25, the limits should be fixed until we have a specific reason for changing them.

The appropriate size of the subgroups will vary from application to application, but we can offer some general advice. Because we are collecting data over time, we do not need to gather large data sets at any particular time. For continuous data, subgroups of 4 or 5 tend to work well. For attribute data, we would like to have samples large enough that we average 5 or more occurrences (defects or defectives) per subgroup. For example, in the situation where we are monitoring visual defects in car finishes, suppose that we average two defects per car. For control chart purposes we would like to plot defects per three cars to get an appropriate sample size.

A key success factor in using control charts is the method of rational subgrouping of the data. This is especially true for variables data. The logical way in which we choose the items to form a given subgroup is important because the width of the limits on the chart is based on the variation within subgroups (Table 5.1) and is intended to represent common-cause variation. Therefore, we have essentially defined sources of variation that occur within subgroups as common cause, and those that only occur between subgroups as special cause. This decision should be made in a logical, well-thought-out manner.

Typically, rational subgrouping is accomplished by grouping items that occurred at approximately the same point in time. For example, in plotting cycle time of approving credit applications, we may select four applications per day and consider each day as a unique subgroup. This subgrouping scheme implies that variation that occurs within a given day is common cause, whereas variation that occurs from day to day is due to special causes. The user must determine if this makes logical sense in light of the nature of the application process.

The distinction between common and special causes is real, but it is also subject to interpretation. For example, in charting commuting time to work, most people would consider a thunderstorm that snarled traffic a special cause. But in an area where thunderstorms occur frequently, individuals might consider this common-cause variation. Give careful thought to the sources of variation that should be considered special causes and be sure that these do not show up within our subgroups. Wheeler and Chambers (1992, Section 5.6) provide an excellent example that illustrates these concepts.

TABLE 5.1 Control Chart Formulas and Constants

Sample Size n^*	A_2	D_3	D_4	A_3	B_3	B_4	d_2
2	1.88	0	3.267	2.659	0	3.267	1.128
3	1.023	0	2.574	1.954	0	2.568	1.693
4	0.729	0	2.282	1.628	0	2.266	2.059
5	0.577	0	2.114	1.427	0	2.089	2.326
6	0.483	0	2.004	1.287	0.03	1.97	2.534
7	0.419	0.076	1.924	1.182	0.118	1.882	2.704
8	0.373	0.136	1.864	1.099	0.185	1.815	2.847
9	0.337	0.184	1.816	1.032	0.239	1.761	2.97
10	0.308	0.223	1.777	0.975	0.284	1.716	3.078

*n refers to the number of items per subgroup, not the number of subgroups.

(continued on next page)

TABLE 5.1   Control Chart Formulas and Constants (*continued*)

	Formulas	
Chart	Center Line	Control Limits
X-bar and R	$\bar{\bar{x}}$, or x double bar: average of $\bar{x}$'s	UCL = $\bar{\bar{x}} + A_2\bar{R}$ LCL = $\bar{\bar{x}} - A_2\bar{R}$
	$\bar{R}$: average of R's	UCL = $D_4\bar{R}$ LCL = $D_3\bar{R}$
X-bar and S[†]	$\bar{\bar{x}}$, or x double bar: average of $\bar{x}$'s	UCL = $\bar{\bar{x}} + A_3\bar{S}$ LCL = $\bar{\bar{x}} - A_3\bar{S}$
	$\bar{S}$: average of S's	UCL = $B_4\bar{S}$ LCL = $B_3\bar{S}$
Individuals and Moving Range	$\bar{x}$: average of individuals $\overline{MR}$: average of moving ranges	UCL = $\bar{x} + 2.66\overline{MR}$ LCL = $\bar{x} - 2.66\overline{MR}$ UCL = $3.267\overline{MR}$ LCL = 0 (None)
p[‡]	$\bar{p}$: average of p's	UCL = $\bar{p} + 3S_p$ LCL = $\bar{p} - 3S_p$ $S_p = \sqrt{\dfrac{\bar{p}(1-\bar{p})}{n}}$
np	$n\bar{p}$	UCL = $n \times$ (UCL for p chart) LCL = $n \times$ (LCL for p chart)
c	$\bar{c}$: average of c's	UCL = $\bar{c} + 3\sqrt{\bar{c}}$ LCL = $\bar{c} - 3\sqrt{\bar{c}}$
u[§]	$\bar{u}$: $\bar{u} = \dfrac{\text{Sum of units} \times \text{defects}}{\text{Sum of units}}$	UCL = $\bar{u} + 3\sqrt{\dfrac{\bar{u}}{n}}$ LCL = $\bar{u} - 3\sqrt{\dfrac{\bar{u}}{n}}$

[†]S (sample standard deviation) of each subgroup is:

$$\sqrt{\dfrac{\text{sum of}(x - \bar{x})^2}{n-1}}$$

[‡]Limits on p chart may be of unequal width, depending on sample size of each subgroup.

[§]Limits on u chart may be of unequal width, based on area of opportunity for each point.

FIGURE 5.11   Chart With Too Much White Space

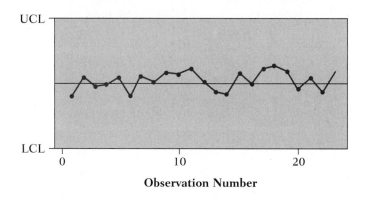

Systematic differences should not appear within the subgroups. For example, due to current market conditions, we may consider branch-to-branch sales volume differences as common-cause variation. If there are consistent differences between branches, however, we would not want to include these branches in the same subgroups. This consistent branch-to-branch difference will cause all the subgroup ranges (or standard deviations) to be large, resulting in a large average range. The large average range will cause the limits on the X-bar chart to be artificially wide, and the resulting X-bar chart will typically have all the points tightly bunched around the center line, with few if any points approaching the limits. This situation is sometimes referred to as "too much white space on the chart" (Figure 5.11). A better approach in this case would be to chart the different branches on separate charts, or at least to randomly sample from the branches so that this systematic difference does not show up in each subgroup.

Step 4: Gather the initial data. We may be able to use historical data, especially if we are using an individuals chart. When using historical data, the subgrouping scheme used to collect these data must be consistent with the approach chosen for future data. New data should reflect the normal behavior of the process, so no special actions should be taken to ensure that the data "look good." Data should be collected long enough to incorporate the most likely sources of variation, such as shift changes or end of month financial closings. Typically, 20 to 25 subgroups of data are needed to start the chart. Chart limits should not be calculated without at least 10 subgroups.

Step 5: Plot the data on the chart. Table 5.1 shows the calculations required for each type of chart. Note that the control charts for continuous data use tabled constants that depend on subgroup size, whereas the charts for discrete data do not. The bar over a variable, such as $\bar{p}$, indicates the average of all these values, the subgroup proportion defective in this case. For continuous data, the width of the limits on the X-bar (or individuals) chart is based on the average measure of variation (range or standard deviation). This is why the choice of items to go into individual subgroups is so important. Note also that for continuous data we have two charts: one for the average level and one for the variation. For discrete data, we only have one, for the average level of occurrences. As we shall see in Chapter 9, this is due to the underlying statistical distributions used by these charts. For these distributions the variation is a function of the average; hence, we can monitor both average and variation in one chart.

Note also that several constants are used that are dependent on subgroup sample size. This is why Table 5.1 has different values for A_2, D_3, and so on for each subgroup size. For subgroup sizes greater than 10, see the tables in Wadsworth, Stevens, and Godfrey (1986). As previously noted, all these limits represent ±3 standard deviations of whatever is being plotted. The formulas with tabulated constants were originally developed by Walter Shewhart in the 1920s, prior to widespread access to computer software. Note also that in some cases there is no lower limit on the range or standard deviation chart. Therefore, it is possible to get zero values, even when only common-cause variation is present.

These limits are all referred to as "3-sigma" or 3 standard deviation limits (the Greek letter sigma, σ, is used to represent the standard deviation), because the limits are all equivalent to the overall measure of average or variation ±3 standard deviations of whatever you are plotting. For example, the limits on the X-bar chart are the overall average ±3 standard deviations of the subgroup average. Because raw data are subject to variation, any calculation based on raw data, such as an average or range, will also be subject to variation, or uncertainty. This variation can be quantified with a standard deviation, just as a standard deviation can be calculated for raw data. Quantifying the standard deviations of calculated measures of sample data is part of the field of sampling distributions, which will be addressed in greater detail in Chapter 9.

The use of 3-sigma limits was originally selected by Shewhart because he found this to be a good economic balance between narrower limits, which might result in frequent false alarms (indications of a special cause when there really is none), and wider limits, which might result in missing some special causes. Theoretically, depending on the exact distribution of the data, we would only expect about three false alarms per 1000 data points plotted on the chart.

The ranges on the individuals and moving range chart are artificial ranges in that they are simply the ranges of consecutive observations. For example, if we plot sales figures, the first moving range is the difference between the first and second sales figure (largest minus smallest). The second moving range is the range of the second and third sales figure, and so on. We will have one less moving range than sales figure, and most sales figures are used twice, hence the term *moving range*. When using individuals and moving range charts, we are defining the sources of variation that occur from one data point to the next as common cause. Additional sources of variation, such as year-end closings, would be considered special causes.

The subgroup ranges for X-bar and R charts are the largest subgroup value minus the smallest. The subgroup standard deviation is calculated using the formula in Table 5.1. As previously noted, the standard deviation uses all the data and is a more precise measure of variation than the range. Because the standard deviation calculates the deviation of each data point from the overall average, it can be thought of as a type of average deviation of individual data points from the average. Squaring the deviations avoids having them average out to zero and results in the standard deviation not being exactly the same as the average deviation. The larger the standard deviation, the more variation in the data.

For the vast majority of data sets, most of the data points (60% to 75%) will be within 1 standard deviation of the average. The vast majority of the data (90% to 98%) typically are within 2 standard deviations of the average, and almost all (99% to 99.7%) are within 3 standard deviations. For example, if the average net sales for 25 sales regions this month is $1 million, with a standard deviation of $100,000, we would expect most regions to have sales between $900,000 and $1.1 million, the vast majority to have sales between $800,000 and $1.2 million, and virtually all between $700,000 and $1.3 million.

The software used to plot the chart should select scales wide enough to include the limits, as well as the most extreme values to be plotted. The limits should be drawn on the graph, and the data values ($\bar{x}$, R, p, c, and so on) plotted. For ease of interpretation, lines connecting consecutive values are drawn.

Step 6: Interpret the chart and reevaluate the original sampling plan. The chart allows the process to "talk to us." Look for any noteworthy behavior that could be telling us something important about the state of the process. These include trends, cycles, individual "outliers" (extremely high or low values), and so on. The center line and control limits on the chart provide benchmarks for interpreting the data. If the data are subgrouped in a rational manner, the control limits represent the approximate width of variation that could be explained or predicted by the common-cause variation in the system. If no special causes are present, the data values will tend to vary randomly within the limits. A point beyond the limits is evidence that a special cause has had an impact at least at that point. In general, use the problem-solving strategy to eliminate special causes prior to applying the process improvement strategy to improve the process fundamentally. Similarly, nonrandom behavior, such as a trend of points steadily increasing, provides evidence of a special cause, even if none exceed the control limits. Rules for detecting these types of special-cause signals will be discussed shortly.

Figure 5.12 represents safety data for a multinational corporation. It records the number of injuries, as defined by the Occupational Safety and Hazards Administration (OSHA), per 100,000 hours worked. Because we are counting the number of occurrences (injuries) and cannot count nonoccurrences, and because occurrences are rare, these data are best treated as defects, or Poisson data. The number of employees, and therefore the number of hours worked, was fairly stable over this time period (it varied less than 20%), so a c chart would normally be used. But the standard formula for reporting this type of data is to "unitize" it— that is, report injuries per 100,000 hours worked—so a u chart was used. A c chart of these data would look the same but would have a different scale.

The safety performance definitely improved over time, and this improvement was the result of a major safety initiative that provided considerable training worldwide and focused on unsafe behaviors, even if they did not result in an injury. The control limits had to be recalculated several times because there was evidence that the system was operating at a new level. These improvements could not

FIGURE 5.12
u Chart of OSHA Incident Rate (Safety Data)

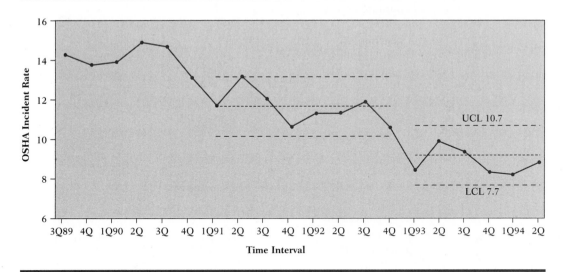

reasonably be attributed to the common-cause variation in the system because they exceeded the control limits. This is an example where a special cause actually represents an improvement to the process.

A key point to be made about interpreting control charts is the fact that random variation within the limits cannot really be "explained." For example, if management had been looking only at tables of OSHA injury rates comparing this quarter with previous quarters or previous years, they might conclude that a special cause had occurred each quarter. After the second quarter of 1993, suppose management asked the Safety Director: "Why did the injury rate go up this quarter?" Because this injury rate is the result of all the sources of common-cause variation in the system (work procedures, safety equipment, employee behavior, management attitudes, and so on), there is no single answer to this question. We might as well ask why a flipped coin came up tails this time when it came up heads last time. Unfortunately, these questions are asked frequently by business leaders. Even worse, employees waste a tremendous amount of time attempting to provide an answer and may resort to making up excuses. Of course, if this quarter's injury rate exceeds the control limits, it is a perfectly reasonable question to ask why. There is a reason, and we should be able to find it.

Sometimes the subgrouping scheme used is not appropriate. For example, if most of the points are outside the limits, this could indicate that important sources of common-cause variation were excluded from the subgroups (for continuous data) or that we used an inappropriate type of chart (for attribute data). Conversely, if all the points are hugging the central line, some systematic variation was included within continuous subgroups or an inappropriate type of attribute chart was used. If we see evidence of special causes, the data points involved

FIGURE 5.13   Additional Rules for Detecting Special Causes

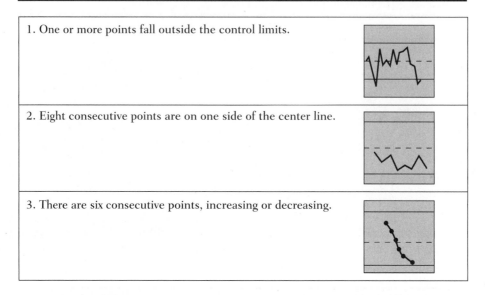

1. One or more points fall outside the control limits.

2. Eight consecutive points are on one side of the center line.

3. There are six consecutive points, increasing or decreasing.

should be removed and the limits recalculated to represent only the common-cause variation in the system.

In addition to points outside the control limits, there are other symptoms of special causes. These "run rules" are typically used to detect more subtle special causes that do not result in any points outside the control limits. For example, 8 consecutive points above (or below) the center line provides sufficient statistical evidence of a special cause. Six or more consecutive points steadily increasing (or decreasing), whether or not they cross the center line, is another indication. Figure 5.13 illustrates three commonly used run rules for detecting special causes. These rules of thumb are intended to help calibrate the human eye when it searches for patterns in the plot. The use of the one point outside (3-sigma) control limits and the "8 in a row" rule provide good ability to detect special causes and do not require calculation of additional limits. Most software packages will calculate whatever rules are desired, however. More sophisticated rules, using several zones within the control limits, are discussed in Brassard and Ritter (1994).

Step 7: Maintain the chart over time to improve or control the process. Once limits that truly represent only common-cause variation are identified, we should lock in these limits until we have evidence that the system has changed permanently. In general, do not update the limits as each new data point is taken. A gradual increase in the variable being charted would result in a gradual increase in the limits, and we might miss this trend. The philosophy is to force the process to follow the limits rather than force the limits to follow the process. In the OSHA safety data, the system was consciously changed to lower injury rate, and the limits were recalculated to reflect the new process level.

The chart will help us rapidly detect special causes over time so that we can identify and eliminate the root causes. This will prevent the process from deteriorating from the levels we are attempting to maintain. Recall that elimination of special causes in itself will not improve the fundamental capability of the process, it will only maintain it. By reviewing the charts periodically, however, we continue to learn about the process and are able to more effectively intervene to truly improve performance.

Sometimes very few, if any, special causes show up over time after our improvement efforts. In these cases we may decide to reduce the frequency of sampling or eliminate the chart altogether and focus our energies elsewhere. The control chart should only be continued as long as it adds value.

Variations   We have only covered the basic charts in this section. Other charts may be used by people with more advanced statistical training. One of these is the exponentially weighted moving average (EWMA), which is a weighted average of all the data, with the most recent data weighted most heavily. This chart is commonly used in engineering and financial applications. A similar chart is the cumulative sum, or "cusum," which is a cumulative measure of deviation from target. These charts tend to be more effective at detecting small shifts in the process average but less effective in detecting individual points affected by special causes or in detecting unusual patterns, such as seasonality in monthly data. For more information on these charts, see Wadsworth et al. (1986).

Tips
- Control limits represent the process—what it is currently capable of doing—and do not necessarily represent what is "good" or "bad." The specification limits should represent the needs of the customer. A process can be very stable but consistently perform at an unacceptable level. Capability analysis compares the process with customer needs to see if the current process is capable of meeting those needs.
- The individuals chart is generic and can be applied to virtually any situation. When in doubt as to the correct chart to use, try the individuals chart.

Histogram

Purpose   Histograms provide a view of the overall distribution (average level, variation, and shape) that a process is producing. The histogram is a bar graph that depicts how frequently different values occurred in the data. This graph is typically applied in the Evaluate Process Capability step.

Benefits   Histograms reveal the average level, the degree of variation, and the shape of the distribution. They can detect abnormalities in the data that should be investigated, such as outliers, the existence of two separate distributions, values that appear too frequently or infrequently, and so on. When plotted versus process standards (specifications), the histogram graphically portrays process capability.

FIGURE 5.14 Histogram of Age of Card Holders

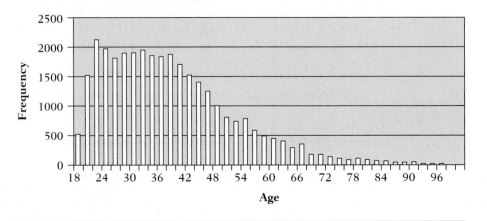

Limitations The major limitation of the histogram is the fact that it can produce misleading results if the underlying process is unstable because it hides the time dimension. In addition, the shape of the graph can be dependent on the number of bars used.

Examples A histogram of yield in the resin case study revealed that two distinct groups were mixed in the data. This led to further investigation and, eventually, improvement. Figure 5.14 is a histogram that shows the age distribution of credit card customers for a retail general merchandise store. This information is helpful for understanding current customer demographics. The obvious cut-off on the left side of the distribution is due to the fact that card holders must be 21 years old (18 under special circumstances). Figure 5.15 shows the distribution of the cycle time of order to receipt for a specialized financial service. Note the bimodal nature of this histogram, which led the business to investigate and find the source of multiple processes (different processes depending on amount of money involved).

Procedure Histograms are typically graphed using software such as Excel, Minitab, or JMP. Once process stability has been determined and a reasonable amount of data (typically at least 50–100 data points) have been collected, a histogram can be plotted. The histogram is a plot of the frequency with which each value or range occurred. Therefore, we must create ranges if we have continuous data. For example, credit card balances might be categorized using these ranges: $0–$99, $100–$199, $200–$299, and so on. Then the number of data values that fall into each of these ranges can be calculated. Typically, strive for 10 to 15 ranges, which should be based on round numbers. Most computer software programs will automatically calculate "reasonable" categories but will allow the user to select different categories if they wish. After calculating the number of values in each range, plot a bar graph where the horizontal axis represents the ranges and the height of the bar represents the number of values in this range.

FIGURE 5.15 Order-to-Receipt Cycle Time

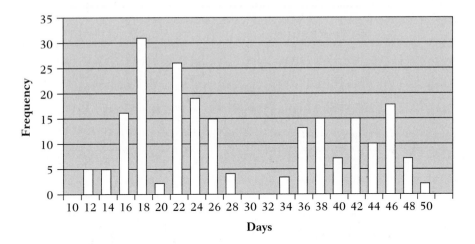

The standards can also be plotted on the histogram. When interpreting the histogram, we look at average level, degree of variation, and shape. If the standards are also noted on the graph, we can quickly see if we have a problem with the average, the variation, or both.

The shape of the histogram reveals the underlying distribution of the process. We are often interested in whether the process is producing the typical bell-shaped, or normal, distribution. This becomes important when performing statistical inference (see Chapters 8 and 9) because many of the tools used assume a normal distribution. Many measures of time—time to pay invoices, time to process checks or loan applications—tend not to have a normal distribution because it is impossible to have a negative time but possible to have very large times. These distributions tend to be steeper on one side than on the other. This is called a skew. Figure 5.16 shows potential histogram shapes.

Variations With small data sets one may choose to plot each data point without any grouping. If a value occurs more than once, the dots can be stacked on top of each other, resulting in a histogram shape. This is called a dot plot, for obvious reasons. Another variation is back-to-back histograms, or placing two histograms vertically, back to back, to enable easier comparison of the two distributions.

Tips
- It is important to determine process stability prior to interpreting a histogram. If the process is unstable, the histogram can be confusing or even misleading because it hides the time dimension.
- The shape of the histogram is affected by the number of bars used. If the shape looks atypical for the type of data being plotted, vary the number of bars and replot the histogram.

FIGURE 5.16
Potential Histogram Shapes

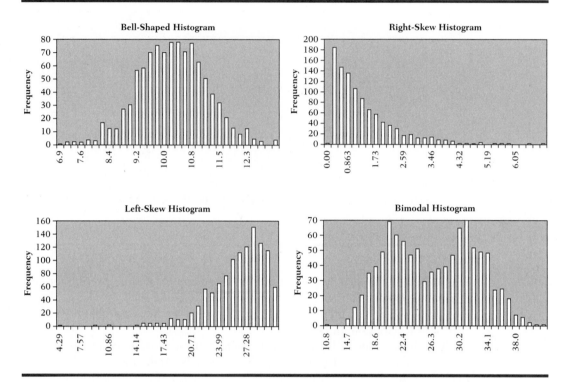

Pareto Chart

Purpose   A Pareto chart depicts the frequency with which different issues have occurred, drawing attention to the most important issues. Pareto charts are often used to plot data from checksheets. These charts are typically used in the Analyze Common-Cause Variation and Document the Problem stages.

Benefits   Having a graph of actual data helps ensure a rational versus emotional approach to improvement. The graph also helps focus the team's energy on the critical few versus the important many causes.

Limitations   The Pareto chart is only used with discrete data when documenting the type or cause of an issue or problem. These charts do not solve problems but only identify them.

Examples   A Pareto chart was used in the case study involving the telephone waiting time for a bank to plot the frequency of occurrence of various causes for a long waiting time (over five rings). In the soccer team study a Pareto chart plotted the frequency with which different causes of goals-allowed occurred.

FIGURE 5.17

Pareto Chart of Reasons for Being Able to Break Computer Passwords

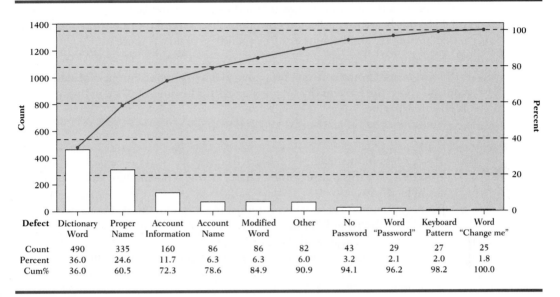

Defect	Dictionary Word	Proper Name	Account Information	Account Name	Modified Word	Other	No Password	Word "Password"	Keyboard Pattern	Word "Change me"
Count	490	335	160	86	86	82	43	29	27	25
Percent	36.0	24.6	11.7	6.3	6.3	6.0	3.2	2.1	2.0	1.8
Cum%	36.0	60.5	72.3	78.6	84.9	90.9	94.1	96.2	98.2	100.0

In Figure 5.17 a chart of computer passwords guessed by the Security Administrator during a security evaluation is plotted on a Pareto chart (Bradt, personal communication 1998). The count represents the number of times the password was guessed, using a variety of techniques—searching dictionary words, searching proper names, and so on. This chart highlights some typical mistakes users make when selecting passwords.

Procedure   Constructing a Pareto chart is quite simple and does not really require a computer, although a computer will tend to produce a better looking plot. Excel, Minitab, JMP, and many other packages can create Pareto charts. We simply add up the number of times each cause has occurred and plot these frequencies on a bar graph. The only uniqueness is that the bars are plotted in descending order of frequency; that is, the largest bar is first, then the second largest, and so on to visually prioritize the causes that are most important to address. If a large number of causes are identified, group the lowest frequency items in a "miscellaneous" or "all other" category to reduce the size of the chart.

Typically, the height of the bars decreases rapidly, indicating that a few key causes are producing most of the problems observed. (The chart received its name from the Pareto principle, which states that 80% of the problems are due to 20% of the causes.) These are the causes to focus improvement efforts on. If the largest bar is too big to address directly, break it down into finer detail and produce a second Pareto chart of the elements of this cause. For example, if "disputes" is the most frequent reason for reduced payment of invoices, a Pareto chart of all the specific causes of disputes could be created.

In some cases, it makes sense to plot total cost for each cause rather than frequency. This might be done if the costs associated with one cause are significantly different from those associated with another, such as medical claims when a hospital stay for heart surgery could cost more than several routine surgeries. In this case charting the total claims cost for each type of stay is more useful.

Variations   Some software programs plot a line representing cumulative percentage above the bars on the top of the chart. This line is called an ogive, and it is shown in Figure 5.17.

Tips
- If a process in unstable, the relative frequency of various causes may change over time. For this reason Pareto charts are most meaningful when applied to a stable process.
- Employ common sense when interpreting a Pareto chart. Focus on a cause that is not the largest bar on the chart if it can be dealt with more quickly or easily than the cause with the largest bar.
- If the miscellaneous category is fairly large relative to the others, too many items have been grouped together. Break some of these out as separate categories.

Run Chart (Time Plot)

Purpose   A run chart can help to quickly evaluate the performance of the process over time, with special emphasis on diagnosing the degree of stability. It is typically applied in the Assess Process Stability stage.

Benefits   The run chart helps identify trends, cycles, shifts, and unusual data points. It is one of the most versatile tools: it can be applied to any type of data, makes no assumptions, and requires no calculations other than an overall average.

Limitations   Because the run chart does not have limits representing the expected range of variation in the data, take care not to jump to conclusions about outlying points or even trends. The human eye is very effective at detecting patterns, and it is possible to see patterns where none exist.

Examples   Run charts were applied in several previous case studies, including the net realized revenue case, where they were key in solving the problem. Run charts have wide applicability and are particularly valuable when looking at financial data, such as earnings, that have only one data point at a time. The run chart in Figure 5.18 shows the proportion of calls to a computer support help desk that were not answered within 30 seconds (continuous data of time to answer were not available in this case). This plot reveals at least one spe-

FIGURE 5.18   Run Chart for Proportion of Calls Answered in More Than 30 Seconds

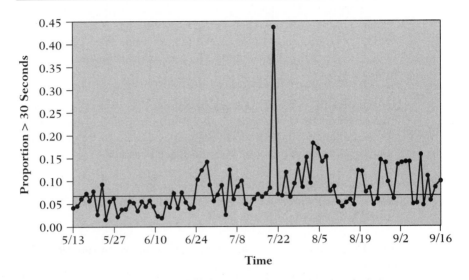

cial cause—the extreme high point—as well as a gradual increase over time (Sweat, personal communication, 1997).

Procedure   A run chart is a plot of data over time and may be thought of as a scatter plot where the horizontal scale is time. The variable plotted over time may be raw data, averages, some measure of variation (such as standard deviation or range), percent defective, or any other statistic of interest. If we put statistically based limits on the run chart, it becomes a control chart. Creating the chart is quite simple: plot the variable of interest on the vertical axis and time on the horizontal axis. Typically a center reference line is added to represent the average (or median, i.e., 50th percentile) of the data as an aid in interpretation. Then look for any obvious patterns, trends, or cycles.

Variations   For more rigorous interpretations, control chart run rules that do not involve control limits can be used. These include the 8 in a row above (below) the average and 6 in a row steadily increasing (decreasing) rules. It is also possible to count the number of times the center line is crossed. Of course, the run chart can also be converted to a control chart by calculating the appropriate limits.

Tips
- A run chart is a good alternative if you are uncertain of the appropriate type of control chart to use.
- It is a good idea to look at a run chart prior to calculating summary statistics or performing more detailed analysis of data. This will reveal situations where summary statistics are misleading due to a lack of stability.

Scatter Plot

Purpose   The scatter plot is used to identify relationships between variables. The plot reveals whether the variables appear to be related and, if so, indicates the strength and type of the relationship. The scatter plot is typically applied in the Study Cause-and-Effect Relationships and Identify Potential Root Causes stages.

Benefits   The scatter plot helps identify which input or process variables have an important relationship to an output variable. In addition to showing the overall relationship between two variables, the scatter plot draws attention to any outlying points, which may indicate special causes.

Limitations   Scatter plots reveal correlation, but they do not imply causation. In other words, seeing that two variables are related does not prove that one directly causes the other.

Examples   Scatter plots were used in the resin variation case study to identify a possible relationship between output and feed ratio, and in the advertising case in Chapter 2 to look for a relationship between change in advertising and change in sales. Another example is given in Figure 5.19, which shows the amount of payment made by credit card customers versus their total balance. This plot shows a lot of variation from customer to customer but does suggest that higher balances lead to higher payments.

FIGURE 5.19   Payment Amount Versus Credit Card Balance

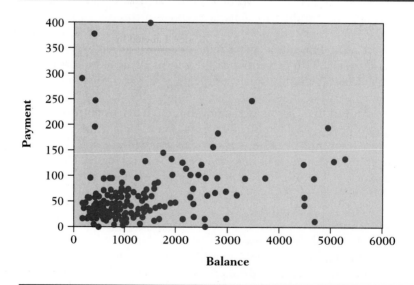

Procedure   Excel is good for scatter plots, although any statistical package will plot them. A scatter plot shows data points on two scales: a horizontal (x) scale and a vertical (y) scale. The placement of the point on the chart represents its value for the two scales, which represent two variables, such as advertising and sales. Typically, the scales are set up so that the horizontal scale represents the "cause" variable and the vertical scale represents the variable "affected" by the cause variable. This determination is made using subject matter knowledge. For example, we would expect advertising to cause sales to vary rather than the other way around. Remember, though, that the plot itself does not prove causality.

A scatter plot should have 50 or more data points, if possible. The scatter plot requires paired observations—for example, sales figures for the same months as the advertising figures—that are represented by two columns on a spreadsheet. Once the data have been plotted, any noteworthy relationship, or even isolated points that may be unique, are candidates for further study. As always, subject matter knowledge must be employed to properly interpret the plotted data. Figure 5.21 shows possible relationships that we might see in the plot. Note that real data may not conform to any of these relationships.

Variations   A somewhat more complicated but useful plot can be made by labeling the plotted points according to the value of a third variable, creating a "labeled" scatter plot. This helps determine the degree to which a third variable is influencing the relationship. Figure 5.20 shows a plot of paper towel strength versus absorbency, labeled by manufacturing process. The product development team was originally puzzled over the scatter plot of strength versus absorbency because it suggested that these variables have a positive relationship. This contradicts the fundamental theory of paper making, which indicates that strength and absorbency have a negative relationship. Once the team realized that the data

FIGURE 5.20   Scatter Plot of Strength Versus Absorbency, Labeled by Manufacturing Process

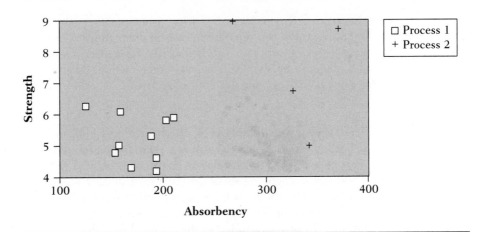

Correlation Coefficient   The strength of the relationship seen in a scatter plot can be quantified using a correlation coefficient, which is an index number that quantifies the degree of linear association. Nonlinear relationships will not be well quantified by the correlation coefficient, hence it is always a good idea to plot the data before doing numerical calculations (see Chapter 6).

Correlation coefficients range between −1 (a perfect negative linear association) and +1 (a perfect positive linear association). A correlation coefficient of zero indicates no linear relationship. In a negative relationship one variable goes up and the other goes down. In other words, a line through the data would have a negative slope. A positive relationship means just the opposite. Figure 5.21 shows examples of various relationships. The correlation coefficient for the weak positive and weak negative relationships in Pair A are about .4 and −.4, respectively. The strong positive and strong negative in Pair B are about .9 and −.9. The curved and no relationship in Pair C are each about 0. Note that the curved relationship appears noteworthy, but the correlation coefficient is 0 because the relationship is not linear. As a rule of thumb, correlation coefficients greater than about .7 (or less than −.7) are typically considered noteworthy.

The formula for the correlation coefficient between two variables, say, x and y, is:

$$r = \frac{\text{sum of} \left[(x - \bar{x})(y - \bar{y}) \right]}{\sqrt{\left[\text{sum of} \left(x - \bar{x} \right)^2 \right] \left[\text{sum of} \left(y - \bar{y} \right)^2 \right]}}$$

contained two different manufacturing processes, the labeled scatter plot was made. This revealed two separate negative relationships caused by differences in the manufacturing process. In other words, for a given manufacturing process, strength and absorbency have a negative relationship. By changing the manufacturing process, we can increase both properties.

Tips
- When there is uncertainty as to whether the plot indicates a real relationship or not, quantify the strength of the relationship using a correlation coefficient. (See the box on correlation coefficients above.)
- A reasonable degree of variation must exist in each variable to detect a relationship. If the variables have been tightly controlled for business purposes, minimal variation will show in the plot, reducing the chance of detecting any patterns. Insufficient range in the sampled data does not prove that there is no relationship.

FIGURE 5.21 Potential Scatter Plot Relationships

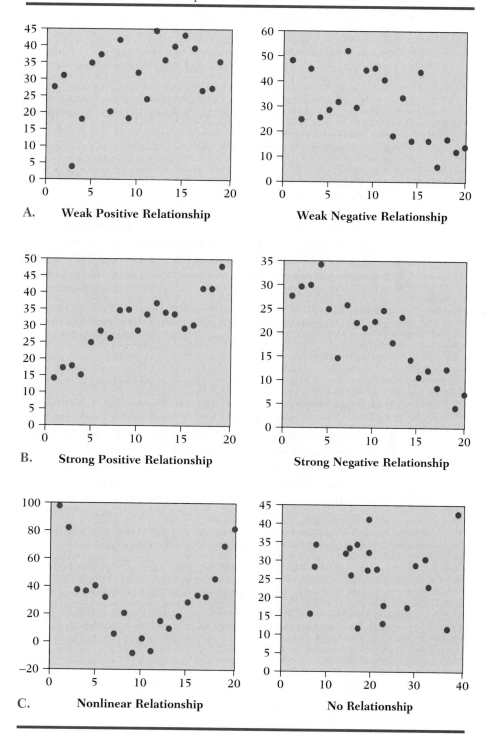

Stratification

Purpose   Stratification is used to identify key variables that have an impact on the output variable being analyzed. This helps us get below the symptoms and identify root causes. Stratification is typically applied in the Analyze Common-Cause Variation and Identify Potential Root Causes steps.

Benefits   Stratifying the data and plotting it by strata quickly determines whether a variable suspected of having an impact on an output is important or not. By stratifying we can also reduce a data set that includes both "apples and oranges" to subsets comprised of only one conceptual population.

Limitations   We must come up with the original means of stratifying through subject matter knowledge.

Examples   The resin case study detected an abnormal histogram pattern in output. When the data were stratified, the two separate batches, which had been combined, each had its own distribution pattern. Similarly, the surprising results of the second experiment in the advertising case study, where the "stimulus–response" curve did not match the theory, led investigators to identify multiple segments in the population and attempt to stratify the data to obtain individual stimulus–response curves for each segment. Other examples are stratifying customers by amount of business, income levels, geographic location, or male versus female for targeted marketing.

Procedure   Stratification is nothing more than breaking down the data into logical categories to detect differences. For example, calculating statistics on the heights of adults in the United States would not be very useful. It is well known that there is a systematic difference between male and female heights. A more logical approach would be to stratify by sex and analyze adult male heights separately from adult female heights. This is what clothing manufacturers do on a regular basis. Various statistical tools, such as scatter plots, can benefit from stratification. For example, when plotting the percentage of defaulted loans versus loan amount, we might want to label the points from each branch with a different symbol, creating a labeled scatter plot. This would reveal branch-to-branch differences as well as the relationship, if any, between loan amount and defaults.

To stratify the data, identify the potential variables included in the data set that could cause two or more distinct populations. Subject matter knowledge is required here. Then segregate the original data set into two or more groups (strata) according to the variable identified and compare these individual distributions with histograms, other graphs, or calculated statistics to see if there is a noteworthy difference. If there is not, keep the data combined. If there is a noteworthy difference, we may choose to keep the groups separate for further analysis.

Variations   In most cases data are stratified to compare one strata to another. Of course, there can be more than two strata. For example, we might look at five different regional branches of a pharmaceutical supply business. Sometimes strata are analyzed separately rather than against each other. An example might be a clothing manufacturer that designs and markets men's, women's, and children's clothing. Data gathered on people's physical dimensions to ensure that their off-the-rack clothing fits as many men or women as possible would be most useful if kept separate and stratified by sex and by adult/child.

Tip   You can never do too much stratification. If the variable you used to stratify turns out not to be important, cross it off the list and move on to other variables.

5.5   Knowledge-Based Tools

Knowledge-based tools are intended to organize and analyze the knowledge we already have. They are therefore highly dependent on the knowledge of the people applying the tool. Having the right people "in the room" when the tool is applied is essential. Knowledge-based tools are particularly useful when applied by cross-functional teams because we benefit from the knowledge of the entire team.

Affinity Diagram

Purpose   An affinity diagram organizes a collection of ideas into "natural" groupings, so that they can be more effectively addressed. The affinity diagram is particularly valuable after a brainstorming exercise that has produced a large number of ideas. It is typically applied in the Choose Best Solutions stage when the ideas generated by brainstorming become the "data" to be analyzed.

Benefits   In addition to organizing the ideas into a "workable" number, this grouping process also helps to identify the uniqueness, or essence, of each group. The process is designed to ensure that everyone's creativity comes into play and that no one person dominates the decision making.

Limitations   Like all knowledge-based tools, the affinity diagram is limited by the knowledge of those who participated in its use. It is also possible to lose details of individual ideas once they have been grouped.

Examples   Brassard and Ritter (1994) depicted an affinity diagram developed by Goodyear to organize issues team members had identified surrounding implementation of the business plan. In this particular case the ideas to be organized were related to identifying problems rather than to prioritizing solutions. The following issues were identified:

- Reconciliation with corporate resource allocation
- Functional groups not trusting each other
- Poor definition of prioritization for market introduction

- Insufficient team approach to new product development and introduction
- Unrealistic goals create blue-sky attitude
- Complexity driven by customer demands requires added investment
- Lack of integration of support group plans
- New government regulations
- Sales forecast is not accurate
- Production capacity
- Possibility of economic downturn
- Faster pace of product introductions stretches resources
- Capital availability limits opportunities
- Them/us perception
- Group members not making individual commitment to success of the plan
- Fighting daily problems ("fighting alligators versus draining the swamp")
- Ownership of plan does not cross functional lines
- Plan is not linked to unit financial goals
- Rewards do not compensate team playing
- Communication between functional groups is difficult
- Group is not focal point for conflict resolution
- Production capability to support changing requirements

Clearly, it would be difficult, if not impossible, to deal with each of these issues independently. The first task was to identify similar ideas and combine them. The ideas were grouped together and labeled, producing the affinity diagram in Figure 5.22.

When these groups were examined, some were found to be similar in nature and could be grouped themselves. In the second round additional groupings were made, which produced the affinity diagram shown in Figure 5.23. This affinity diagram suggests that there are fundamentally only four macro issues surrounding implementation of the business plan:

- Our business planning approach must be improved
- The group could function more effectively
- Limited resources are a challenge
- External factors affect implementation

Four subteams were formed, and each attacked one of the major issues. Discussing the issues, addressing them, and even reporting on progress are all greatly simplified by this organization.

Procedure   We typically begin by gathering a large number of ideas through brainstorming, but that need not be the case. We can create an affinity diagram using any collection of ideas. The best method for handling this is to write ideas on individual Post-It notes that can be moved easily from spot to spot. State the idea in a concise manner (with at least a noun and a verb), and place the notes on a large visible work area, such as a wall, flipchart pad, or table, in a haphazard fashion. It is important that team members be able to see all of the ideas.

The team is then asked to look over the ideas and think about which ones might naturally go together—for whatever reason makes sense to each individual. There are no predetermined categories. This enables creative or "out of the box"

FIGURE 5.22 Business Plan Issues

Plan Not Integrated

> Reconciliation with corporate resource allocation
> Ownership of plan does not cross functional lines

No Strong Commitment to the Group

> Functional groups not trusting each other
> Group members not making individual commitment to success of the plan
> Rewards do no compensate team playing

Lack of Time and Resources

> Complexity driven by customer demands requires added investment
> Capital availability limits opportunities

External Factors Affect Implementation

> New government regulations
> Possibility of economic downturn

Fast New Product Introductions Stretch Resources

> Lack of integration of support group plans
> Faster pace of product introductions stretches resources

Capacity May Not Meet Needs

> Sales forecast is not accurate
> Production capacity
> Production capability to support changing requirements

Planning Approach Not Standardized

> Poor definition of prioritization for market introduction
> Fighting daily problems (alligator/swamp)

Communication Issues Within the Group

> Insufficient team approach to new product development and introduction
> Them/us perception
> Communication between functional groups is difficult
> Group is not focal point for conflict resolution

Means Not Clearly Defined

> Unrealistic goals create blue-sky attitude
> Plan is not linked to unit financial goals

FIGURE 5.23
Macro Issues of Business Plan

External Factors Affect Implementation

New government regulations
Possibility of economic downturn

The Group Could Function More Effectively

No Strong Commitment to the Group

Functional groups not trusting each other
Group members not making individual commitment to success of the plan
Rewards do not compensate team playing

Communication Issues Within the Group

Insufficient team approach to new product development and introduction
Them/us perception
Communication between functional groups is difficult
Group is not focal point for conflict resolution

Limited Resources Are a Challenge

Capacity May Not Meet Needs

Sales forecast is not accurate
Production capacity
Production capability to support changing requirements

Fast New Product Introductions Stretch Resources

Lack of integration of support group plans
Faster pace of product introductions stretches resources

Lack of Time and Resources

Complexity driven by customer demands requires added investment
Capital availability limits opportunities

Our Business Planning Approach Must Be Improved

Plan Not Integrated

Reconciliation with corporate resource allocation
Ownership of plan does not cross functional lines

Planning Approach Not Standardized

Poor definition of prioritization for market introduction
Fighting daily problems (alligator/swamp)

Means Not Clearly Defined

Unrealistic goals create blue-sky attitude
Plan is not linked to unit financial goals

Source: From M. Brassard & D. Ritter, *Memory Jogger II*, 1994. Copyright © 1994. Reprinted by permission of GOAL/QPC.

thinking and prevents the team from falling into historical ruts. After reviewing the ideas the team organizes them into groups, one note at a time. The rules for the organizing process are as follows:

- No talking or nonverbal signals are allowed.
- Everyone is allowed to move notes simultaneously.
- You may move a note someone else has already moved, but prior to doing so, pause to think about why he or she moved it there.

It is also suggested that people use their opposite hand (left hand if the person is right handed) to help stimulate both sides of their brain.

If a note is being fought over—that is, being moved back and forth between groups—it is usually a good idea to make another copy of it so the idea appears in both groups. There are no magic numbers of groups, nor items per group, although 5 to 10 groupings are common. It is possible to have groups of one individual note, or a "miscellaneous" grouping, which is sometimes called an "orphan pile." The organizing continues until movement of notes comes to a stop, or at least slows down to a trickle. Typically, this takes only a few minutes.

Moving notes simultaneously and working in silence both speed up the process and improve the quality of the results. If discussion takes place over each movement of a note, the process tends to bog down into a lengthy debate. Having everyone work simultaneously also greatly speeds up the process. Another advantage of working in silence is that it levels the playing field; it equalizes any organizational hierarchy, dominant personalities versus bashful contributors, and so on. There will be time to discuss the results, and even change them, but this comes later.

Once the items have been organized, we evaluate the organization and name the groups by asking three questions of each group, one group at a time:

- Do all the items in this group belong together?
- What is the uniqueness or essence of this group?
- Do any orphan pile notes belong in this group?

It is important to ask the questions in this order. The team can be slowed down by debating the essence if some notes in the group do not belong there. Similarly, team members may suggest that items in other groups belong here. To make the process effective and maintain focus, ask these questions sequentially for one group at a time.

Answers to these questions should be developed by team discussion and consensus. Any notes that are determined not to belong in a particular group should be put in the orphan pile unless there is clear and immediate agreement that they belong to some specific other group. To ensure clarity, the essence of each group should be stated as a sentence or phrase rather than as one word. These become the "header cards" or "parents" of the group and are typically written in a different color and placed at the top of the group. Sometimes an individual note is selected as the header card. If there are a large number of header cards, the team may wish to organize them, just as they did the individual ideas. This results in a multilevel affinity diagram such as the business plan diagram in Figure 5.23.

Variations   If the group is larger than 10 people, it may make sense to work in waves. Have the team break up into groups of 8 or less and take turns moving the notes. The first group may work for 5 minutes, the second for 5 minutes, and so on. It is important to ensure that each group has an opportunity to influence the final diagram.

Tips

- Because this process is new for many people and involves "rules," it is a good idea to use a trained facilitator who does not participate in organizing the ideas. The facilitator is neutral and can focus completely on how the process is working. (The role of the facilitator in effective teamwork is discussed in Appendix A.)
- If there are potential relationships among the header cards, it is often helpful to do an interrelationship digraph (ID)—that is, identify cause-and-effect relationships—of the headers. This helps determine which headers may have the greatest overall impact and takes into account how the headers are related to one another. (IDs are discussed in more detail on page 199.)

Brainstorming

Purpose   Brainstorming is used in many contexts to rapidly generate a diverse list of ideas or potential root causes. In the application of statistical thinking it is typically applied in the Identify Potential Root Causes step.

Benefits   The use of a formal methodology maximizes the openness of input from the team by preventing criticism, premature rejection, or organizational hierarchy from interfering with the process.

Limitations   If ideas are not rigorously evaluated after brainstorming, the ideas can reflect people's biases rather than the best current knowledge of the situation.

Examples   Brainstorming was used by the baby wipe team to come up with their original list of potential causes for poor flushability. It was also used by the telephone response team to come up with categories for a checksheet, which was subsequently used to document the major causes of customer waiting. It is very effective any time a team wants to get all the ideas documented prior to deciding which to follow up on.

Procedure   Brainstorming typically begins in a formalized fashion by agreeing on the specific question or issue the team is brainstorming. The issue needs to be concisely worded so all team members are clear on its meaning. Team members think about the issue for a few minutes and write down their ideas. Next, each person states one idea. These ideas are written down by a "scribe," preferably in plain view of the entire team, and should not be modified or reworded in any way unless agreed to by the person who suggested it originally. (The role of the scribe in effective teamwork is discussed in Appendix A.)

Each idea must be a concise statement; no long speeches are allowed. In addition, no discussion, disagreement, or modification of an idea is allowed at this stage. Discussion of ideas at this stage can bog down the team in long debates, discourage team members from suggesting controversial ideas, and lead to domination of the session by team members who have higher rank in the organization. Allowing everyone their turn without interruption levels the playing field, encourages team members to be creative, and greatly speeds the process. Questions of clarification are allowed to ensure that other team members understand the idea presented.

Once all members have given an idea, the process is repeated as members give second, third, or more ideas. Having all the ideas visible prevents replication of ideas and enhances creativity by enabling one idea to spark others. As the process continues, some members will run out of new ideas. When this occurs, they may pass on their turn. When only a few ideas are left, members may switch to free flow and call them out.

Once all the ideas have been documented, the team discusses them to further clarify ideas, suggest consolidation of ideas, or even eliminate some. This discussion is not intended to result in a lengthy debate, but only to allow for obvious points that could not be made during the formal brainstorming session. The output of the brainstorming session is rarely complete in itself. Almost always this information is input to some other process, such as an affinity diagram, multivoting, or an interrelationship digraph.

Variations   With a small team (four or less) it may make sense to begin the process in free flow format. You must be careful that no one dominates the session, however.

Tip   Because brainstorming is such a disciplined process, it often helps to have a formal facilitator, who is neutral and does not suggest ideas, run the session. As teams get more experience with the process, they can self-facilitate without needing an outside facilitator.

Cause-and-Effect Diagram

Purpose   Cause-and-effect diagrams document the relationship between observed effects, and their causes, in decreasing levels of detail. Cause-and-effect diagrams are also known as "fishbone" or "Ishikawa" diagrams. They are typically used in the Study Cause-and-Effect Relationships step.

Benefits   This diagram serves as a useful reference document to help the team address an issue in a disciplined manner, identify holes in the team's knowledge, and even as a training or problem-solving tool for others who work on the same process.

Limitations   The use of predetermined categories of causes may limit thinking. In addition, the completeness of the diagram will be limited by the knowledge of those who participated in its development.

Examples   A cause-and-effect diagram was used in the resin study to depict the different types of causes of resin variation. One was also used by the bank telephone response team to help determine the causes of waiting to include in the checksheet used to gather data. Figure 5.24 shows a cause-and-effect diagram developed by a team attempting to reduce the time it takes to make corporate tax payments. This served as a base of reference for the team as they attacked the problem.

Procedure   Statistical or graphical software can be used to create and store electronic versions of cause-and-effect diagrams. The cause-and-effect diagram begins with a careful statement of the effect, which is typically the problem or issue the team is trying to address. Major categories of causes are then identified and agreed upon. These form the major branches of the diagram, or fishbone. The logic in a cause-and-effect diagram is just the reverse of that of an affinity diagram. To create an affinity diagram, we begin with individual ideas and group them into larger categories that are only named later. To create a cause-and-effect diagram, we begin with predefined categories and then identify ideas at greater levels of detail.

The traditional categories used are people, materials, technology, methods (procedures), and measurement, with environment often added as a sixth category. Depending on the particular application, other categories may be more useful. If the team does not have good ideas for the categories, the traditional categories provide a useful starting point because they force broad thinking. Teams sometimes focus only on people or technology, but the most important root causes are often procedures (or lack thereof) for doing the work, poor raw materials (including databases), or problems with the measurement system.

Once the major categories are chosen, subcauses for the effect within each category are identified and included on the graph. For example, if technology is a category for the effect "computer system crashes," specific aspects of technology that could cause the system to crash would be identified: operating system, servers, applications, and so on. This process is typically continued for two or three levels of detail, resulting in a final diagram in the form of a fishbone, such as the one shown in Figure 5.24.

Variations   It is possible to creatively come up with cause categories by creating an affinity diagram. This chart can then be recast in the form of a cause-and-effect diagram.

Tip   It is very useful to review the final diagram with a wide diversity of people involved with the process to ensure that it accurately depicts our best understanding of the relationships between the causes and the effect. As with any knowledge-based tool, it should be developed by those most knowledgeable about the process, and updated as our knowledge increases over time.

FIGURE 5.24

Cause-and-Effect Diagram for High Cycle Time in Paying Corporate Property Taxes

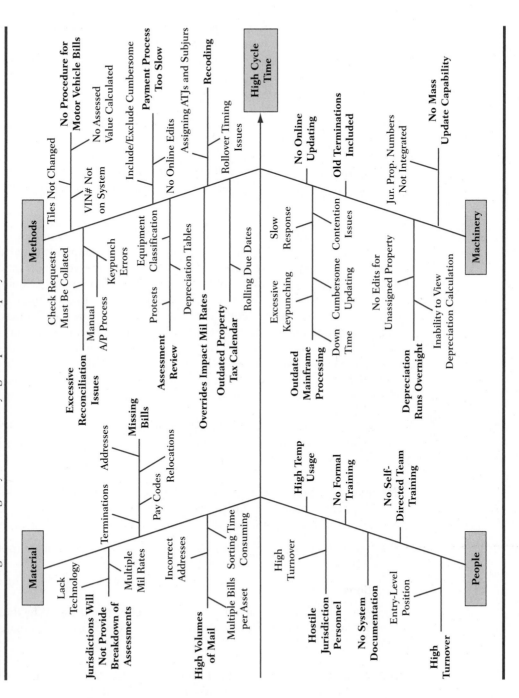

Five Whys

Purpose   The Five Whys are used to probe beneath the symptoms of a problem to get down to the root causes. The intent is to ask "Why" a symptom has occurred, and then why the cause occurred, and then why the cause of the cause occurred, and so on. Often, we have to dig down through as many as five causes to get to the root cause. This technique is typically used in the Identify Potential Root Causes step.

Benefits   Helps the team discover the true root causes that must be dealt with to permanently address the issue at hand.

Limitations   This technique is limited by the team's ability to accurately answer the Why? question several times and to know the appropriate point at which to stop.

Examples   Safety data are particularly prone to evaluating symptoms versus root causes. For example, "Worker violated safety rules" is commonly listed as the cause of safety incidents. If we want to prevent reoccurrence, an important question to ask is, Why did the worker violate the rule? Was it ignorance (lack of training), time pressure, ambiguity in the rule, or something else? Each of these causes of the cause would lead to different actions to prevent reoccurrence. For example, if it was lack of training, why was the worker not trained? Because we don't have a good training system in place. Why don't we have a good training system in place? Because management doesn't see the value in doing it. Why don't they see the value in doing it? Because they have never been educated about the real cost-benefits of training. Now we know what must be done to prevent reoccurrence over the long term.

Although many of the cases presented in this text used Five Whys thinking, it is not documented in a formal sequence of questions and answers. This is typical in practice.

Procedure   The procedure is quite simple. We simply continue to ask "Why?" the problem occurred until we get to a root cause. It is safe to conclude that we have described a root cause when the cause meets one of these criteria:

- Responding to the cause would prevent reoccurrence of the problem (not just resolve the current symptoms).
- The sources of this cause are obvious (Why is it humid in Florida in the summer?).
- The sources of this cause are general and difficult to address (Why do managers in the United States tend to be more short-term focused than managers in Europe or Asia?).

Tip   Finding the root cause may take less than five "Whys" or more. Five is simply a typical number. Continue to ask "Why?" until the root cause is discovered.

Flowchart

Purpose   The purpose of a flowchart is to graphically depict the flow of a process. Such a picture is particularly important when the actual process cannot be physically seen, such as the flow of an electronic order. Developing a flowchart is typically the first tool used in the Understand the Process step of the process improvement strategy.

Benefits   This chart helps everyone grasp the big picture of the overall process and puts individual steps in proper context. The chart also ensures that everyone has the same understanding of how the process actually flows.

Limitations   If the flowchart is made at too high a level, it can be very limited in value. If it is at too low a level, it can be confusing because of all the detail.

Examples   There are many types of flowcharts. Most of the flowcharts in the case studies in this text are macro-level flowcharts. That is, they depict the process at a macro level. These include the net realized revenue case, the baby wipe case, and the resin variation case. Often, a key step from the macro flowchart is selected to develop a more detailed flowchart of this one step. These are called top-down flowcharts. This type of chart is explained in the section on variations.

Procedure   Flowcharts are typically created with graphical rather than statistical software. To make a macro-level flowchart, the team must identify the major value-added steps in the process. At this point details of each step should be ignored. Once identified, the major steps are listed sequentially, either going left to right or top to bottom on the page. Of course, numerous software packages produce flowcharts quite easily.

Unfortunately, most processes are not as simple as 1, 2, 3, done. Most processes involve some decision points (e.g., "If the ordered item is not is stock, then . . ."), or some flow back, that is, moving back in the process. For this reason arrows are used to indicate the direction of flow, and a variety of symbols indicate the type of action done at each step. Although there is no official standard, these symbols are commonly used in practice:

Oval	Beginning and end of process
Rectangle (box)	Value-adding activity in the process
Diamond	A decision point
Circle with letter (or number) inside	A break in the chart (letter/number indicates where to go next)

Using these symbols, we can develop a chart, step by step, that accurately portrays the flow of a process.

Variations   Another approach is to compare the actual process with a flowchart depicting what the process should be if there were no non-value-added work. This helps identify potential areas of improvement. The examples in Figures 5.25 and 5.26 (from Salamone, personal communication, 1997) compare the

FIGURE 5.25
Mailing Incomplete Patent Applications

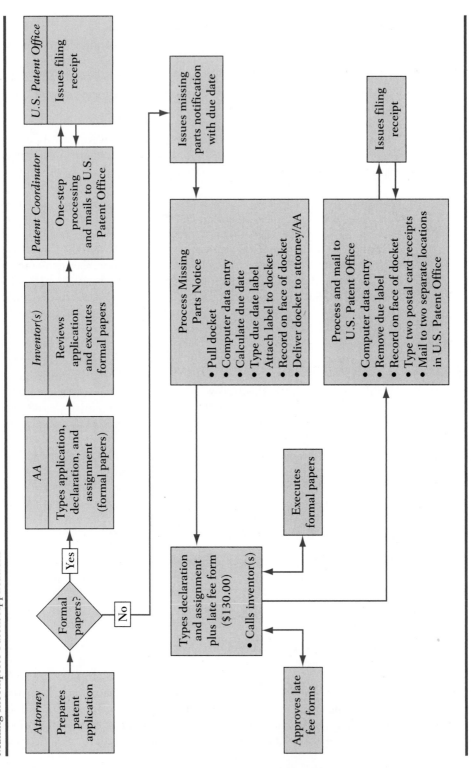

FIGURE 5.26   Mailing Complete Patent Applications

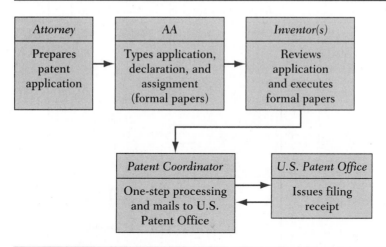

current process for patent filings in a research and development organization with what it would look like if there were no incomplete filings.

Other versions of flowcharts include a deployment flowchart, which lists process steps vertically and organizational units across the top of the page. The step is then placed in the appropriate column to indicate which organizational unit is accountable for that step. This type of chart is particularly useful for managing large, cross-functional processes. The deployment flowchart in Figure 5.27 shows the process for resolving customer complaints used by a specific business at 3M Corporation (from Britz, Emerling, Hare, Hoerl, & Shade, 1996).

Another type of flowchart is the top-down flowchart. This is a macro-level flowchart showing the major activities of the process beginning on the left and continuing to the right. Detailed steps for each element in the macro flowchart are shown below these activities. This chart combines the information from macro and detailed charts. Figure 5.28 is a top-down flowchart for the process of giving off-list prices (OLP) to customers. (Off-list pricing is below the list price.) This process can be broken down into the four major steps shown across the top of the chart.

Tips
- Start out by documenting the process as it currently is rather than as it should be. Prior to deciding on changes, we must know the current state of reality.
- Ask several team members to develop flowcharts independently and compare the results. Typically, there is much more variation than anticipated, which helps the team see the need to cooperate.

FIGURE 5.27   Process for Resolving Customer Complaints

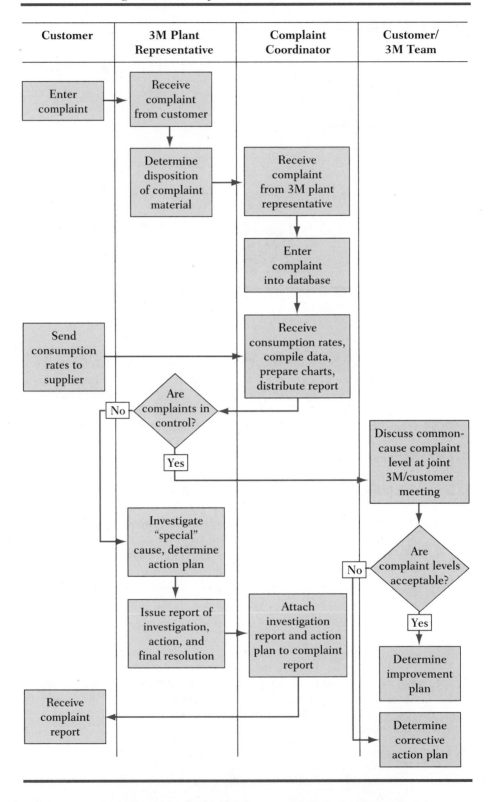

FIGURE 5.28   Top-Down Flowchart for Off-List Pricing

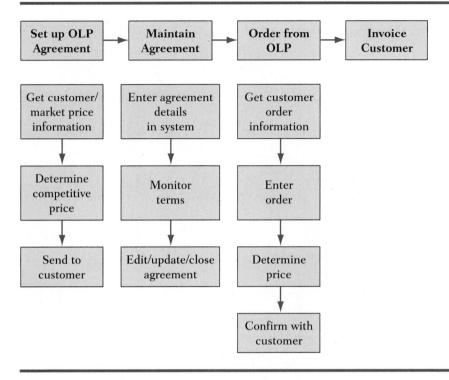

Interrelationship Digraph

Purpose   The purpose of an interrelationship digraph, or ID, is to evaluate the cause-and-effect relationships between issues to identify which are the "drivers" and which are the "effects." The drivers will have high leverage in the system because they are causes of other issues identified. Improving them first will have a positive impact on the other issues. The ID is typically used in the Choose Best Solutions step.

Benefits   The ID helps prioritize issues and pinpoints where to implement solutions first. It also provides a picture that reveals the interrelationship of the issues.

Limitations   The ID only looks at relationships between two variables at a time, so it is difficult to see complex relationships involving several issues. It also does not quantify the strength of the relationships or identify whether they are positive or negative.

Example   Figure 5.29 shows a completed ID that details the reasons statistical thinking is not used more often by managers of a large corporation. As you can see, these digraphs can become quite complex.

FIGURE 5.29   Reasons Statistical Thinking Is Not Used More

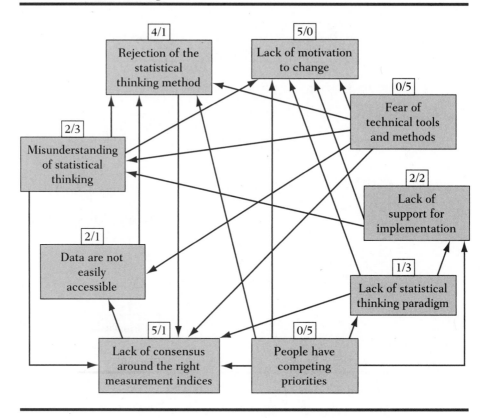

Procedure   The ID begins by arranging the individual items (issues, root causes, and so on) in a circle, typically on a flipchart or wall chart (Figure 5.30). It is important that everyone can see the items and the resulting digraph. The team then evaluates the relationship between every possible pair of individual items. This is typically done by starting with the item at the top of the digraph and comparing it with the item to its right, and then the next item to the right, and so on, in a clockwise fashion (Figure 5.31) (Hare, personal communication, 1996). Once all items have been evaluated against the item at the top, we move to the item to its right and evaluate its relationship with each remaining item, in the same clockwise fashion. This process continues until we have evaluated all possible pairs (Figure 5.32). We do not need to evaluate pairs twice. When we get to the next to the last item, there is only one other item left to evaluate it with, the last. All other pairs involving the next to the last item will have already been evaluated.

When evaluating each pair, the question we are answering is: Does item A tend to cause item B, does item B tend to cause item A, or is their relationship negligible? If A causes B, we draw an arrow going from A to B. If B causes A, we draw it the other way. If the relationship is negligible, no line is drawn. Note that

FIGURE 5.30 First Step in Creating an Interrelationship Digraph

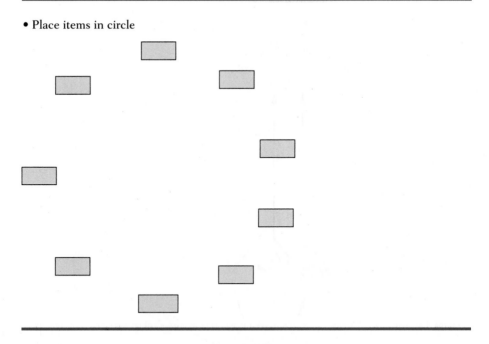

• Place items in circle

FIGURE 5.31 Second Step in Creating an Interrelationship Digraph

• One item at a time
• Arrows indicate primary cause
• Arrows one way only
• May have no arrow

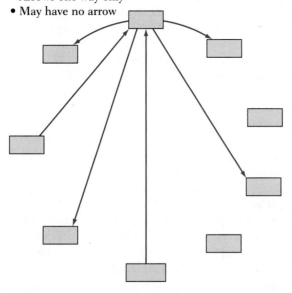

FIGURE 5.32   Completed Interrelationship Digraph

- Complete the circle
- Label "In"/"Out"
- High number of "Ins" are effects
- High number of "Outs" are drivers (root causes)

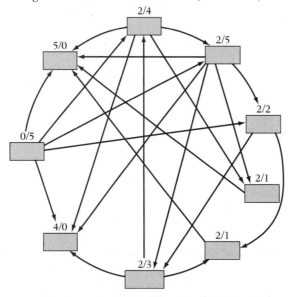

we force a decision one way or the other. It is not acceptable to say they cause each other. A determination must be made of which is more likely to cause the other. Forcing a one-way arrow ensures that we will get meaningful information concerning which items are the drivers of the others.

Once all arrows have been drawn, count the number of "in" and "out" arrows for each item and write these numbers on the digraph. The items with the most "outs" are drivers, or leverage points. If we address these issues, we will indirectly address many of the other issues as well. These are the issues we generally address first. The items with the most "ins" are effects, or symptoms. They will be addressed indirectly if we work on the other items. Addressing them first would not have any positive impact on the others.

Variations   It is possible to be much more creative with this chart. Here are some examples: draw a picture of an organization and begin drawing arrows to reflect which groups influence others; draw arrows between several priorities of a business to note interrelationships; or create any meaningful diagram on which to draw arrows of influence or interrelationship. The possibilities are limitless. It is also possible to weight the strength of each arrow.

Tips

- For the ID to work the items being evaluated must have some interrelationships; that is, they must relate somehow. If the items are totally independent, some other tool would be more appropriate.

- Limit the ID to 15 items or less. If there are more items than this, refine the categories using a tool such as an affinity diagram.

Is–Is Not Analysis

Purpose   Is–Is Not analysis is a rigorous process for carefully documenting an issue so that true root causes can be identified. Specifically, it notes when, where, and how an issue occurred, as well as when, where, and how it did not occur, but could have. This is typically done in the Identify Potential Root Causes stage.

Benefits   This analysis documents the issue in significant detail, and comparing when an issue did or did not occur can provide important clues about potential root causes. This analysis provides a logical basis for evaluating suggested root causes by pointing out discrepancies between when, where, and how the issue occurred versus when it did not.

Limitations   This thought process works particularly well for special-cause variation because there is something unique about special-cause observations. It may not work with common-cause variation, however. For example, if the number of late shipments this month is unacceptably high but within statistical control limits, this may be due to common-cause variation and not to any change in the process. If this is the case, there is no root cause, and searching for a single reason for the high late shipments may be in vain.

Example   In the flushability complaints case study an informal Is–Is Not analysis was done of poor flushability to help determine the root causes. The problem was identified as being unique to a particular time frame and to one production facility. It could have occurred in all production facilities, and it could have occurred at other time periods—but it didn't. This information was very helpful because potential root causes that applied to facilities not involved in the problem, such as common fibers or chemicals, could be eliminated.

In Figure 5.33 we show a more formal Is–Is Not analysis for not supplying customers in a timely fashion in a distribution process. Note that this typically is done in some type of spreadsheet, such as Excel.

Procedure   Begin an Is–Is Not analysis by carefully describing the problem or issue. Next, note the key questions to be asked. These typically include What? (the type of issue), Where? (location or region of occurrence), When? (the time dimension), and Who? (people involved). Depending on the specific application, other questions may also be appropriate, such as How? (mode of failure, for example) or Why? (root cause, if known).

FIGURE 5.33
Is–Is Not Analysis for Late Delivery of Items in a Distribution Process

	Is	Is Not	Potential Explanations	Actions
Where	Northeast United States	South, Midwest, or West	Different warehouses	Review data on individual warehouses
What	Out of stock on variety of parts	Late delivery on stocked parts	Delivery process is OK, but inventory management is not	Check inventory management system
When	Problem started around the beginning of October	Prior to that time	New computer system installed at the end of September, efforts to reduce general inventory levels began in October	Check for differences in inventory management algorithms between computer systems, document any change in targeted inventory levels.
Who	Involves all order takers	N/A	N/A	N/A

Once the questions are determined, we answer them for where (when, who, etc.) the issue *is*. This documents the observed facts about the issue. We then answer them in terms of where (when, who, etc.) the issue *is not*—but could be. These questions force us to think beyond the observed facts and provide the basis for identifying root causes. By comparing these two sets of answers, we document clues that could explain the occurrence of the issue in some areas, but not others. These clues are placed in the next column of the spreadsheet and help us identify potential root causes. They also provide a framework with which to evaluate suggested root causes. For example, if someone suggested that the root cause of late deliveries in the distribution process was some new order takers who were hired, it would be obvious from the spreadsheet that this is unlikely because late deliveries were occurring with all order takers. This process helps us to both generate and test hypotheses about root causes of the issue.

Next, we determine appropriate actions to follow up on the potential root causes identified and place these in the next column of the spreadsheet. The spreadsheet then becomes not only documentation but also a planning tool, which can be reviewed periodically to track progress on follow-up.

Variations   We do not need to limit ourselves to the questions of How, When Where, and Who. We can ask whatever set of questions makes most sense for our particular situation. See Kepner and Tregoe (1979) for more details.

Tips
- This process works very well but can be time consuming. Reserve this approach for problems that are particularly difficult to solve.
- Verify that there is a true difference (special cause) prior to choosing an Is–Is Not analysis. This approach generally does not work with common-cause variation.

Multivoting

Purpose   Multivoting is used to prioritize a large group of suggestions so that teams can follow up on those most likely to be fruitful first. Multivoting is generally used in the Choose Best Solutions step.

Benefits   Multivoting is a quick means of testing for general consensus, hence it can be a tremendous time saver. It is also more democratic in that everyone gets an equal vote.

Limitations   Because of its democratic nature, it is possible to arrive at the most popular—rather than the most likely—solutions. Defer to those with the best knowledge of the problem.

Example   The flushability team used multivoting to select the most likely causes of poor flushability in the baby wipe case. It was quickly revealed that the team considered four potential causes most probable, and each was subsequently followed up by a subteam. Multivoting saved considerable time here because the team was at a standstill trying to agree on a single most probable cause. Each expert had his or her own opinion, and it was impossible to come to that level of consensus. By multivoting, there was general consensus on the top four potential causes, and the team agreed to follow up on all four.

Procedure   In the most basic form of multivoting, the team identifies, clarifies, and lists the ideas or suggestions on which to vote. Next, each team member prioritizes the items from most important (or likely) to least important. Typically, the most important item is given the largest number, and the least important the smallest. Team members write down their votes and take turns showing them to the rest of the team. The votes are added for each item and displayed, typically as a Pareto chart. By looking at the Pareto chart the team can see if a few items clearly stick out as the consensus of the team (the largest bars on the Pareto) or if no real consensus is emerging. The team should then openly discuss the output and make sure that they are all comfortable with it. It is always possible that an important item was not rated highly because of some type of misunderstanding, and an open discussion by the team can help bring out these oversights. If the team has general consensus, they can begin plans to follow up on the prioritized items. If no general consensus is emerging, it usually indicates that people have different understandings of the issue or are approaching it from very different perspectives. In this case, it probably makes sense to take a step back and discuss the fundamental issue the team is addressing, what the team's objectives and priorities are, and so on.

Variations   Two popular variations are the "N over 3" method, and the "dot" method. In the N over 3 method, you do not vote on each item, but only on the subset each voter deems most important. If there are a large number of items, it often makes sense for each person to select one-third of the items to vote for (hence the term "N over 3," where N is the total number of items). Obviously, each person will select the third that seem most important to him/her. It is not necessary that each person choose the same third, since this method essentially assigns a vote of zero to items not selected.

The dot method gives each person a fixed number of votes, which they can allocate however they see fit, with no restrictions. For example, given 10 votes, one person could give all 10 to one item and none to the others, whereas another person might give 2 votes each to five items. There are no set rules for number of votes, but each team member is typically given 1 vote per item being voted on, that is, 20 if there are 20 items. This works best using sticky dots, which can be directly placed on flipcharts, one dot per vote (hence the term "dot" method).

Tips
- The vote is intended to capture the most probable causes, not the most popular. Be sure the team only votes on the order in which to follow up ideas and does not try to vote on the "answer."
- Prioritized items should always be followed up to ensure that they are the real root causes.

- Only people with the right subject matter expertise should vote.
- Team members should write down their votes prior to disclosing them to avoid being swayed by hearing how others voted.

5.6 Summary

1. Various types of tools are used in the process improvement and problem-solving strategies.
2. Data collection tools are used to obtain the most appropriate data for analysis.
3. Sampling concepts can help us obtain the right quantity and quality of data.
4. Data analysis tools are used for analyzing quantitative (number) data.
5. Knowledge-based tools are used for analyzing qualitative (idea) data. The results of using these tools is dependent on participation of people with the right knowledge.

5.7 Project Update

Now that you have learned about the tools themselves, begin applying them to your project in the context of either the process improvement or problem-solving strategy. Obtain data over time, perhaps with a checksheet or survey, and diagnose stability with a run or control chart. Based on the results of this analysis, you should begin applying one of the strategies. Remember that it typically takes several tools used in conjunction, and perhaps several cycles through the entire strategy, to make the desired level of improvements. At this point you should be able to apply all the basic tools discussed in this chapter to the extent that they are appropriate for your project. Describe the specific tools you will apply to your project. In the chapters that follow you will be exposed to more formal statistical tools and methods. Some of these may also apply to your project and may allow more thorough analysis of the data you have collected.

EXERCISES

1. In what situations might you prefer to use a checksheet versus a survey, and vice versa?
2. Develop a checksheet to be used in an Internet book and music business to note the location of damage on boxes of product coming from a warehouse.
3. Develop a flowchart of the process you typically use to get ready for school or work when you wake up in the morning. Choose one step in this flowchart and develop a more detailed flowchart of how you do that single step.
4. Comment on the appropriateness of these sampling schemes:

   - A project team is working on improving the number of errors in placing orders from customers. The team samples 50 orders from the last month to estimate the current level of errors made on the customer purchase order (PO) number.

There is only one PO per order in this business. They selected the specific orders sampled by selecting the 50 largest orders in terms of dollar amount.

- An Internet auto leasing business is evaluating customer satisfaction. They perform an electronic survey of customers by randomly choosing customers from five major metropolitan areas.
- To address a concern over billing accuracy, an improvement team at an insurance company meets at 4 P.M. every day for 2 weeks to physically check accuracy of all invoices processed from 4 P.M. to 5 P.M. that day (about 150 per day).
- A human resources team is reviewing accuracy of their records. Because they have such a large database, they only look at records of every 50th person on a list arranged in alphabetical order.
- To monitor delivery cycle time on pharmaceutical orders, a health maintenance organization reviews cycle time of all the pharmaceutical shipments received in its Chicago facility last month (about 20) because that facility had the most detailed records.
- A bank includes a customer satisfaction survey with its monthly statements. The bank received about 400 responses out of a total of 55,000 customers who received that monthly statement.

5. The data in Table 5.2 list country code and the order to remittance (OTR) time for hardware/software installations for the last 76 installations (from first to last). OTR is the time it takes from an order being placed until the system is installed and we receive payment (remittance). Because this company does business internationally, it also notes the country of installation using a country code. This code is given in the first column.

   Does the OTR time appear to be stable? If you were to use a control chart to evaluate stability, which chart would you use? Why? What can you learn about the distribution of the installation process? Does it appear that country has an impact on installation time? Brainstorm a list of reasons one might find differences between countries and organize these into logical categories using one of the knowledge-based tools from this chapter. If you were actually on a team that was going to investigate these categories, would it make more sense to prioritize them with an interrelationship digraph (ID) or with multivoting? Why?

6. A software development organization is concerned about recent losses of key technical talent, often to Internet startups. The following data show the number of people who have resigned over the past 12 months, by organizational unit, labeled A–R. The organizational units are approximately equal in size and represent different technical specialties of the software developers.

Unit	Resignations	Unit	Resignations	Unit	Resignations
A	7	G	14	M	2
B	3	H	5	N	7
C	9	I	8	O	24
D	7	J	1	P	6
E	1	K	7	Q	3
F	2	L	5	R	1

TABLE 5.2 Country Code and OTR Cycle Time for Software System Installation

Country Code	Cycle Time	Country Code	Cycle Time
1	20	5	29
1	24	6	40
1	46	7	157
1	26	8	19
14	38	5	24
1	15	1	81
1	15	7	53
17	23	7	26
1	31	1	28
1	31	1	34
6	64	1	34
5	29	7	50
5	44	1	52
1	32	1	19
1	15	1	44
7	11	14	150
7	14	7	29
1	89	17	25
17	41	6	79
7	41	17	13
1	36	6	32
8	43	7	61
17	21	8	42
8	28	8	46
7	18	7	88
8	47	14	24
6	26	7	7
6	47	1	33
5	9	5	129
7	42	17	41
5	5	17	43
6	27	14	42
6	27	14	42
1	33	7	53
7	44	7	53
1	21	7	48
1	22	5	21
1	50	1	19

Place the data in an appropriate graph to show which units appear to have the largest problems. How else, other than by organizational unit, might the data be stratified to look for patterns in resignations? What additional data would you want to analyze?

7. Prepare a cause-and-effect diagram for why you might perform well in this class (the effect). Consider the branches of people, materials, methods, technology (machine), environment, and measurement. Conversely, consider a class for which your performance did not meet your expectations. Ask "Why?" five times to determine root causes for this disappointment.

8. A financial organization, General Financing, makes bids to guarantee municipal bonds. This means that they will (for a fee) guarantee the bond so that the municipality (which may have a poor bond rating and therefore be forced to pay high interest rates on the bond) can get an excellent (AAA) bond rating. In the bidding process, several organizations may bid for the bond, and other factors—such as bond rating—being equal, the municipality will typically select the lowest bid. The two biggest factors influencing General Financing's profitability are "hit rate," or percent of the time GF gets the bid, and "money left on the table," or the difference between GF's bid and the next-lowest when GF did get the bid. Obviously, it does little good to get the bid if one has to bid so low that one loses money. A perfect bid is one that gets the bid, but does so by being just barely below the next-lowest bid—that is, GF got the bid without "leaving anything on the table."

    Recently, General Financing seems to have been losing a lot of bids that they had anticipated winning. Although the hit rate has deteriorated, the money left on the table seems to be stable. Data reveal that of their four regional branches, those in the Northeast and South have had noticeably lower hit rates. By plotting the data over time, GF finds that over the past several years the hit rate in these regions has been reasonable for the first three quarters, only to plummet in the fourth quarter. Within both of these regions, however, some bid coordinators (who have overall responsibility for a given bid) have not done worse on their fourth-quarter bids. Assuming the above conclusions are accurate, conduct an Is–Is Not analysis to pinpoint root causes of the decreased hit rate. If you are not familiar with municipal bond markets, you will obviously have to speculate on potential explanations for differences. Once you have completed your analysis, evaluate the following potential causes using the Is–Is Not analysis.

    - "The decrease in hit rate is due to a new competitor who entered the bond market around the beginning of this year. They're killing us throughout the country."
    - "The problem in the Northeast is due to the inexperience of the new regional VP. She just doesn't have enough experience in this business."
    - "Every fourth quarter, our competition is more determined to get bids to make their annual quota and presents lower bids than normal, just to get the business. Unless you have a lot of experience and really understand the bidding system well, you won't be able to appropriately factor this into your bidding."
    - "There has been a lot of expansion in the Northeast and South over the past 3 or 4 years. Most of the bid coordinators are fairly new in these regions, with only a few experienced coordinators left. This is why these regions are having problems."
    - "We implemented a new computer bidding system 3 years ago. It is intended to help the coordinators by gathering all the relevant background information

on the bond and suggesting a bid. I think this new system has actually made things worse and is the root cause of the problem."

Which, if any, of these explanations is plausible in light of the information presented above? Could any combination of these suggestions explain the drop in hit rate? What additional data might you need to verify any of the suggestions? What follow-up actions could be taken to improve the situation?

9. We have just made significant improvements in the consumer loan business of our bank. We now wish to monitor the process over time to ensure that we maintain the improvements. What type of control chart should we use for each of the following process measurements?

- The accuracy of the monthly payment calculations (principal amount, interest amount, etc.), which are hand checked weekly on each loan.
- The total cycle time of processing a loan application, where we sample five loans completed each day to measure their cycle time.
- The number of errors, typographical or otherwise, in the official document that explains all the terms of the loan.
- Our monthly profit from consumer loans.
- The percentage of loans that result in a default each month. This is calculated by taking the number of defaults this month and dividing it by the total number of outstanding loans.
- The number of complaints our bank receives each month pertaining to our consumer loan business.

10. Organize a team, and brainstorm ways in which your learning of statistical thinking could be improved. The ideas should include specific ways this course could be improved as well as improvements to the text itself, your study habits, and even things you could do outside of this course. Once you have a reasonable number of ideas (25 or more), construct an affinity diagram of these ideas. What are the major categories (headers) of ideas? Now that you have the headers, multivote on the relative importance of each header in terms of its ability to enhance your understanding. In other words, which headers should you focus on to improve your understanding of statistical thinking fastest? After multivoting, perform an ID to prioritize the ideas based on which ideas are leverage points. Which might drive others? Did you get a similar result using both multivoting and the ID? Why do you think this is the case?

11. An Internet financial services company wishes to market a new service, home refinancing loans, to its customers. It currently has about 80,000 customers for a broad range of financial services from consumer loans, to traditional banking, to Internet bill paying, to home mortgages. The company wants to send out an electronic survey to appropriate customers to determine if there is a sufficient market for refinancing loans to make the initiative profitable. Customers will be asked, among other things, how receptive they would be to several types of refinancing loans on a scale of 1–10.

How many customers should they survey? Should they sample from all customers or focus on a subgroup? How could they obtain a random sample, given that the customers are not all in the same database but are segregated by type of service (loans versus savings accounts, etc.)? What advice could you give them about addressing nonresponse bias?

12. The data in Table 5.3 represent the goodwill, total assets, and amortization period for the last 52 acquisitions of a major multinational conglomerate. Goodwill is the amount of money over and above the fair market value of a company (typically measured by market capitalization—stock price × number of shares outstanding) that would be required to acquire the company. In general, one must pay more than the fair market value to acquire a company, to provide an incentive for the current owners to sell. The amortization period is the time period after the purchase over which the goodwill is written off the books. Typically, we want to write off large amounts of goodwill over a long time period to avoid a major hit (negative impact) to our bottom line in any one year.

  Does it appear that the goodwill is dependent on the total assets of the company? Does it appear that the amortization period is dependent on the amount of goodwill?

  Calculate the goodwill as a proportion of total assets (goodwill divided by assets). This eliminates the impact of size of the company and allows us to compare goodwill across major and minor acquisitions. Construct a control chart of this ratio and determine if the process appears to be stable. What type of control chart makes sense here? How would we interpret stability or lack of stability for the acquisition process—that is, what does it mean in practical terms?

13. Respond to this hypothetical scenario. What are your thoughts about the validity of Ron's conclusions? What questions should Ron ask? What would be a better approach? The data you will need appear in Table 5.4.

Statistical Thinking in Sales*

Ron Hagler, V.P. of Sales for Selit Corporation, wanted to see his regional sales managers right away. His staff assistant, Bonnie Teller, just compiled the past 5 years of quarterly sales data for the regions under his authority, and he was not happy with the results. "Marsha, tell the regional managers I need to speak with them this afternoon. Everyone must attend. Oh, and get us a conference room."

  Marsha Underwood had been Ron's secretary for almost a decade. She knew by the tone in his voice that he meant business. She immediately checked the availability of the isolated conference rooms until she found one. The meeting was set for 2 P.M. At 1:55 P.M. the regional managers filed into the room. The only time they were all called to a meeting was when Ron was not happy.

  Ron wasted no time. "Bonnie just finished the fourth-quarter sales report. New England sales were fantastic. Steve, you not only improved 17.6% over last quarter, but you also increased sales a whooping 20.6% over last year. I don't know how you do it!" Steve smiled. His philosophy was to

---

* From Britz et al. (1997). The authors would like to thank Don Wheeler for contributing the original data set.

TABLE 5.3 Goodwill Data (in millions, except amortization, which is in years)

Goodwill	Total Assets	Amortization	Goodwill	Total Assets	Amortization
9.5	3.8	10	32.1	18.8	15
5.2	4.5	5	36.3	7.2	30
0.8	1	10	1200	800	40
3.4	1.9	15	16.2	9.2	10
2.9	0	10	0.6	0.2	5
18	8	12	41	57	15
0.35	0.115	5	334	105	25
1	1.4	10	13	0.7	5
24.1	14.2	10	5.2	2.9	10
9	7.2	10	7.3	10.4	15
1.7	1.3	10	2.5	1.6	20
31	36.6	20	3.9	2.6	20
13.7	11.4	10	0.8	0.6	20
956	752	40	16.7	9.7	20
454	324	20	4.6	3.6	15
1266	601	40	29.3	6.4	10
2.2	1	10	8.1	2	10
2.1	0.3	10	5	14.8	7
1.9	0.8	15	43	13.8	10
1.2	1	10	14.6	9.7	10
46.4	28.8	15	6.1	5.1	7
422.4	281.9	30	27	18.4	10
181	83	30	11.9	10.2	10
1.4	1.2	5	186.5	336.6	30
11.3	8.5	15	20.3	18.1	10
13.5	10.7	10	2.3	1.8	10

end the year with a bang by getting the customers to stockpile units. First-quarter sales were always sluggish, therefore, but a decline in sales at the beginning of the year always went unnoticed.

Ron continued. "Terry, Southwestern sales were also superb. You showed an 11.7% increase over last quarter and an 11.8% increase over last year." Terry also smiled. She wasn't sure how she did so well, but she sure wasn't going to change anything.

"Jan, the Northwestern sales were up 17.2% from last quarter, but down 8.2% from last year. You need to find out what you did last year to make your sales go through the roof. Even so, your performance this quarter was good." Jan tried to hide his puzzled look. Sales for the Northwest were declining. In November a new Smith Brothers' store opened. It was the first big order he had received in a long time.

Ron was now ready to deal with the problem regions. "Leslie, North Central sales were down 5.5% from last quarter, but up 4.7% from last year. I don't understand how your sales vary so much. Do you need more

TABLE 5.4
Sales by Region

	1991					1992			
Region	1st	2nd	3rd	4th	Region	1st	2nd	3rd	4th
NE	924	928	956	1222	NE	748	962	983	1024
SW	1056	1048	1129	1073	SW	1157	1146	1064	1213
NW	1412	1280	1129	1181	NW	1149	1248	1103	1021
NC	431	470	439	431	NC	471	496	506	573
MA	539	558	591	556	MA	540	590	606	643
SC	397	391	414	407	SC	415	442	384	448

	1993					1994			
Region	1st	2nd	3rd	4th	Region	1st	2nd	3rd	4th
NE	991	978	1040	1295	NE	765	1008	1038	952
SW	1088	1322	1256	1132	SW	1352	1353	1466	1196
NW	1085	1125	910	999	NW	883	851	997	878
NC	403	440	371	405	NC	466	536	551	670
MA	657	602	596	640	MA	691	723	701	802
SC	441	366	470	426	SC	445	455	363	462

	1995			
Region	1st	2nd	3rd	4th
NE	1041	1020	976	1148
SW	1330	1003	1197	1337
NW	939	834	688	806
NC	588	699	743	702
MA	749	762	807	781
SC	420	454	447	359

incentive?" Leslie looked down. She had been working very hard the past 5 years and had acquired numerous new accounts. In fact, in 1994 she received a bonus for acquiring the most new business.

"Kim, the Mid-Atlantic sales were down 3.2% from last quarter and 2.6% from last year. I'm very disappointed in your performance. You were once my best sales representative. I had high expectations for you. Now, I can only hope that your first-quarter results show some sign of life." Kim felt her face get red. She knew she sold more units this year than in the previous year. What does Ron know anyway? He's just an empty suit.

As he turned to Dave, Ron felt a surge of adrenaline. "Dave, South Central sales are the worst of all! Sales are down 19.7% from last quarter and down 22.3% from last year. How can you explain this? Do you value your job? I want to see a dramatic improvement in this quarter's results or else!" Dave felt numb. This was a tough region with a lot of competition. Sure, accounts were lost over the years, but those lost were always replaced with new ones. How could he be doing so badly?

REFERENCES

Britz, G., Emerling, D., Hare, L., Hoerl, R., & Shade, J. (1996). *Statistical thinking*. Milwaukee, WI: American Society for Quality Control Statistics Division. (Available from the Quality Information Center, American Society for Quality, P.O. Box 3005, Milwaukee, WI 53201-3005; telephone 800-248-1946. Publication number S07-07.)

Britz, G., Emerling, D., Hare, L., Hoerl, R., & Shade, J. (1997, June). How to teach others to apply statistical thinking. *Quality Progress, 30,* 67–79.

Brassard, M., & Ritter, D. (1994). *The memory jogger II*. Methuen, MA: GOAL/QPC.

Can you trust these polls? (1995, July). *Readers' Digest*, pp. 49–54.

Fienberg, S. E. (1971, January). Randomness and social affairs: The 1970 draft lottery. *Science*, 255–261.

Imai, M. (1986). *Kaizen, the key to Japan's competitive success*. New York: Random House Business Division.

Kepner, C. H., & Tregoe, B. B. (1979). *Problem analysis and decision making*. Princeton, NJ: Princeton Research Press.

Mager, R. (1988). *Making instruction work*. Belmont, CA: Lake Publishing.

Parten, M. (1950). *Surveys, polls, and samples*. New York: Harper.

Wadsworth, H., Stevens, K., & Godfrey, B. (1986). *Modern methods for quality control and improvement*. New York: Wiley.

Wheeler, D., & Chambers, D. (1992). *Understanding statistical process control* (2nd ed.). Knoxville, TN: SPC Press.

PART III

Formal Statistical Methods

Introduction to Minitab

Overview

Prior to Chapter 5, we presented an introduction to Microsoft Excel. This spreadsheet package is commonly used for a variety of graphs; it also contains some basic statistical tools. For more rigorous statistical procedures, such as multiple regression and design of experiments, however, we will need to rely on more powerful statistical software packages. One of the most commonly used, and easiest to learn, is Minitab. We provide an introduction to Minitab here because this is one of the packages you might use to perform multiple regression and to design experiments, topics addressed in Chapters 6 and 7. More detailed instruction on the use of Minitab can be obtained from Minitab Incorporated, which offers several guides and manuals.

What Is Minitab?

You may already be familiar with Minitab. It combines some aspects of a spreadsheet package with the power of a full-scale statistical analysis package. It is compatible with Microsoft products, and data from an Excel file can be copied directly into Minitab. Minitab graphs can be used in PowerPoint presentations. Minitab is basically split into two portions, a session window at the top and a spreadsheet at the bottom. Commands are given via a set of pull-down menu boxes, similar to Excel. Using the data contained in the spreadsheet portion of Minitab, you can produce spreadsheet calculations, graphs, or statistical analyses. The numerical output is presented in the session window, and graphs are presented as individual "windows" that can be shrunk, enlarged, copied, or closed, as with standard

FIGURE 1   Basic Minitab Window

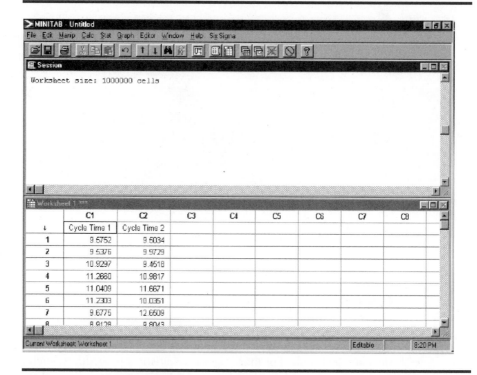

Microsoft Windows tools. Figure 1 shows a typical Minitab window with data in the spreadsheet portion.

Data Storage

Data can be entered into Minitab either manually, by copying data from another compatible application, or by directly opening an Excel data file. Minitab has trouble with text in Excel files, so it is often easier to cut and paste Excel files into Minitab. As with Excel, data are stored in a series of columns. To enter data, one moves the cursor into the cell (a given row of a given column) desired and types in the appropriate number. Text may also be entered in cells, but these cells cannot be statistically analyzed. Such a cell can be used as the horizontal axis of a graph, however. For example, we might wish to make a box plot of cycle time to close the books by three divisions of a conglomerate. The division name could be the horizontal (*x*) axis, with cycle times plotted on the vertical (*y*) axis. By saving the worksheet, the data will be electronically stored exactly as they were when they were saved, and the entire project, including graphs and numerical output from the output window, can also be saved. Minitab is not a database, and it is not particularly efficient for storing large amounts of data.

FIGURE 2   Minitab Menus

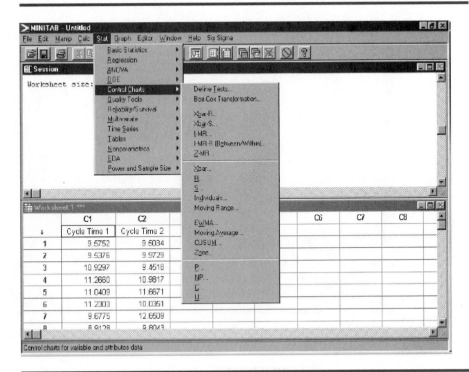

Statistical Calculations and Graphs

Minitab has standard spreadsheet capabilities, which enable the user to create new variables from existing variables in the spreadsheet. However, its real advantages are for performing statistical calculations and creating statistical graphics. For example, if we have monthly income in one column and monthly expenses in another, we might want to calculate monthly profit (income minus expenses). Once the data have been entered in Minitab, we initiate a statistical graph or numerical procedure by selecting one of the menus at the top of the window: Edit, Manip (manipulate), Calc, Stat, and Graph, among others. The Edit, Manip, and Calc menus allow for editing of the spreadsheet and calculation of new variables. These provide functions similar to Excel. The Graph menu provides several options for statistical graphics, such as box plots or histograms. The Stat menu is employed for formal statistical methods such as regression (Chapter 6), experimental design (Chapter 7), confidence intervals (Chapter 8), and hypothesis testing (Chapter 8).

Figure 2 shows the more detailed statistical menu that is obtained by selecting the Stat menu. Note that numerous options are available to us. Each of these options contains a unique submenu. For example, in Figure 2 we have not only selected the Stat menu but then have also selected the Control Charts submenu.

FIGURE 3 Minitab Dialogue Box for Xbar-R Chart

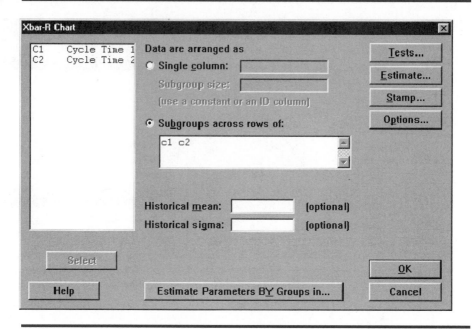

Numerous options are available here, including the standard control charts discussed in Chapter 5. The use of submenus creates a natural tree structure of commands that goes from general (Stat) to specific (Xbar-R).

If we select Xbar-R, a dialogue box appears in which we can choose the data we wish it to use for the Xbar-R chart, how the data are stored (all in one column or subgroups in separate columns), and what type of subgrouping to use. An example of this dialogue box can be seen in Figure 3. Most Minitab commands use similar dialogue boxes. In this case we have collected two cycle time measurements at a time, and columns 1 and 2 form logical subgroups. In other words, row 1 of columns 1 and 2 comprises the first subgroup, row 2 of these columns composes the second subgroup, and so on. This is reflected in how the dialogue box in Figure 3 is filled out.

Figure 4 shows how Minitab presents the results. The session window reminds us of what it has been asked to do, and the Xbar-R chart shows up as a unique window, which can be enlarged, shrunk, copied, saved, or closed. For commands that produce more numerical output, such as regression, we would see the details of the numerical output in the session window. In general, numerical output is presented in the session window and graphs appear as individual windows.

FIGURE 4   Xbar-R Output and Graph

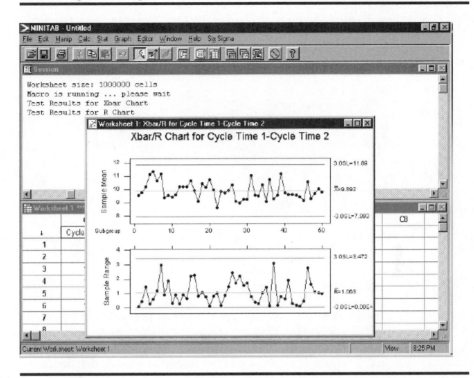

Summary

Minitab is one of the most popular statistical analysis packages. It combines the utility of a spreadsheet package with more formal statistical analysis tools. We recommend Minitab primarily for regression, experimental design, confidence intervals, and hypothesis testing. It is perfectly capable of creating basic graphs as well, however. Minitab should be capable of performing all of the data-based tools discussed in this text. At the time of this writing, the knowledge-based tools, such as affinity diagram or interrelationship digraph, are the only tools not included in Minitab. As with any software package, developing proficiency requires repeated use and experimentation.

Introduction to JMP

Overview

Another more powerful statistical software package is JMP (pronounced "jump"). JMP can be used to perform statistical analyses such as multiple regression and design of experiments, which are addressed in Chapters 6 and 7. As with Microsoft Excel and Minitab, we do not intend to provide detailed, command-by-command instructions. Rather, we introduce the software, explain how it works, and discuss when it would be most useful in applying statistical thinking tools. More detailed instruction on the use of JMP can be obtained from SAS Incorporated, which offers several guides and manuals.

What Is JMP?

Some of you may already be familiar with JMP. It combines some aspects of a spreadsheet package with the power of a full-scale statistical analysis package. It is compatible with Microsoft products and can read data from Excel files directly. JMP graphs can be copied into a PowerPoint presentation. Like Minitab, JMP uses a spreadsheet to view and manage the data and pop-up windows to display the output of analyses and graphics. Commands are given via a set of pull-down menu boxes, similar to Excel and Minitab. These commands allow the user to produce spreadsheet calculations, graphs, or statistical analyses from the data contained in the spreadsheet portion of JMP. The numerical output and graphics are presented together as individual windows, which can be shrunk, enlarged, copied, or closed, as with standard Microsoft Windows tools. JMP output windows are designed to be very interactive, allowing the user to add or delete detail with the click of a button. Figure 1 shows a typical JMP window with data in the spreadsheet portion.

FIGURE 1
Typical JMP Window

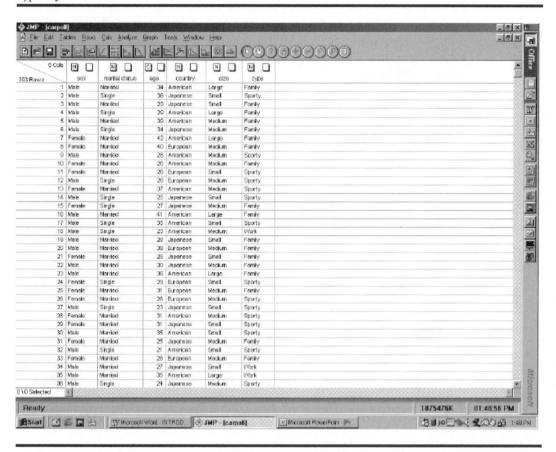

Data Storage

As with Excel and Minitab, data can be entered in JMP either manually, by copying data from another compatible application, or by directly opening an Excel data file. Data are stored in a series of columns. To enter data, one moves the cursor into the cell (a given row of a given column) desired and types in the appropriate number. Text may also be entered in cells, but then these cells cannot be statistically analyzed. They can be used as the horizontal axis of a graph, however. For example, we might wish to make a box plot of cycle time to close the books by three divisions of a conglomerate. The division name could be the horizontal (x) axis, and the cycle times would be plotted on the vertical (y) axis. By saving the worksheet, the data will be electronically stored exactly as they were when they were saved. Like Excel and Minitab, JMP is not intended to be used as a database, and it is not particularly efficient for storage of large amounts of data.

Statistical Calculations and Graphs

JMP has standard spreadsheet capabilities, allowing the user to create new variables from existing variables in the spreadsheet. However, its real advantages are for performing statistical calculations and creating statistical graphics. For example, if we have monthly income in one column and monthly expenses in another, we might want to calculate monthly profit (income minus expenses). Once the data have been entered into JMP, we initiate a statistical graph or numerical procedure by selecting one of the menus at the top of the window: Edit, Tables, Rows, Cols (columns), Analyze, Graphs, and Tools, among others. The Edit, Tables, Rows, and Cols menus allow you to edit the spreadsheet and calculate new variables, similar in function to Excel. The Graph menu provides several options for statistical graphics, such as Pareto charts or control charts.

Some common graphs, such as histograms and box plots, are included under the Analyze menu. The Analyze menu is typically used for formal statistical methods, such as regression (Chapter 6), confidence intervals (Chapter 8), and hypothesis testing (Chapter 8). The Table menu is used for experimental design (Chapter 7). For analysis of individual variables, such as confidence interval for the average of a population, you would select the "Distribution of Y" submenu under Analyze. For analysis of one *y* and one *x* variable (e.g., a hypothesis test comparing two or more averages), select the "Fit Y by X" submenu. For more than one *x* variable, select the "Fit Model" submenu.

Figure 2 shows the more detailed submenu that is obtained by selecting the Analyze menu. Selection of any of these options opens a dialogue box to guide us through the analysis. In Figure 3 we have selected "Distribution of Y," and the resulting dialogue box is shown. The type of data (continuous, ordinal, and nominal) is indicated in the spreadsheet and shown in the dialogue box. The data type determines what analyses are possible for these data. Data types are discussed in de-

FIGURE 2   Analyze Menu

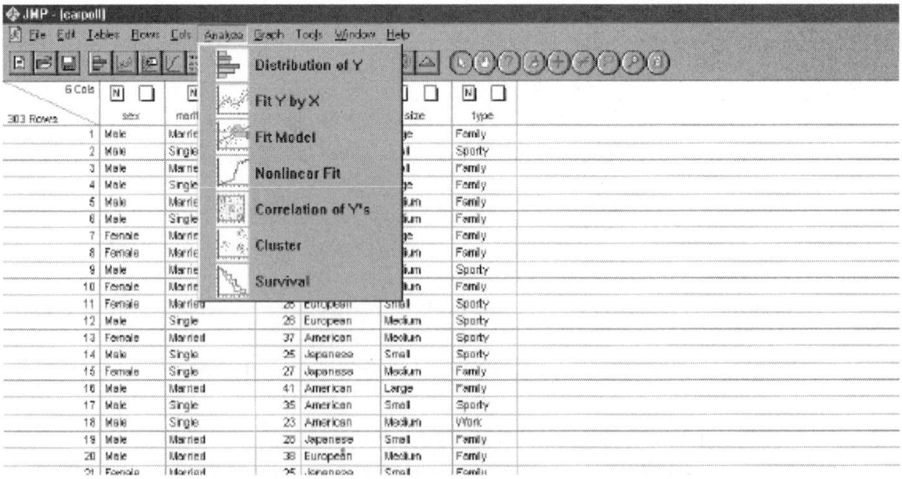

FIGURE 3   "Distribution of Y" Dialogue Box

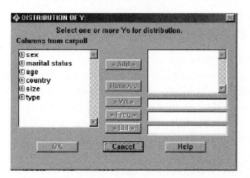

tail in Chapter 9. If we select the variable "age," we obtain the graphical output shown in Figure 4. This includes a histogram and a box plot. JMP output is designed to be interactive. If we click on the pointer button next to the word "age," additional options, including performing a hypothesis test on the average value, can be accessed. Most of the output in JMP works the same way; we select these buttons to obtain additional graphs and analyses.

FIGURE 4   Output from "Distribution of Y" Command

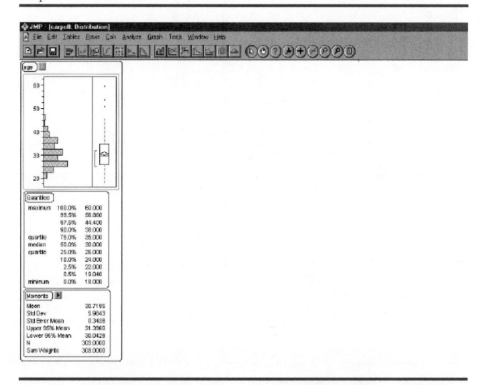

JMP Tools

An interesting feature of JMP, which makes it even more interactive, is the tools. The specific tools and their purposes are as follows:

- "Arrow" tool—general-purpose selecting tool
- "Hand" ("Grabber") tool—manipulates or grabs items to move them
- "Question" tool—obtains a help screen for any topic selected
- "Brush" tool—highlights data points in a rectangular area in a graph
- "Cross-hair" tool—produces a ruler to line up or measure things
- "Scissors" tool—selects irregular areas for a cut-and-paste
- "Lasso" tool—similar to the brush tool but for irregular areas
- "Magnifier" tool—zooms in on areas of interest
- "Annotate" tool—places text on a graph

The arrow tool and the hand tool are shown in Figure 5, and other tools are used in a similar manner. We start with the output obtained from Analyze, "Dis-

FIGURE 5
Use of Arrow and Hand Tools

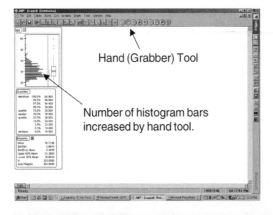

Arrow Tool

"Quantiles" shrunk by arrow tool.

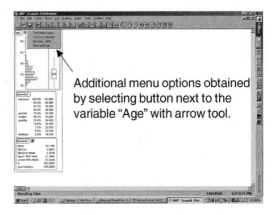

Additional menu options obtained by selecting button next to the variable "Age" with arrow tool.

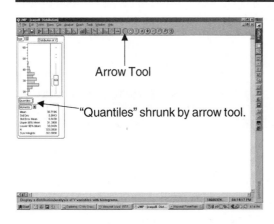

Hand (Grabber) Tool

Number of histogram bars increased by hand tool.

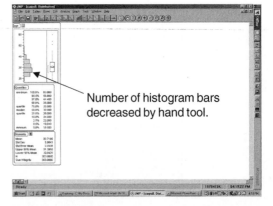

Number of histogram bars decreased by hand tool.

tribution of Y" (see Figure 4), where we looked at the distribution of age. JMP uses the arrow tool as the default tool. If we click on "Quantiles" with the arrow, this report is shrunk. Clicking again will make the analysis reappear. As previously noted, clicking on the button next to the word "age" produces additional menu choices. Next we go up to the hand icon to select the hand tool. If we take the hand over to the histogram of age and click on the histogram while holding down the left mouse button and moving the hand to the right, it increases the number of bars in the histogram. Moving the hand back to the left reduces the number of bars in the histogram.

Summary

JMP is a popular, powerful, and versatile statistical analysis package. Like Minitab, it combines the utility of a spreadsheet package with more formal statistical analysis tools. We recommend JMP primarily for regression, experimental design, confidence intervals, and hypothesis testing. It is perfectly capable of creating basic graphs as well, however. JMP should be capable of performing all of the data-based tools discussed in this text. At the time of this writing, the knowledge-based tools, such as affinity diagram or interrelationship digraph, are the only tools not included in JMP. As with any software package, developing proficiency requires repeated use and experimentation.

6

Building and Using Models

All models are wrong, but some are useful.

George E. P. Box

6.1 Overview

Model building is one of the methods used in the process improvement strategy. Model building is a broad topic, and in this chapter we will develop a basic understanding of the types and uses of models and develop functional capability to apply basic regression analysis to real problems. More extensive study and applications experience will be required to develop mastery of the entire field of regression analysis.

The power of statistical thinking is in developing process knowledge that can be used to manage and improve processes. The most effective way to create process knowledge is to develop a model that describes the behavior of the process. Webster's dictionary defines a model as "a generalized, hypothetical description, often based on an analogy, used in analyzing and explaining something." In this chapter we learn how to integrate our strategies and tools (see Chapters 4 and 5) and enhance them by building models. Model development is an iterative process in which we move back and forth between hypotheses about what process variables are important and data that confirm or deny these hypotheses.

Our overall strategy is as follows: We build a model that relates the process outputs (y's) to process variables (x's). We then use the model to (1) better understand the process, (2) control the process, (3) predict future process performance, (4) measure the effects of process changes, and (5) manage and improve the process. We build a model by collecting data. The process outputs will display variation. Using statistical techniques, in this case least squares regression analysis, we analyze the variation to discover the relationship between the process variables (*x*'s) and the process outputs (*y*'s). These relationships are summarized in the form of a model. In reviewing our strategy we see that we have used all three elements of statistical thinking: process, variation, and data.

Our goal is to build useful models. George Box points out that "all models are wrong, but some are useful" (1979, p. 202). All models are wrong because we never know the true state of nature. Fortunately, a model can be useful without exactly reproducing the phenomenon being studied. A useful model is simple and parsimonious, yet enables us to manage the process effectively, predict future process performance, and predict how process performance will change if the process operations are changed.

There are two basic strategies for developing process models: analyzing existing process data and proactively experimenting with the process. In this chapter we discuss the first approach. This strategy uses the statistical techniques of correlation and regression analysis. The second strategy involves experimenting with the process itself in a systematic, planned manner. This strategy uses the technique of statistical design of experiments. (Studying correlations between variables using scatter plots was discussed in Chapter 5.) Data collected without the aid of a statistical design have many limitations (see Section 6.8). It is easier to appreciate these limitations after one has an understanding of regression analysis (discussed in this chapter) and design of experiments (discussed in Chapter 7).

In this chapter we provide some examples of business models and discuss types and uses of models. This is followed by a discussion of the use of regression analysis of process data to construct process models. The chapter concludes with a summary of key points, a project update, and exercises.

6.2 Examples of Business Models

Discussing a variety of models is helpful in understanding the value and benefits of models and seeing the variety of situations in which models are used. Box, Hunter, and Hunter (1979, p. 540) call our attention to a useful model. In Yellowstone National Park, the park rangers need to predict when the Old Faithful geyser will erupt because this is what tourists come to see. Through data analysis the rangers have determined that if the previous eruption lasted for 2 minutes or less, the next eruption will be in 45 to 55 minutes. If the previous eruption lasted 3 to 5 minutes, the next eruption will be in 70 to 85 minutes. The accuracy of this model and its predictions directly affect the satisfaction of the tourists (i.e., customer satisfaction).

One situation we all can relate to is weight gain of humans and the variables that affect weight gain. Two key variables affect weight gain: the number of calories we consume and the amount of exercise we do. In most situations increasing caloric intake increases our weight, whereas increasing exercise decreases our weight. From a modeling viewpoint, we would say "weight gain is a function of caloric intake and exercise." This is represented mathematically as:

Weight gain = f (caloric intake and exercise)

where *f* is some unknown mathematical function. This is a crude model at this point. We don't know the form of the function *f*, the magnitude and nature of the effects of caloric intake and exercise on weight gain, or whether these two variables affect weight gain independently of each other. The model is useful, however, because it tells us two key variables to pay attention to: caloric intake and exercise.

We call the small subset of variables that have a major effect on the process "key drivers." Caloric intake and exercise are key drivers of (have a major influence on) weight gain of humans. There are no doubt other key drivers of weight gain (e.g., heredity). Our experience is that processes typically have five or six key drivers and perhaps a few other variables that have smaller effects.

Another example of multiple drivers was the Chapter 2 case study for Anheuser-Busch (Ackoff, 1978) investigating the effects of advertising on sales. Three variables were studied: percent change in advertising level, amount spent on sales effort, and amount spent on sales material. The purpose of the study was to determine a functional relationship between sales and percent change in advertising level, which turned out to have a complicated mathematical form.

In another advertising study, Montgomery and Peck (1992) report data on advertising and sales of 30 restaurants that could be described by this linear relationship:

Sales (\$) = 50,976 + 7.92 (Advertising \$)

This model was developed for advertising expenses ranging from approximately \$3,000 to \$20,000 and suggests that a restaurant would have sales of \$50,976 with no advertising and that sales would increase \$7.92 for every dollar spent in advertising. A straight-line relationship is the simplest model possible to describe the relationship between two variables.

Davis (1994) describes a model used by a major credit card company to predict the probability that a delinquent account will pay its bill. Among other things, Davis found that the outcome of the collection strategy depended on the type of account, the balance owed, and the number of months it was overdue. Collection strategies developed using the model produced an annual savings of \$37 million *and* improved customer satisfaction. An unanticipated result found was that in some instances the best strategy is "to do nothing."

Oil companies typically develop blending models that describe the quality of the gasoline (octane number) as a function of the proportion of the types of components that make up the gasoline. A four-component linear blending model might look like this:

Octane = $b_1x_1 + b_2x_2 + b_3x_3 + b_4x_4$

where x_1, x_2, x_3, and x_4 are the fractions of the different components in the gasoline and *b* is the coefficient derived from the data. The coefficients (*b*'s) in the model describe the blending behavior of the components and enable us to predict (calculate) the octane of a blend given the volume fraction of the different components (*x*'s) in the blend.

For example, one blending model developed by Snee (1973) looked like this:

Octane $= 103.2x_1 + 104.5x_2 + 104.0x_3 + 97.9x_4$

where $x_1 =$ Light FCC
$x_2 =$ Alkylate
$x_3 =$ Butane, and
$x_4 =$ Reformate

This model predicts that a gasoline consisting of 20% light FCC, 30% alkylate, 6% butane, and 44% reformate would have an octane of 101.3.

$103.2(.2) + 104.5(.3) + 104.0(.06) + 97.9(.44) = 101.3$

Blending models generally involve 10 to 15 gasoline components and often involve nonlinear blending terms (e.g., x_1x_2, x_5x_6).

Blending models, together with cost and manufacturing data, are typically used as inputs to linear and nonlinear programming algorithms to develop minimum cost blending strategies. They help make management decisions such as ways to operate a process or run a business to maximize profit or minimize cost. When the objective function to be maximized (profit) or minimized (cost) and the constraints are all linear, we use a linear programming algorithm. Nonlinear programming algorithms are used when nonlinear equations are involved in either constraints or the objective function (see Hillier & Lieberman, 1995).

Another area where models can be helpful is in predicting the workforce needed to handle the workload in call centers and telephone repair centers, for example. In both instances it is advantageous to be able to accurately predict the workload so that the right size workforce is scheduled: too few workers produces unsatisfied customers, too many workers unnecessarily increases costs. Similar analyses are needed to "size" Web sites (determine the appropriate number of servers or routers).

Studies have shown that telephone repair needs vary with the day of the week (Monday has the highest load), the season of the year, rainfall, and temperature. Models that predict repair loads from these variables are very effective in managing workloads and maintaining customer satisfaction. In addition to helping understand how the variables affect the repair load, this model can be used for daily resource management, seasonal planning, and risk assessment of the effects of large rainstorms.

The models discussed in the previous paragraphs were developed based on empirical relations derived from data. Models can also be based on theoretical relationships such as engineering laws, econometric theories, and chemical reaction kinetics. For example, in the case of chemical reactions in which material A reacts to produce product B, and it is known that first-order kinetics apply, the rate of formation of product B at any point in time (x) is given by:

Rate $= b_1(1 - e^{-b_2x})$

where the natural number e is about 2.72. The coefficients b_1 and b_2 are unique to materials A and B and are derived from data. This model is called a theoretical or mechanistic model because it is based on theoretical considerations regarding the nature of chemical reactions. The coefficient b_1 is the concentration of material B at the end of the reaction and b_2 is the rate of the reaction.

Models have different amounts of complexity and different uses. In the next section we discuss both in greater detail.

6.3   Types and Uses of Models

A key distinction to grasp is between empirical models and theoretical models. Empirical models are based on data and are developed by studying plots of the relationship between the process outputs (y) and one or more predictor variables (x's). The observed relationships are subsequently described and quantified by a mathematical equation. The resulting equation is a model for the relationship. A linear relationship would be described by the equation:

$$y = b_0 + b_1 x$$

where y is the process output of interest, x is the predictor variable, and b_0 and b_1 describe the intercept and slope of the line. Linear relationships are typically identified by plotting y versus x, hence the form of the model is identified empirically from the data. The linear model for the relationship between restaurant sales and advertising discussed earlier was developed in this manner.

Empirical models can take a number of forms and often include curvilinear terms (squares $x_2 x_2$ and cross products $x_1 x_2$), and exponential terms (e^x). Empirical models and their formulation will be discussed further in Section 6.4.

Theoretical models are based on known theoretical relationships and mechanistic understanding. The chemical reaction rate equation discussed in Section 6.2 is an example of a theoretical model.

$$\text{Rate} = b_1(1 - e^{-b_2 x})$$

Theoretical models are difficult to develop, and can be very complex. In a business setting, economic theory is often used to decide which variables should be considered and what form of mathematical relationship should be used in the model. Theoretical knowledge should be incorporated in the empirical modeling process wherever it is available, but the need for a theoretical model often depends on the use of the model. As Box, Hunter, and Hunter (1978) point out, the empirical model used by the park rangers to predict the eruptions of Old Faithful is good enough for their purposes. Geologists, however, might be interested in the hydrological mechanisms that govern the geyser's eruptions and study rate of pressure build up, water temperature, and so on. Such studies might improve predictions of the time of eruption. In most instances an empirical model will be sufficient. If deep understanding is needed, then the consideration of fundamental theoretical laws is in order. Theoretical models generally do a better job of predicting outside the region of the data (extrapolation) than do empirical models.

Another type of model often used in the analysis of time-related data, such as econometric data, is the *time series model*. Time series models typically predict the value for a given period ((a day, a month, or a year) as a function of the values of the variable observed on previous days, months, or years. The basic assumption is that there is a predictable pattern in the past behavior of the time series, and that the process that produced the data will not change going forward. We can thus use the time series model to predict the behavior of the process in the future. Time series models use autocorrelation functions and are beyond the scope of this book. For more information on this topic, we recommend Box and Jenkins (1976), Cryer (1986), and Vandaele (1983).

Uses of Models

There are four main uses of models:

- Predict future process performance
- Measure the effects of process changes
- Control, manage, and improve the process
- Deepen understanding of the process

Models that forecast future sales, revenues, or earnings are examples of using models to make predictions. A model can also help us determine the effects of potential process changes without disturbing the existing process. Using a model we can analyze the predicted process performance for various conditions of interest. It is also often important to predict how the process performance will change given specified changes in the levels of the predictor variables. For example, "If I double my advertising expense, how will my sales volume change?" The third use of models helps us control and manage the process to produce consistent products. When the process is off target, the model can tell us our options: which variables can be changed by how much to get the process back on target.

Finally, as we construct and use the model, we deepen our understanding of the process and the variables that drive its performance. The modeling process, if done properly, forces us to look systematically at all parts of the process. At some point even the best models fail. The detection of the failure and subsequent enhancement of the model adds to our knowledge of the process.

One strategy for developing models is to observe the process "as is" by collecting process data. When we have the data in hand, we need a technique to analyze the data and build the model, and *regression analysis* is the technique we use. Regression analysis is sometimes referred to as a mathematical French curve because it helps us smooth out variation and identify predictive relationships in many dimensions (i.e., many predictor variables). For many years engineers and draftspeople used a flexible straightedge, called a French curve, to draw lines and curves through clouds of points in one (y versus x) and sometimes two dimensions (y versus x_1 and x_2). Today, of course, this is done with computer software. There are a number of ways to build models, both formal and informal, but we focus in the following sections on regression analysis because it is so widely used. (Chapter 7 explains how to use regression analysis to analyze the results of designed experiments.)

FIGURE 6.1   Process Diagram

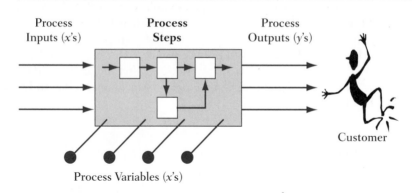

6.4   The Regression Modeling Process

When building process models using regression analysis we must consider (1) the number of predictor variables (x's), (2) the nature of the relationship between the response (y) and the predictor variables (x's), (3) a procedure for calculating the coefficients in the model, and (4) an overall method for building the model. These considerations are the subject of this section.

Schematically, the process can be represented as shown in Figure 6.1. Process inputs (x's) move through a series of process steps influenced by process variables (additional x's) to produce process outputs (y's), which are then sent to an internal or external customer. Variation in the process inputs and process variables cause variation in the process output measures (y's). In the case of our weight gain example, caloric intake is an input variable (x), amount of exercise is a process variable (x), and weight gain is a process output variable (y). Amount of caloric intake and exercise have effects on (cause) weight gain.

The overall objective is to use process data to build a model that helps us understand how the process works and to predict process performance (y) from various predictor variables (x's), which include process inputs and process operating variables. In the case of one predictor variable (x) in which the relationship between the response (y) and x can be described by a straight line, the linear model would have the following form:

$$y = b_0 + b_1 x$$

where b_0 and b_1 are coefficients to be estimated from the data by a technique known as least squares. In this model, b_1 is the slope of the line relating y and x and b_0 is the intercept—the value of y where $x = 0$ (Figure 6.2). This is the same form of an equation for a straight line used in algebra: $y = mx + b$ where m is the slope (m is the same as b_1) and b is the intercept (b is the same as b_0). If the relationship between y and x is not straight but curved, a model that contains a quadratic term might be used. For example,

FIGURE 6.2 Straight-Line Model

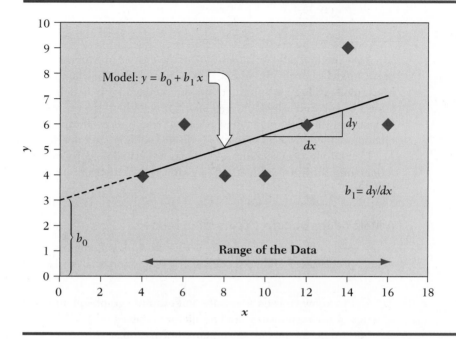

$$y = b_0 + b_1x + b_{11}x^2$$

We will restrict our discussion to linear and quadratic models because these models are widely used in practice and illustrate the key aspects of building and using models. We acknowledge that many other forms of curvilinear models are also encountered in practice.

Multiple Predictor Variables

Few processes involve a single predictor variable. When we include multiple predictor variables, the linear model for p process variables is

$$y = b_0 + b_1x_1 + b_2x_2 + \cdots + b_px_p$$

and the quadratic model becomes

$$\begin{aligned} y = {} & b_0 + b_1x_1 + b_2x_2 + \cdots + b_px_p \\ & + b_{12}x_1x_2 + b_{13}x_1x_3 + \cdots \\ & + b_{11}x_1x_1 + b_{22}x_2x_2 + \cdots \end{aligned}$$

In the case of three predictor variables, the linear model is

$$y = b_0 + b_1x_1 + b_2x_2 + b_3x_3$$

and the quadratic model is

$$y = b_0 + b_1x_1 + b_2x_2 + b_3x_3$$
$$+ b_{12}x_1x_2 + b_{13}x_1x_3 + b_{23}x_2x_3$$
$$+ b_{11}x_1x_1 + b_{22}x_2x_2 + b_{33}x_3x_3$$

As before, the coefficients (b's) in the models are estimated from process data using the method of least squares. As in the case of one predictor variable, the quadratic model is formed by adding curvilinear terms—for example, x_1x_2, x_1x_1—to the linear model. The cross product terms in the quadratic models (x_1x_2, x_2x_3) describe a form of curvilinearity known as *interaction*. When two variables interact, the effect of one variable (x_1) is dependent on the level of another variable (x_2). This characteristic of process variables is discussed in greater detail in Chapter 7.

In the following sections we discuss a methodology for building a regression model and illustrate the methodology in the situation of one and three predictor variables.

A Method for Building Regression Models

A key aspect of statistical thinking is that all work is done in a system of interconnected processes. Building models using regression analysis is a process with five key steps:

1. Get to know your data. Create and examine graphical and analytical summaries of the response (y) and predictor variables (x's).
2. Formulate the model: linear, curvilinear, and so on.
3. Fit the model to the data.
4. Check the fit of the model.
5. Report and use the model as appropriate.

Modeling begins after the problem has been properly formulated, the use of the model has been determined, the process variables are understood, and the data have been collected. This five-step process is summarized in detail in Table 6.1 and shown graphically in Figure 6.3. Model building is not a linear, single-pass process; many iterations and recycle loops may be needed to develop a useful model.

Least Squares

In step three of the methodology, "fit the model to the data," we use a technique known as least squares. Least squares calculations can be done in Excel, JMP, or Minitab, so we will focus on the interpretation of the results of these calculations and how these models can be used to better understand and manage the process being studied. To understand least squares, let's study an example involving one predictor variable (x) and one response (y). For simplicity we will assume that we have five observations:

Observation	x	y
1	8	1
2	10	6
3	12	4
4	14	6
5	16	10

TABLE 6.1 Method for Building Regression Models

---

1. *Get to know your data*
   - Examine the summary statistics: mean, standard deviation, minimum, maximum, for all x's and y's.
   - Examine the correlation matrix.
   - When data are collected sequentially in time, do a time plot for each x and y.
   - Plot y versus each of the x's.
   - Construct scatter plots for all pairs of x's or at least those x's that have high correlation coefficients in the correlation matrix.
   - Examine all plots for outlier or atypical points.

2. *Formulate the model*
   - Use subject matter knowledge whenever possible to guide the selection of the model form.
   - Study the form of the relationship between y and each of the x's in the plots to determine whether linear relationships exist or whether curvilinear terms need to be added to the model.

3. *Fit the model to the data*
   - Calculate the regression coefficients in the model and the associated regression statistics.
   - Examine the regression results for significant variables (key drivers).
   - Assess the fit of the model using the adjusted R-squared statistic.
   - Study the regression coefficient variance inflation factors (VIFs) to determine whether correlations among predictor variables (multicollinearity) is a problem.

4. *Check the fit of the model*
   - Construct plots of the residuals (observed minus predicted) to identify any abnormal patterns or atypical data points:

     Residuals versus predicted values
     Residuals versus predictor variables (x's)
     Residuals versus time or observation sequence
     Normal probability plot of residuals

   - Nonrandom patterns in the residuals indicate that the model is not adequate.

5. *Report and use the model*
   - Use the model as appropriate.
   - Create and circulate a document that summarizes the model, the data, and the assumptions used to create it.
   - Establish a procedure to continually check the performance of the model over time to detect any deviations from the assumptions used to construct the model.

These observations are plotted in Figure 6.4, showing a linear relationship between y and x. This relationship can be described by the equation

$$y = b_0 + b_1 x + e$$

where b_1 is the slope of the straight line, b_0 is the y-intercept of the line (i.e., the value of y when $x = 0$), and e is the error or "residual" representing the difference between the observed value of y and the value of y predicted by the straight line $(b_0 + b_1 x)$,

FIGURE 6.3 Regression Analysis Method

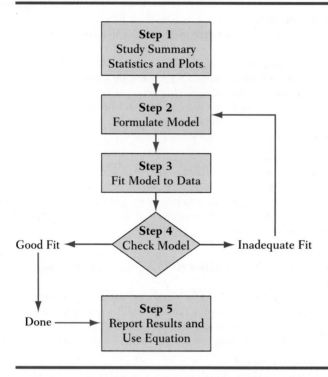

FIGURE 6.4 One-Variable Example, Response Versus Predictor Variable

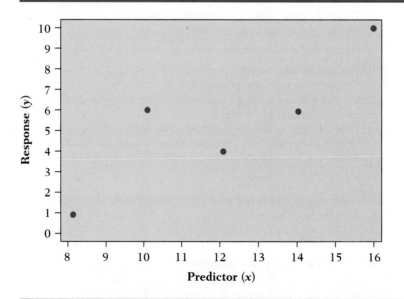

$e = y - (b_0 + b_1 x)$
$\quad = $ observed y − predicted y

The least squares technique finds b_0 and b_1 so that the sum of the squares of the residuals (e) is the smallest; hence the name least squares.

The residuals of the model defined by the least squares coefficients will be a mixture (approximately 50–50) of positive and negative values. If all the residuals are positive, the observed value of y is uniformly larger than the predicted value of y and the equation passes *below* the data. Conversely, if all the residuals are negative, the line passes *above* the data.

Table 6.2 shows residuals for models based on three different pairs of coefficients. The three models and the data are shown graphically in Figure 6.5. The line defined by Model 1's set of coefficients, $b_0 = -2.4$, $b_1 = 0.9$, passes above the data, and Table 6.2 shows that all the residuals are negative. The Model 2 set of coefficients, $b_0 = -5.0$, $b_1 = 1.0$, passes through the data, but not as closely as Model 3's set of coefficients, $b_0 = -5.4$, $b_1 = 0.9$, which are the coefficients calculated using the least squares technique. Notice in Table 6.2 that the residual sum of squares for the least squares coefficients is the smallest of the three models. The least squares technique produces the coefficients that result in the minimum residual sum of squares. Table 6.3 shows how the model components (b_0, $b_1 x$, e) combine to create each value of y.

Three assumptions must be satisfied to use least squares regression to make statistical inferences regarding the variables in the regression model. The residuals (e) are assumed to

- Be independent of each other
- Have a homogeneous variance
- Follow a normal distribution

The independence assumption is usually checked by studying how the data were collected and determining if any observation might influence the value (outcome)

TABLE 6.2 Least Squares Illustration

Observation	x	y	Residuals (e)		
			Model 1	*Model 2*	*Model 3*
1	8	1	−3.8	−2	−0.8
2	10	6	−0.6	1	2.4
3	12	4	−4.4	−3	−1.4
4	14	6	−4.2	−3	−1.2
5	16	10	−2.0	−1	1.0
Residual sum of squares			55.8	24.0	10.8

Model 1	$b_0 = -2.4$	$b_1 = 0.9$
Model 2	$b_0 = -5.0$	$b_1 = 1.0$
Model 3	$b_0 = -5.4$	$b_1 = 0.9$ (Least Squares Model)

FIGURE 6.5 Fitting by Least Squares

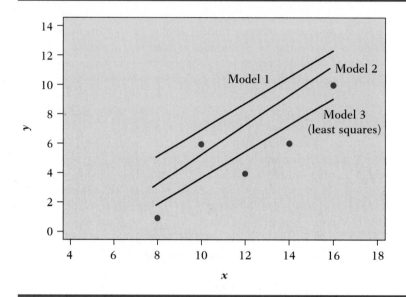

TABLE 6.3 Least Squares Illustration of Straight-Line Model Components

Data			Model Component	
x	*y*	b_0	$b_1 x$	Residual (*e*)
8	1	−5.4	0.9(8) = 7.2	−0.8
10	6	−5.4	0.9(10) = 9.0	2.4
12	4	−5.4	0.9(12) = 10.8	−1.4
14	6	−5.4	0.9(14) = 12.6	−1.2
16	10	−5.4	0.9(16) = 14.4	1.0

Note that $y = b_0 + b_1 x + e$ for all values of y.

of another observation. One example of nonindependent data is cumulative daily (monthly) sales. Each cumulative sales value is dependent on the values of the sales on previous days (months). The assumptions that the residuals have a homogeneous variance and are normally distributed are checked by analyzing plots of the residuals of the fitted models (see Section 6.5). It is assumed that all residuals have a common variance for all values in the region spanned by the values of the *x*'s. The normal distribution assumption is required for doing tests of significance of the regression results. These tests and the normal distribution, the familiar bell-shaped curve, are both discussed further in Chapters 8 and 9.

Although other regression methods exist for fitting the model to the data, the least squares method is by far the most popular. We restrict our discussion to least squares in this text and point out its limitations where appropriate. For a detailed discussion of the formulas used, we recommend Draper and Smith (1998) and Montgomery and Peck (1992).

6.5   Building Models With One Predictor Variable

To understand the art and science of building models, it is helpful to begin with the study of the relationship between two variables: a predictor variable, x, and a response variable, y. Table 6.4 shows monthly data on the customer call abandon rate (the percentage of customers who hang up before they are connected) and time to answer for a call center that markets credit cards. The goal is to reduce call abandon rate. It was felt that there was a relationship between time to answer

TABLE 6.4   Call Center Data

Month	Time to Answer	Abandon Rate	FITS1	RESI1
1	12	22	20.4384	1.5616
2	8	19	16.7277	2.2723
3	9	25	17.6554	7.3446
4	9	26	17.6554	8.3446
5	18	27	26.0044	0.9956
6	17	30	25.0767	4.9233
7	17	30	25.0767	4.9233
8	20	33	27.8597	5.1403
9	22	34	29.7151	4.2849
10	24	34	31.5704	2.4296
11	25	31	32.4981	−1.4981
12	30	39	37.1364	1.8636
13	23	47	30.6427	16.3573
14	16	29	24.1491	4.8509
15	16	30	24.1491	5.8509
16	17	30	25.0767	4.9233
17	16	17	24.1491	−7.1491
18	12	19	20.4384	−1.4384
19	9	14	17.6554	−3.6554
20	10	21	18.5831	2.4169
21	23	23	30.6427	−7.6427
22	33	40	39.9194	0.0806
23	33	36	39.9194	−3.9194
24	24	27	31.5704	−4.5704
25	24	24	31.5704	−7.5704
26	10	11	18.5831	−7.5831
27	10	14	18.5831	−4.5831
28	15	12	23.2214	−11.2214
29	16	13	24.1491	−11.1491
30	10	12	18.5831	−6.5831

Variables: Time to Answer and Abandon Rate. *Units:* Seconds and %.

(x) and call abandon rate (y); that is, increasing the time to answer increases the abandon rate because people get tired of waiting and hang up. We are working with monthly data because they were the only data available. Monthly data ignore many sources of variation that may occur within the month. This is one of the hazards of working with available data rather than collecting the data required to satisfy the needs of the study.

Step 1. Get to Know Your Data

The first step in the model building process is to get to know your data from an analytical and graphical viewpoint. The summary statistics for these data are

	Average	Range
Time to Answer (x)	17.6	8–33
Abandon Rate (y)	25.6	11–47

Figures 6.6 and 6.7 show how these variables vary over the 30 months of observation. Both variables show peaks for observation ranges 10 to 20 and 20 to 30, suggesting some correlation might be present.

Figure 6.8 is a scatter plot of abandon rate versus time to answer, which shows a positive linear relationship with one outlying observation (#13, visible in Figure 6.6). As time to answer increases, the abandon call rate increases. This makes sense and is consistent with our hypothesis. We conclude that we have a linear relationship in the time to answer range of 8–33 seconds. The correlation coefficient for this relationship is 0.724, suggesting the strong relationship shown in the figure.

FIGURE 6.6   Call Center Data, Abandon Rate Versus Time

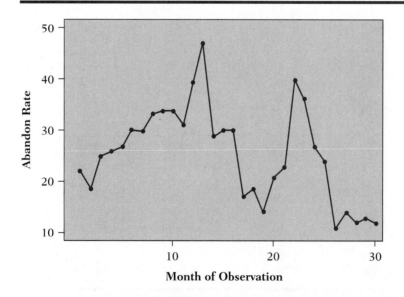

FIGURE 6.7 Call Center Data, Time to Answer Versus Time

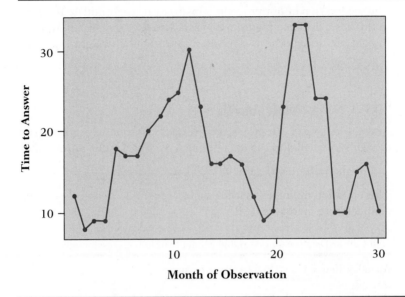

Month of Observation

FIGURE 6.8 Call Center Data, Abandon Rate Versus Time to Answer

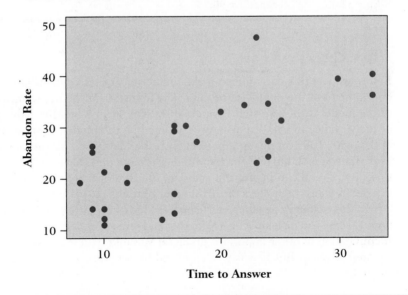

Time to Answer

When there is only one y variable and one x variable involved, in many instances plots of y versus x and of x and y versus time tell essentially the whole story. A regression model summarizes and quantifies what you see in a scatter plot.

Step 2. Formulate the Model

The second step of our procedure instructs us to formulate the model. This is easy in this case, because Figure 6.8 shows a straight-line relation, which makes sense in light of our hypothesis. Hence, we postulate a straight-line model

$$y = b_0 + b_1 x$$

Step 3. Fit the Model to the Data

In the third step of our procedure, we use least squares regression analysis to fit the straight-line model to the data (Table 6.5). This results in the following equation:

Abandon Rate = 9.31 + 0.928 (Time to Answer)

This equation indicates that for each 1-second increase in time to answer (x), the abandon rate increases 0.928%.

TABLE 6.5 Call Center Data for One-Variable Model (Minitab regression analysis printout)

Abandon Rate = 9.31 + 0.928 (Time to Answer)

Predictor	Coef	SD Coef	t-Ratio
Constant	9.306	3.170	2.94
Time to	0.9277	0.1670	5.56

R-Sq = 52.4%   R-Sq(adj) = 50.7%

Step 4. Check the Model Fit

Before we use this equation to make predictions or guide decisions, we need to check the fit of the model to the data. The first statistic we use to measure the fit of the model is the adjusted R-squared statistic, which measures the percentage of the variance in the response (y) explained by the model. It varies from 0% to 100%, with 100% implying a perfect fit to the data (i.e., all residuals are zero). This statistic is "adjusted" for the number of nonconstant terms in the model (in this case, 1) and can be used to compare the fit of a number of different models to a set of data—regardless of the number of terms in the models.

In this example the adjusted R-squared statistic is 50.7%, which gives us some hope that we can develop a useful model but leaves a lot of room for improvement. Identification of other important predictor variables will increase the adjusted R-squared statistic. But that is getting ahead of our story. We will return to this issue later.

Residual Analysis   The fit of the model is also checked by analyzing the residuals from the fitted model. A residual is the difference between the observed value of y and the value of y predicted by the model.

Residual = y Observed – y Predicted

There is one residual for each observation, and in this example we have 30 observations so there are 30 residuals. The residuals (Resi1) and predicted y values (Fits1) for the linear model are summarized in Table 6.4.

Analysis of the residuals is an effective method to assess the fit of the model and to determine whether the model is useful. Study a variety of residual plots, and look for any patterns or trends. Generally speaking, the absence of patterns and trends (i.e., random scatter of points) indicates that the model is adequate because the residuals will be made up of random variation only. All systematic behavior in y has been accounted for in the model, leaving nothing but "noise," or random variation, in the residuals. If our model in not adequate, however, then the residuals will contain both random variation and systematic patterns or trends.

Analysis of residuals can help us perform the following important checks on the adequacy of the model: determine the adequacy of the mathematical form of the model, decide whether important variables may be missing from the model, check the correctness of the model assumptions, and determine whether any abnormal or atypical observations in the data set may be unduly influencing the results.

The modeling process is not complete until the residuals have been thoroughly analyzed. Residuals are typically analyzed by constructing the following four scatter plots (a histogram of the residuals may be viewed as well):

- Residuals versus Predicted Values
- Residuals versus Predictor Variables (x)
- Residuals versus Normal Probability Scale
- Residuals versus Observation Sequence

These plots are not independent of each other and should be interpreted as a set. It is not uncommon for one or more of these model inadequacies to show up in more than one plot. Caution must be exercised in determining the root causes of the patterns and trends in the residuals to avoid errors. It is also important to recognize that there is a lot of judgment used in the interpretation of residual plots. The ability to properly interpret these plots grows with experience.

1. Residuals versus Predicted Values. The plot of the residuals versus predicted values should be "boring" and show no pattern or trend if the model gives an adequate fit to the data. No relation is observed in Figure 6.9, and we conclude that the model fit is adequate so far.

When a nonrandom pattern is observed, consider changing the form of the model. One pattern commonly observed is an increasing variation in the residuals as the predicted values increase (a V or megaphone shape). This indicates that the homogeneous variance assumption for the residuals is not satisfied by the data. The V-shaped pattern suggests that the variation increases with the average. In this case, a transformation such as a log (i.e., analyze log y in place of y) or weighted regression (Draper & Smith, 1998) should be used. The log transformation is usually helpful if the ratio of maximum y to minimum y is greater than 10. Further details on transformations are discussed in Chapter 9.

A curved relationship between the residuals and predicted values suggests that curvilinear terms (e.g., x^2, x_1x_2) should be included in the model. Further

FIGURE 6.9   Call Center Data, One-Variable Model Residuals Versus Fitted Values

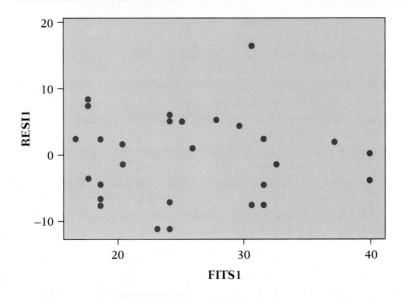

insight to this situation is provided by examining plots of the residuals versus the predictor variables (x's). A transformation of y, as noted in the next section, will often remove the curvilinearity and the need for adding a curved term to the model.

2. Residuals versus Predictor Variables (x's). The plot of the residuals versus the predictor variables (x's) should also be a random scatter. In Figure 6.10 we see that a random scatter exists for this model and conclude that the linear model is adequate. With only one x variable, this plot is equivalent to Figure 6.9.

If the pattern in the plot of the residuals versus the predictor variables (x's) is not random, it often shows some curvature. Such a pattern suggests curvilinear terms be added to the model or that either or both y and x variables be transformed (i.e., log, square root, reciprocal) prior to fitting the regression equation.

In the previous section we noted that the log transformation can be used to make the variations in the residuals homogeneous throughout the range of the predicted values, thereby satisfying an assumption of least squares. Reexpression of relationships using log transformations and other types of transformations (e.g., square root, reciprocal) can also straighten out curved relationships. Many relationships observed in nature are exponential (e.g., growth in compound interest) and can be described by the model $y = b_0(10^{b_1 x})$. Taking logs (base 10) of both sides of the equation produces the linear relation:

$$\log y = \log b_0 + b_1 x$$

Hence, we have a straight-line relationship between $\log y$ and x.

FIGURE 6.10   Call Center Data, One-Variable Model Residuals Versus Time to Answer

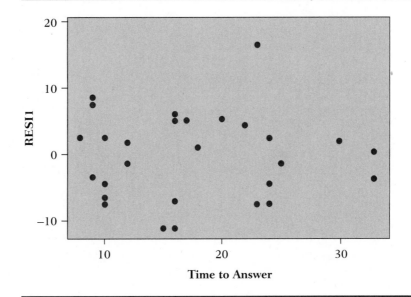

We conclude that transforming *y*, or *x*, or both can straighten out curved relationships. Transforming *y*, particularly through the use of a log transformation, can also produce residuals with a homogeneous variance. See Chapter 9 for further details on transformations.

3. Residuals versus Normal Probability Scale. The normal probability plot of the residuals is constructed to check on the normal distribution assumption of the least squares fitting process. In the normal probability plot, the residuals are ordered by size and plotted versus the normal probability scale. If the plot forms approximately a straight line, the assumption that the residuals follow a normal distribution is reasonably satisfied.

The normal distribution assumption is needed to make tests of significance (hypothesis tests) on the regression coefficients and the regression model and, especially, to calculate prediction intervals for additional data (Chapter 8). In general terms, a test of significance helps us determine whether an observed effect is real (important) or can be attributed to random chance variation (not important). Tests of significance for regression coefficients tell us which individual variables (*x*'s) are important and which variables can be ignored. The *F*-ratio for the overall fit of the model tells us whether there is any correlation between the model, as a whole, and the data. Further details on tests of significance are provided in Chapters 8 and 9.

In Figure 6.11 we see that for this example the normal probability plot of the residuals is essentially a straight line. We conclude that the normal distribution assumption is reasonably satisfied by these data. No real data will ever follow a

FIGURE 6.11   Call Center Data, One-Variable Model Residuals (Normal Probability Plot)

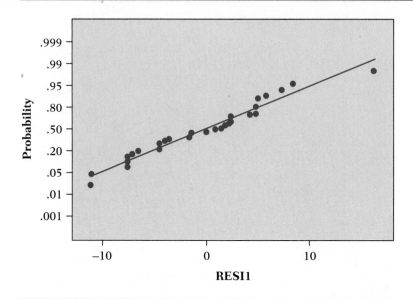

normal (or any other) distribution exactly, so we are not looking for a perfect line in this plot, just a general linear trend.

The presence of a curvilinear relationship in the normal probability plot of the residuals suggests that the normal distribution assumption is not satisfied. The curvilinearity may also be due to the relationship between y and one or more of the x variables not being modeled properly. For example, we may need to add an x^2 or x_1x_2 term to the model. In this situation, as noted earlier, it may also be appropriate to consider transformations for y and one or more of the predictor variables (x's).

4. Residuals versus Observation Sequence (Run Chart). This plot should show no trends if the model is adequate. In Figure 6.12 we see a negative trend in the residuals over time. This suggests that there are one or more variables not currently in the model that have changed during the time spanned by these data. Predictability of the model would be improved if this variable or variables could be added to the model. These variables are the source of the instability.

Outlier Residuals   Outlier residuals result when one or more observations are not adequately fit by the model. When outliers exist, they typically show up in several, if not all, of the residual plots. Outliers suggest that the observation is due to a special cause (measurement error, keying mistake, and so on) or that the model is inadequate (missing variables or missing model terms).

In this example the outlier (observation #13) that we saw in Figure 6.8 appears as an outlier in all residual plots except the normal probability plot. We should in-

FIGURE 6.12   Call Center Data, One-Variable Model Residuals Versus Time

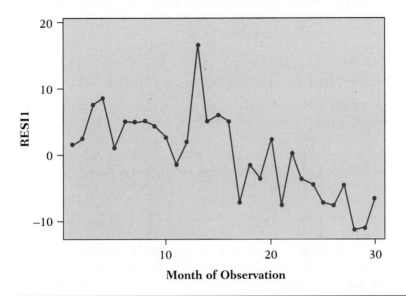

vestigate this point to determine the source of the large residual (special cause) or how the model might be inadequate (e.g., add another variable to the model).

The patterns and trends found in residual plots and some likely problem sources are summarized here. Outliers will show up in any or all of these plots. If the model is adequate, the residuals will show no patterns or trends.

Plot	Pattern	Likely Source of Problem
Sequence	Shifts and trends	Missing variables
Normal probability	Nonlinear	Residuals not normal Multiple data groups Transformation needed
Predicted values	V-shaped Curved	Transformation needed Curved terms needed Transformation needed
Predictors (x)	Curved	Curved terms needed New variables (x) needed Transformation needed

Step 5. Report and Use the Model

We conclude at this point that there is a linear relationship between time to answer and abandon call rate. This model explains 50.7% of the variance in the abandon call rate. The trend in the plot of the residuals versus observation sequence (Figure 6.12) suggests that there are other important variables not in the model. This is not unexpected, as it is unlikely that a phenomenon as complicated as abandoned calls could be explained by a single variable. We will continue our analysis in the next section as we learn about multiple regression analysis.

Extrapolation Can Be Like Skating on Thin Ice

One of the uses for a model is prediction of future observations. Assume, for the sake of argument, that the model just constructed for call abandon rate did fit the data (i.e., no peculiar patterns were found in the residuals) and we wanted to use the results for prediction. Under what circumstances is it appropriate to use the model for prediction? From a theoretical point of view an empirical model should be used only to predict within the region of the x variable data. In our example, the range in time to answer is 8–33 seconds. We have no idea how well the model will predict outside this range because we have no data outside this range and no hypothesis to guide us as to what the nature of the relationship should be.

When a lake is frozen near the shore but not in the center, you might be able to skate on the ice near the shore. But closer to the center of the lake the ice is thinner, and skating becomes more dangerous. And so it is with extrapolation: The further you move outside the range of the data, the less accurate your predictions become and the more "danger" you encounter making decisions on model extrapolations. Extrapolate with care, and be aware of how far you are from the shoreline (i.e., the region of the data).

6.6   Building Models With Several Predictor Variables

Most processes encountered in practice have several predictor variables that can be used to predict measures of process performance. For example, in Chapter 2 we discussed a study that showed beer sales were affected by amount (\$) of advertising, level of sales effort, and amount (\$) spent on sales materials (Ackoff, 1978). Davis (1994) found that bill collection success depended on type of account, balance owed, and number of months overdue. The techniques used to build models involving multiple predictor variables are similar to, but more complicated than, those used for one predictor variable.

We will use the call center data again to illustrate the development of a model involving multiple predictor variables. When analysis of the residuals of the one predictor variable model indicated that the model was not adequate, the investigators searched for some other variables that might be related to abandon call rate. Two were identified: percentage of calls answered by the virtual response unit (VRU), an automated answering system, and the number of customer calls. Using the VRU should enable more calls to be answered at a given point in time, thereby reducing the abandon rate. It was also conjectured that as the call volume increased, the ability to provide a timely response would decrease and drive the abandon rate up. The addition of these two new variables results in a total of three predictor variables and one response:

x_1 = time to answer
x_2 = % answered by VRU × 10
x_3 = call volume
y = call abandon rate

The same procedure is used to build this model as for the single predictor variable model discussed previously (see Figure 6.3 and Table 6.1). The data are shown in Table 6.6.

TABLE 6.6 Call Center Data

Month	Call Volume	Time to Answer	VRU	Abandon Rate	FITS1	RESI1
1	135	12	142	22	26.0955	–4.0955
2	150	8	156	19	21.9757	–2.97574
3	109	9	187	25	20.1204	4.87959
4	109	9	178	26	20.7166	5.28341
5	128	18	189	27	28.0203	–1.02027
6	125	17	148	30	29.8294	0.17059
7	124	17	155	30	29.3496	0.65039
8	141	20	200	33	29.2179	3.78206
9	133	22	175	34	32.4621	1.53787
10	153	24	192	34	33.3751	0.62493
11	155	25	177	31	35.2594	–4.2594
12	294	30	108	39	46.3611	–7.36114
13	295	23	132	47	38.778	8.22197
14	305	16	165	29	30.7437	–1.74369
15	221	16	251	30	23.694	6.30597
16	172	17	261	30	23.1009	6.89905
17	174	16	278	17	21.1486	–4.14855
18	165	12	272	19	17.9671	1.03287
19	177	9	290	14	14.3926	–0.39257
20	183	10	304	21	14.4203	6.57971
21	179	23	282	23	26.9735	–3.9735
22	191	33	323	40	33.0357	6.96435
23	170	33	292	36	34.751	1.24903
24	174	24	289	27	27.2878	–0.28776
25	187	24	281	24	28.0271	–4.02706
26	198	10	290	11	15.5893	–4.58926
27	161	10	309	14	13.7348	0.26523
28	150	15	302	12	18.3137	–6.31373
29	149	16	283	13	20.4147	–7.41472
30	139	10	302	12	13.8442	–1.84416

Variables: Call Volume, Time to Answer, VRU, and Abandon Rate.
Units: Thousands, Seconds, Thousands, % × 10, and %.

TABLE 6.7   Call Center Data

| Variable | Summary Statistics | | | |
	Mean	SD	Minimum	Maximum
Time to Answer	17.6	7.2	8	33
VRU	230	66	108	323
Call Volume	172	50	109	305
Abandon Rate	25.6	9.3	11	47

| | Correlation Matrix | | | |
	Time to Answer	VRU	Call Volume	Abandon Rate
Time to Answer	1.0			
VRU	−0.053	1.0		
Call Volume	−0.108	0.331	1.0	
Abandon Rate	0.724	−0.519	0.360	1.0

Step 1.   Get to Know Your Data

The summary statistics and correlation matrix for these variables are summarized in Table 6.7. The correlation matrix has 1's on the diagonal because the correlation of each variable with itself is a perfect correlation with a coefficient of 1.0. The matrix is symmetrical because the correlation between two variables—say, x_1 and x_2—is the same as the correlation between x_2 and x_1. Correlation coefficients measure linear relationships between pairs of variables, and this matrix summarizes all the correlation coefficients for all pairs of variables. For example, the correlation coefficient between VRU and abandon rate is −0.519, indicating a negative relationship. The correlation between VRU and time to answer is −0.053, indicating little or no correlation. Generally speaking, the correlations between the response y and the predictor variables x should be high, and the correlations among the predictor variables should be low. We will return to this issue when we discuss multicollinearity (correlation between multiple predictor variables) in Section 6.7.

The time plots of the x and y variables for the abandon rate versus time and the time to answer versus time are identical to those shown in the previous example (see Figures 6.6 and 6.7). Of particular note is the step-change increase in VRU (Figure 6.13) and the three high values for call volume (Figure 6.14).

In the scatter plots of abandon rate versus the predictor variables, we see a positive relationship with time to answer ($r = 0.724$), as shown in Figure 6.15; a negative relationship with VRU ($r = -0.519$), as shown in Figure 6.16; and a weak positive relationship with call volume ($r = 0.360$), as shown in Figure 6.17.

FIGURE 6.13   Call Center Data, VRU Versus Time

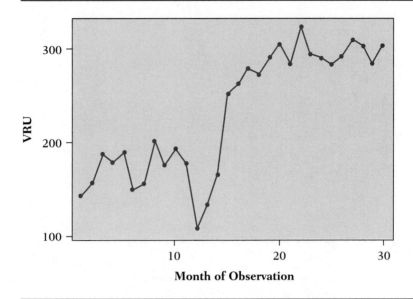

FIGURE 6.14   Call Center Data, Call Volume Versus Time

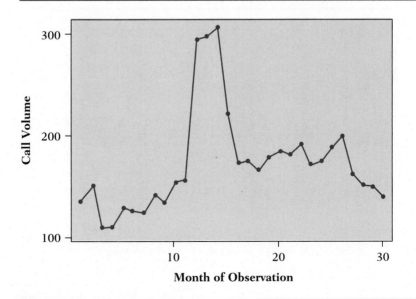

FIGURE 6.15 Call Center Data, Abandon Rate Versus Time to Answer

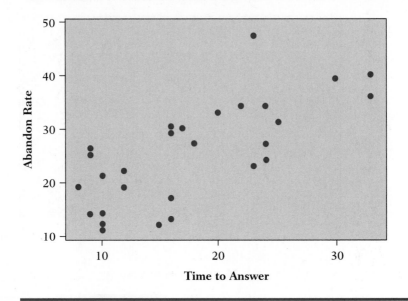

FIGURE 6.16 Call Center Data, Abandon Rate Versus VRU

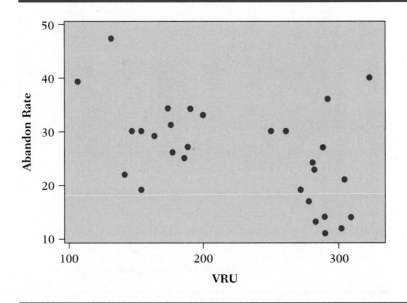

FIGURE 6.17 Call Center Data, Abandon Rate Versus Call Volume

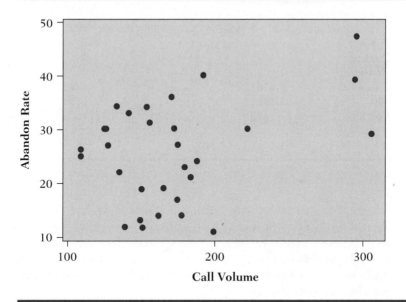

Step 2. Formulate the Model

The relationships between abandon rate and the predictor variables are all linear, suggesting that we should fit a model of the form

$$y = b_0 + b_1 x_1 + b_2 x_2 + b_3 x_3$$

where y = abandon rate, x_1 = time to answer, x_2 = VRU, x_3 = call volume, and b_0, b_1, b_2, and b_3 are coefficients to be estimated from the data.

Step 3. Fit the Model to the Data

Least squares regression is used to calculate the regression coefficients from the data. The results of the least squares calculations are summarized in Table 6.8, which gives the values of the regression coefficients (b's), the standard deviations of the regression coefficients, and the t-ratios. The coefficient standard deviations are a measure of the uncertainty of the regression coefficients. If we were to collect more data, the standard deviations of the regression coefficients would decrease in size and there would be less uncertainty in the values of the regression coefficients. The actual values of the coefficients would change slightly, reflecting the fact that data had been added to the analysis. The t-ratio, obtained by dividing the coefficient by its standard deviation, is a yardstick that can be used to assess the importance of the variable as a predictor of the response.

TABLE 6.8   Call Center Data, Three-Variable Model (Minitab regression analysis printout)

The regression equation is
    Abandon Rate = 23.0 + 0.858 Time to Answer − 0.0662 VRU + 0.0161 Call Volume

Predictor	Coef	SD Coef	t-Ratio	VIF
Constant	23.026	4.817	4.78	
Time to	0.8585	0.1296	6.62	1.1
VRU	−0.06624	0.01341	−4.94	1.0
Call Vol	0.01610	0.01873	0.86	1.1

R-Sq = 76.3%   R-Sq(adj) = 73.6%

Predictor variables with t-ratios of 2.0 or larger are considered important. Formal tests of significance (hypothesis tests) using the t-ratios will be discussed in Chapters 8 and 9. In Table 6.8 we see that time to answer has a strong positive relationship ($t = 6.62$), VRU has a strong negative relationship ($t = −4.94$), and call volume has no detectable relationship ($t = 0.86$) over the range of the data in this study.

Step 4. Check the Model Fit

Before reaching any conclusions or making any decisions on these results, we must check the adequacy of the fit of the model by checking the adjusted R-squared statistic and analyzing the residuals. The adjusted R-squared statistic for the model is 73.6%—considerably higher than the 50.7% for the model using only time to answer as the predictor variable.

The Difference Between R-Squared and Adjusted R-Squared   The R-squared statistic, which measures the percentage of variability (sum of squares) explained by the model and ranges from 0% to 100%, is also used by some as a measure of model fit. For one predictor (x) variable, R-squared is the square of the correlation coefficient between y and x. The R-squared statistic is of limited use as a measure of model fit because it is an increasing function of the number of terms in the model. It will increase even if a term added to the model is not important (when $t < 2.0$). For this reason, we do not recommend its use as a measure of model fit. It is used appropriately in the calculation of the variance inflation factor discussed in Section 6.7, however.

To measure model fit, we recommend the adjusted R-squared statistic, which takes into account the number of terms in the model. The adjusted R-squared statistic only increases if the predictive ability of the model increases.

FIGURE 6.18 Call Center Data, Three-Variable Model Residuals Versus Time

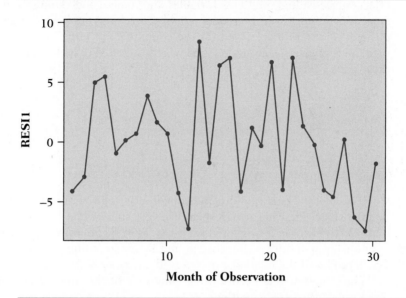

Next we analyze the residuals. No peculiar patterns were observed in plots of the residuals versus either the predicted values or the predictor variables. The normal probability plot of the residuals was approximately a straight line, indicating no significant departure from the normal distribution assumption. The plot of the residuals versus observation sequence was much improved, but a slight curvilinear pattern did appear (Figure 6.18). Observation 13, which was poorly fit by the one predictor model, was much better fit by the three predictor model, although it still had the largest residual.

Step 5. Report and Use the Model

We conclude from the analysis that we have identified two key drivers of abandon rate: time to answer and use of the automatic answering system (VRU). More work needs to be done, however. The adjusted R-squared statistic for the three variable model was 73.6%, and working with monthly data should provide a better fit to the data. The call answering process should be studied to determine what, if any, important variables are missing. The data set could also be updated to include the most recent data, which could extend the utility of the model. If call volume continues to be unimportant, we may decide to eliminate it from the equation. Time to answer may be accounting for the impact of call volume.

6.7 Multicollinearity, Another Model Check

Before interpreting the individual coefficients in a regression model, we should check for a condition called "multicollinearity," which refers to high correlations among the predictor variables. This condition makes it difficult to determine the effects of individual predictor variables and to get good estimates of the coefficients for the terms in the model. Highly correlated predictor variables are very difficult to differentiate (even for the computer!) in terms of their relationship with the response (y). To understand why this is so, let's look at an example using these data:

x_1	x_2	y
1	2	4
2	4	10
3	6	15
4	8	20
5	11	23

If we plot y versus x_1 and x_2 individually, we find very high correlations, 0.994 and 0.981, respectively, which seem like very strong relationships (Figures 6.19 and 6.20). However, if we plot x_1 versus x_2, as in Figure 6.21, we also see a strong relationship ($r = 0.996$). Because x_1 and x_2 are so strongly linked it is not clear which produced the relationship observed: x_1, x_2, or both x_1 and x_2.

The situation gets even murkier when we fit the linear model to the data. Then we get the fitted equation:

$$y = -1.0 + 10.3x_1 - 2.5x_2$$

FIGURE 6.19   Multicollinearity Example, y Versus x_1

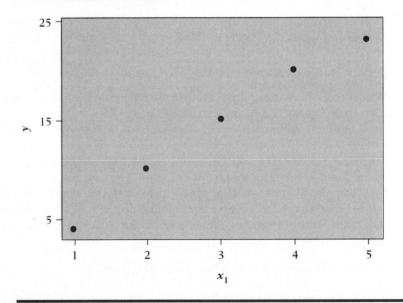

FIGURE 6.20   Multicollinearity Example, y Versus x_2

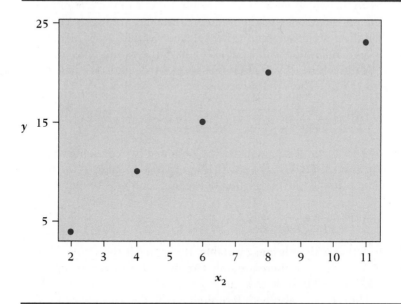

FIGURE 6.21   Multicollinearity Example, x_2 Versus x_1

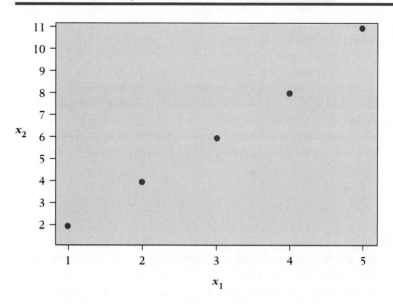

This says that x_1 has the largest effect and x_2 has a *negative effect*, which is contrary to the strong positive relationship we saw in the plot of y versus x_2. This is a common occurrence when predictor variables are highly correlated. In general, when the correlations among the predictor variables are high, the coefficients are too large in absolute value and often have the wrong signs, as we see in this case.

Multicollinearity is often measured by diagnostics called variance inflation factors (VIF). The VIF for variable x is

$$\text{VIF} = \frac{1}{1 - R^2}$$

where R^2 is the R-squared statistic obtained by regressing this x on the other x's in the model. That is, x becomes the response variable in the regression equation. In our example, the correlation between x_1 and x_2 is 0.996, which gives

$$\text{VIF} = \frac{1}{1 - (.996)^2} = 122.0$$

Recall that with only one x variable, R-squared is the square of the correlation coefficient. Because there are only two predictor variables involved in this model, the VIF is 122.0 for both x_1 and x_2.

The VIF can also be interpreted as the degree to which the variances (standard deviations squared) of the regression coefficients have been inflated due to multicollinearity. If VIF = 4, the coefficient variance is four times larger than it would have been if there was no multicollinearity (standard deviation is twice as large). VIF = 1 means that there is no multicollinearity.

As a rule of thumb, when the VIF > 10 we conclude that multicollinearity is a problem and that we should not base our decisions on the magnitude and sign of the regression coefficients. When VIF = 5 or less, we conclude that multicollinearity is not a problem. When VIF is between 5 and 10, we must be concerned about multicollinearity.

In the call center abandon rate data, the VIFs for x_1, x_2 and x_3 were, respectively, 1.0, 1.0, and 1.1, indicating that multicollinearity is not a problem.

When one or more VIF is greater than 10, the model should generally be reformulated (e.g., delete one or more of the correlated predictor variables) or a regression method other than least squares (ridge regression, for example) should be used to analyze the data. Montgomery and Peck (1992) provide a good discussion of the methods to fit regression models when the predictor variables are highly correlated.

6.8   Some Limitations of Using Existing Data

Using existing process data to better understand the process and build models is attractive because the data are readily available and you do not have to interfere with the process. But this approach has some limitations due to the quality of process data (see Chapter 5).

The first limitation is that these data records are frequently incomplete, missing values and omitting predictor variables. In many large processes the data needed for analysis do not exist because the data system was not designed to collect appropriate data. Most data collection systems are designed for purposes other than process improvement and management (e.g., compensation or financial records). For example, cycle time is a key measure for invoice processing, yet many businesses do not routinely record the time required to process invoices, although they do record the amount of payment (financial measure).

The data system may not collect data in the proper frequency. It is not uncommon for the data system to produce monthly data, yet daily, or even hourly, data may be needed to properly understand the process. This was the case with the call abandon rate data discussed earlier in this chapter. When data are collected over time, special causes are often hidden in the data. For example, a stock split may result in both pre- and postsplit prices in the same data set.

Another limitation is that the key predictor variables may have varied over a small range. When a variable is known to be a key driver, it is often held constant so that the process can be better controlled. Variation of a predictor variable over a small range reduces our ability to identify relationships because there is little opportunity for y to be impacted by changes in the predictor. The explained variation will be less, resulting in a low adjusted R-squared.

When the process is observed "as is," the resulting predictor variables are often highly correlated, producing the multicollinearity condition discussed previously. We tend to change several variables simultaneously when we manage processes, and as a result, we often create correlation among the predictor variables.

These limitations suggest that the quality of existing process data is generally low from a model building viewpoint. The probability of producing a useful model is often less than 50%. We recommend learning as much as possible from the existing data, with a particular focus on identifying the key drivers (process variables) that have a major impact on the process. This information may be sufficient to better manage and improve the process. When a better model is needed, statistically designed experiments, such as those discussed in Chapter 7, can be used to collect the high-quality data needed to better understand and model the process.

6.9 Summary

The strategy for better understanding our processes and building models discussed in this chapter is based on observing the process "as is." It uses regression analysis of process data to help us quantify the relationship between the process outputs (y) and the process variables (x's). The results can be used to identify the vital few process variables (key drivers), identify root causes and fix problems permanently, control and manage the process, and predict future process performance. Designed experiments are often necessary to establish cause-and-effect relationships and to identify ways to further improve the process.

Regression analysis is a valuable tool, but it can also be abused. Here are some abuses that commonly occur:

Abuses of Regression Analysis

- Predicting far outside the region of the data (extreme extrapolation)
- Building models from poor-quality data due to poor measurement systems or missing variables
- Ignoring or not understanding the process that generated the data
- Ignoring or not seeking out known scientific and econometric theory regarding the variables
- Ignoring the caveats of the model developer
- Using the model without understanding the context of its development

Effective model building results from adhering to these key ideas:

Building Useful Models

- Be clear about the purpose of the model and how it will be used.
- Make the model useful for its intended purpose.
- Search for and use subject matter knowledge to guide the model development process.
- Recognize and understand the strengths and limitations of the data used to build the model.
- Focus on identification of key drivers.
- Focus on keeping models simple and parsimonious.
- Recognize the iterative nature of model building and move back and forth between hypotheses and data.
- Continually challenge (check) the model over time and revise it as needed.
- Use the model to guide, not dictate, decisions.
- Validate and document the performance of the model.

Regression analysis is an important tool for model building, but it can be a complicated subject, particularly to the beginner. Be mindful of these key ideas as you build process models using regression analysis:

Making Effective Use of Regression Analysis

- Plot all the data, both raw data and residuals.
- Focus on identifying the key drivers.
- Analyze the residuals.
- Be aware that a significant correlation does not guarantee a cause-and-effect relationship.
- Be careful extrapolating beyond the region of the data.
- As you conduct your analysis, keep a record of questions and ideas variable by variable.
- Practice good file management; identify data files and graphs with titles, dates, and the names of those who did the work.
- You can identify significant predictor variables even when the model has low adjusted R-squared statistics. The poor fit may be due to important variables missing from the model or high measurement variation.
- Relate the regression results to the process that generated the data, reporting the results, recommendations, and decisions in the language of the process.

- High correlations among the predictor variables can produce regression coefficients that are too large and have the wrong signs.
- Validate the proposed model whenever possible, and assess how well the model predicts a new sample of data. Identify any process changes that have occurred and that may invalidate the model.
- Be skeptical of your model and its utility. Remember that "all models are wrong, but some are useful."

Paying attention to these key ideas can take you a long way toward constructing useful process models.

6.10 Project Update

At this point in your project you should have collected a reasonable amount of data. Perhaps you analyzed these data using some of the basic tools presented in Chapter 5. Now that you have learned about model building, consider whether this would be an appropriate tool for your project. If it is an appropriate tool but you lack appropriate data, wait to build your models until you have completed Chapter 7, where you will learn more about the best way to collect data.

Try to get a feel for when model building would be helpful and how to go about doing it, and remember to use the diagnostic tools, such as residual analysis. Model building is typically an iterative process; it may take several attempts to find a model that makes sense in light of your subject matter knowledge and also fits the data well.

EXERCISES

1. **Model for Your Process.** Develop a model for a process you are familiar with using existing data on the operation of the process.
2. **Models Used in Your Field of Interest.** Identify two to five models that are useful in your field of interest. Include the model context, process involved, model utility, and variables involved.
3. **Examples of Theoretical Models.** Identify two or three examples of theoretical models, preferably in your field of interest. Discuss the strengths and limitations of these models and under what circumstances empirical models would be better.
4. **Linear and Nonlinear Programming.** Research linear and nonlinear programming algorithms on the Internet or at a library. Determine situations in your field of interest where each would be used. Identify the purpose and value of this approach in each situation.
5. **Time Series Models.** Research time series models on the Internet or at a library. Identify situations in your field of interest where time series models would be useful. Discuss the strengths and limitations of time series models.
6. **Timeliness of Order Delivery.** The order fulfillment process of a major distribution center is having trouble delivering orders on time. It is conjectured that order

TABLE 6.9
Order Fulfillment Response Time

Day	Order Volume	Average Response Time	Day	Order Volume	Average Response Time
1	31	−2	26	20	−5
2	91	9	27	88	19
3	13	−4	28	97	12
4	69	15	29	93	19
5	70	12	30	72	10
6	64	6	31	6	11
7	38	7	32	55	23
8	50	4	33	15	12
9	94	23	34	10	16
10	82	24	35	21	20
11	15	−2	36	88	43
12	42	−4	37	55	23
13	27	−6	38	27	16
14	66	14	39	66	32
15	73	10	40	3	13
16	45	−2	41	35	25
17	21	−12	42	40	15
18	7	−1	43	64	32
19	69	11	44	43	17
20	38	5	45	30	28
21	2	−17	46	73	33
22	99	18	47	46	19
23	36	8	48	82	32
24	82	9	49	35	22
25	58	21	50	2	6

volume is the root cause and that problems occur when order volumes are high. A new computer system has been requested to handle the increased volume of orders. The data in Table 6.9 were collected to support the requested funding for the computer system. A negative response time indicates early delivery.

- Do the data in Table 6.9 support the request? Why or why not? (*Hint:* Be sure to do a *complete analysis* of the data.)
- What recommendations would you make based on your analysis of these data?
- What are some other possible solutions to this problem?

7. **Effect of Variable Range on Ability to Detect Relationships.** Using the order delivery data in Table 6.9, compute the correlation coefficients for the relationship between delivery time and order volume using all the data and only those data for order volume between 0 and 50 orders per day. Construct and compare scatter plots for both sets of data. Be sure to note the differences in the scales of the two plots. How do the correlation coefficients compare? What does this compari-

TABLE 6.10   Antique Clock Selling Prices

Clock	Clock Age	Number of Bidders	Price ($)
1	108	6	729
2	108	14	1055
3	111	7	785
4	111	15	1175
5	113	9	946
6	115	7	744
7	115	12	1080
8	117	11	1024
9	117	13	1152
10	126	10	1336
11	127	13	1235
12	127	7	845
13	132	10	1253
14	137	8	1147
15	137	15	1713
16	137	9	1297
17	143	6	854
18	150	9	1522
19	153	6	1092
20	156	12	1822
21	156	6	1047
22	159	9	1483
23	162	11	1884
24	168	7	1262
25	170	14	2131
26	175	8	1545
27	179	9	1792
28	182	11	1979
29	182	8	1550
30	184	10	2041
31	187	8	1593
32	194	5	1356

son tell you about the effect of the range of a variable on the ability to detect important predictor variables?

8. **Antique Clock Selling Prices.** McClave and Benson (1985) report the data in Table 6.10 on the age of antique clocks and the price paid at auction. The number of bidders participating in each sale is also given.

   ■ Does either the age of the clock or the number of bidders have any effect on the purchase price of the clock? If so, which variables?
   ■ What is the nature of the relationships: positive, negative, curved, strong, weak? Does the nature of the effects make sense compared with what you expect might happen?

- Are the effects of the variables independent of each other?
- What is your best prediction model for the prices of clocks sold at this auction? Under what conditions should your model be used?
- What is your prediction of the selling price of a clock that is 225 years old when 20 people bid for the clock? How confident are you in your prediction?

9. **Measuring the Effect of a Process Change.** The data in Table 6.11 are from a study conducted by an insurance company to determine the effect of changing the process by which insurance claims are approved. The goal was to improve policy-holder satisfaction by speeding up the process and eliminating some non-value-added approval steps in the process. The response measured was the average time required to approve and mail all claims initiated in a week. The new procedure was tested for 12 weeks, and the results were compared to the process performance for the 12 weeks prior to instituting the change.

- What was the average effect of the process change? Did the process average increase or decrease and by how much?
- Analyze these data using the regression model $y = b_0 + b_1x$, where y = time to approve and mail a claim (weekly average), $x = 0$ for the old process, and $x = 1$ for the new process. How does this model measure the effect of the process change? How much did the process performance change on the average? (*Hint:* Compare the values of b_1 and the average of new process performance minus the average of the performance of the old process.)

TABLE 6.11 Insurance Claim Approval Time

Old Process		New Process	
Week	Elapsed Time (days)	Week	Elapsed Time (days)
1	31.7	13	24
2	27	14	25.8
3	33.8	15	31
4	30	16	23.5
5	32.5	17	28.5
6	33.5	18	25.6
7	38.2	19	28.7
8	37.5	20	27.4
9	29	21	28.5
10	31.3	22	25.2
11	38.6	23	24.5
12	39.3	24	23.5

10. **Call Center Study.** This call center operates 24 hours per day, 7 days per week. The call center wanted to increase the percentage of calls answered. The data in Table 6.12 were collected to aid in the analysis of the problem.

TABLE 6.12
Call Center Data

Week	Day	% of Calls Answered in 20 Seconds	Total Calls	Number Answered	Number That Got Busy Signals	Number Abandoned	Hold Time (in seconds)	Workforce (full-time equivalents)
1	1	79	449	293	131	25	381	11
1	2	98	494	487	6	1	413	17
1	3	99	586	579	1	6	375	22
1	4	99	746	743	0	3	427	19
1	5	94	689	517	167	5	433	21
1	6	70	704	312	386	6	514	13
1	7	78	1679	1539	97	43	464	43
2	1	62	821	700	71	50	*	8
2	2	68	1128	506	601	21	526	18
2	3	85	826	573	244	9	442	27
2	4	86	754	556	194	4	416	21
2	5	86	637	470	159	8	377	18
2	6	*	*	*	*	*	*	*
2	7	70	2680	1411	1203	66	463	52
3	1	42	1453	510	883	60	348	12
3	2	89	826	624	197	5	414	30
3	3	87	756	576	174	6	335	18
3	4	68	950	490	446	14	360	19
3	5	72	1001	564	422	15	493	19
3	6	99	521	490	27	4	432	31
3	7	61	2231	1910	207	114	439	53
4	1	44	1559	803	682	74	344	12
4	2	53	1027	675	308	44	437	21
4	3	86	832	599	199	34	374	27
4	4	74	747	580	133	34	412	24
4	5	73	818	601	195	22	372	23
4	6	84	586	445	132	9	364	20
4	7	60	3122	2518	454	150	380	64
5	1	50	2020	851	1127	42	292	18
5	2	69	902	738	127	37	462	31
5	3	79	795	717	53	25	424	25
5	4	49	560	502	0	58	441	20
5	5	87	996	504	451	41	405	20
5	6	67	630	457	157	16	519	28
5	7	52	2615	1692	830	93	496	70
6	1	42	2561	824	1679	58	359	23
6	2	52	1220	693	488	39	471	33
6	3	57	1167	669	447	51	492	25
6	4	46	1348	535	742	71	422	21
6	5	82	484	335	136	13	508	20
6	6	86	798	675	109	14	452	32
6	7	45	4309	2536	1448	325	489	67

(continued on next page)

TABLE 6.12
Call Center Data *(continued)*

Week	Day	% of Calls Answered in 20 Seconds	Total Calls	Number Answered	Number That Got Busy Signals	Number Abandoned	Hold Time (in seconds)	Workforce (full-time equivalents)
7	1	74	2279	1491	752	36	363	37
7	2	62	1124	783	303	38	448	41
7	3	59	1040	729	267	44	323	37
7	4	54	1062	619	401	42	432	34
7	5	37	1156	390	709	57	488	24
7	6	77	664	607	39	18	496	39
7	7	60	3603	2647	704	252	477	74
8	1	88	1861	1567	267	27	354	53
8	2	81	987	718	249	20	466	56
8	3	76	749	685	38	26	455	28
8	4	80	1019	634	372	13	443	45
8	5	42	1312	491	755	66	352	43
8	6	86	719	661	48	10	480	51
8	7	63	3070	2250	738	82	431	70
9	1	93	2117	1856	249	12	322	50
9	2	99	681	676	3	2	377	33
9	3	99.9	736	735	0	1	389	24
9	4	93	929	919	2	8	432	25
9	5	71	1048	879	113	56	443	21
9	6	96	714	700	8	6	437	38
9	7	97	3316	3297	0	19	398	94
10	1	79	2230	1964	180	86	343	52
10	2	98	1047	1043	0	4	448	34
10	3	98	993	985	0	8	405	26
10	4	87	945	924	3	18	454	24
10	5	70	992	834	110	48	438	12
10	6	99	719	712	1	6	439	30
10	7	95	3599	3525	32	42	373	71
11	1	93	2153	1895	240	18	342	63
11	2	47	293	136	151	6	303	9
11	3	77	1049	947	83	19	394	24
11	4	99	955	950	0	5	387	29
11	5	98	905	898	0	7	495	21
11	6	99.9	611	610	0	1	436	31
11	7	90	3558	3501	8	49	403	73
12	1	95	2385	2253	123	9	316	67
12	2	99.9	986	986	0	0	457	38
12	3	99.9	905	901	0	4	426	45
12	4	97	894	872	17	5	442	49
12	5	80	820	791	12	17	400	34
12	6	80	717	676	22	19	388	29
12	7	99	3532	3513	7	12	392	84

(continued on next page)

TABLE 6.12
Call Center Data *(continued)*

Week	Day	% of Calls Answered in 20 Seconds	Total Calls	Number Answered	Number That Got Busy Signals	Number Abandoned	Hold Time (in seconds)	Workforce (full-time equivalents)
13	1	97	2112	2040	61	11	324	72
13	2	97	955	951	0	4	385	38
13	3	97	933	927	0	6	397	33
13	4	96	912	902	3	7	431	27
13	5	99	861	835	24	2	402	23
13	6	99.9	663	656	6	1	440	30
13	7	99	3663	3644	7	12	382	81
14	1	96	2224	2136	79	9	315	63
14	2	99.9	875	871	0	4	415	39
14	3	99	919	914	0	5	437	39
14	4	98	791	780	7	4	469	28
14	5	61	930	796	79	55	449	18
14	6	99.9	631	629	0	2	393	29
14	7	98	3330	3312	1	17	413	82
15	1	98	2247	2191	48	8	369	83
15	2	97	920	898	15	7	448	40
15	3	97	880	874	0	6	378	45
15	4	99.9	889	886	0	3	399	39
15	5	91	913	898	4	11	434	37
15	6	99.9	671	668	0	3	401	32
15	7	98	4647	4607	22	18	359	53
16	1	93	2407	2161	228	18	341	79
16	2	99	1026	1005	19	2	415	39
16	3	99	962	960	0	2	342	39
16	4	99.9	937	930	0	7	372	34
16	5	99.9	903	902	0	1	373	39
16	6	99.9	610	607	0	3	420	33
16	7	96	3310	3170	117	23	376	57
17	1	98	2198	2149	41	8	329	76
17	2	99.9	810	809	0	1	424	37
17	3	99.9	1013	1007	1	5	423	40
17	4	95	967	943	15	9	365	34
17	5	80	937	897	25	15	399	35
17	6	99.9	650	649	0	1	402	35
17	7	96	3736	3392	317	27	321	57
18	1	97	2214	2111	85	18	370	72
18	2	80	1032	943	65	24	418	33
18	3	79	925	866	33	26	411	35
18	4	99	786	765	16	5	429	31
18	5	99.9	848	848	0	0	386	34
18	6	99.9	642	641	0	1	406	33
18	7	98	3440	3272	155	13	362	51

(continued on next page)

TABLE 6.12

Call Center Data (*continued*)

Week	Day	% of Calls Answered in 20 Seconds	Total Calls	Number Answered	Number That Got Busy Signals	Number Abandoned	Hold Time (in seconds)	Workforce (full-time equivalents)
19	1	96	2340	2203	128	9	345	72
19	2	99.9	933	928	1	4	487	32
19	3	98	940	778	157	5	404	34
19	4	89	833	815	4	14	453	33
19	5	92	779	758	16	5	469	35
19	6	96	745	663	68	14	445	31
19	7	81	5726	4703	915	108	357	51
20	1	91	3003	2612	370	21	365	76
20	2	76	1244	1015	209	20	451	33
20	3	99	819	816	1	2	425	34
20	4	99	872	867	0	5	400	35
20	5	99	845	845	0	0	407	38
20	6	99.9	607	603	0	4	377	34
20	7	98	2757	2704	28	25	385	53

*Missing data.

- Draw a schematic of the process identifying input, process, and output variables.
- What are the key drivers of the percentage of calls answered in 20 seconds?
- What would you recommend be done to increase the percentage of calls answered? What are the implications of your recommendations?
- What did you learn about the management of the process over the 20-week period?

11. **Gasoline Mileage of 1997 Automobiles.** *Consumer Reports* published the data in Table 6.13 on the characteristics and measurements of 1997 model year automobiles. Weight distribution measures the car's weight balance, front versus back.

- Which of these variables are predictive of auto gasoline mileage?
- What are the key drivers of gasoline mileage?
- What process are we studying with these data? What are the input, process, and output variables?
- What theoretical relationships can help us predict the variables that will have an important effect on gasoline mileage?
- Are there any abnormal observations or examples of multicollinearity in this data set?

TABLE 6.13
1997 Automobile Gasoline Mileage Data

Make and Model	Drive	Miles/ Gallon	Fuel Type	Tank Capacity	Engine Size	Number of Cylinders	Horse Power	Transmission Type	Speeds	Weight	Weight Distbn
Acura Integra	Front	30	Regular	13.2	1.8	4	142	Manual	5	2665	62
Acura RL	Front	20	Premium	18	3.5	6	210	Auto	4	3670	60
Acura SLX	Rear	19	Regular	22.5	3.2	6	190	Auto	4	4430	51
Acura TL	Front	23	Premium	17.2	2.5	5	176	Auto	4	3278	59
Audi A4	Front	22	Premium	16.4	2.8	6	172	Auto	5	3220	63
BMW 3 Series	Rear	24	Premium	16.4	2.8	6	190	Auto	4	3225	51
BMW 5 Series	Rear	20	Premium	18.5	2.8	6	190	Auto	4	3685	51
Buick LaSabre	Front	20	Regular	18	3.8	6	205	Auto	4	3450	65
Buick Riviera	Front	17	Regular	20	3.8	6	240	Auto	4	3770	62
Cadillac KATRA	Rear	20	Premium	18	3	6	200	Auto	4	3885	54
Chevy Astro	Rear	15	Regular	27	4.3	6	190	Auto	4	4520	53
Chevy Blazer	Rear	17	Regular	19	4.3	6	190	Auto	4	4180	55
Chevy CK 1500	Rear	15	Regular	25	5	6	220	Auto	4	4605	58
Chevy Cavallier	Front	26	Regular	15.2	2.2	4	120	Auto	3	2795	65
Chevy Lumina	Front	22	Regular	16.6	3.1	6	160	Auto	4	3350	64
Chevy Malibu	Front	24	Regular	15	2.4	4	150	Auto	4	3040	64
Chevy Montecarlo	Front	18	Regular	17.1	3.4	6	215	Auto	4	3450	65
Chevy S Series	Rear	17	Regular	20	4.3	6	190	Auto	4	3560	61
Chevy Tahoe	Rear	13	Regular	30	5.7	8	250	Auto	4	5335	52
Chrysler Cirrus	Front	22	Regular	16	2.5	6	168	Auto	4	3145	64
Chrysler Concorde	Front	21	Mid Grade	18	3.5	6	214	Auto	4	3550	64
Chrysler LHS	Front	20	Mid Grade	18	3.5	6	214	Auto	4	3605	64
Chrysler Sebring	Front	22	Regular	15.9	2.5	6	163	Auto	4	3175	64
Chrysler Town Country	Front	18	Regular	20	3.3	6	158	Auto	4	4035	57
Dodge Avenger	Front	22	Regular	15.9	2.5	6	163	Auto	4	3175	64
Dodge Caravan	Front	19	Regular	20	3.3	6	158	Auto	4	3985	59
Dodge Grand Caravan	Front	18	Regular	20	3.3	6	158	Auto	4	4035	57
Dodge Intrepid	Front	20	Regular	18	3.3	6	161	Auto	4	3435	64
Dodge Neon	Front	26	Regular	12.5	2	4	132	Auto	3	2600	64
Dodge Ram 1500	Rear	13	Regular	26	5.2	8	220	Auto	4	4785	57

(continued on next page)

TABLE 6.13
1997 Automobile Gasoline Mileage Data (continued)

Make and Model	Drive	Miles/ Gallon	Fuel Type	Tank Capacity	Engine Size	Number of Cylinders	Horse Power	Transmission Type	Transmission Speeds	Weight	Weight Distbn
Dodge Stralus	Front	20	Regular	16	2.4	4	150	Auto	4	3085	63
Eagle Vision	Front	21	Regular	18	3.5	6	214	Auto	4	3550	64
Ford Aspire	Front	36	Regular	10	1.3	4	63	Manual	5	2140	62
Ford Contour	Front	24	Regular	14.5	2	4	125	Auto	4	2895	64
Ford Crown Victoria	Rear	19	Regular	20	4.6	8	210	Auto	4	4010	55
Ford Escort	Front	28	Regular	12.7	2	4	110	Auto	4	2585	64
Ford Explorer	Rear	17	Regular	22	4	6	160	Auto	4	4440	53
Ford F-150	Rear	16	Regular	25	4.6	8	210	Auto	4	4450	58
Ford Ranger	Rear	18	Regular	20	4	6	160	Auto	4	3680	60
Ford Tarus	Front	21	Regular	16	3	6	200	Auto	4	3515	64
Ford Thunderbird	Rear	20	Regular	18	4.6	8	205	Auto	4	3705	59
Ford Winstar	Front	20	Regular	20	3.8	6	200	Auto	4	4050	61
GMC Jimmy	Rear	17	Regular	19	4.3	6	190	Auto	4	4180	55
GMC Safari	Rear	15	Regular	27	4.3	6	190	Auto	4	4520	53
GMC Sierra CK 1500	Rear	15	Regular	25	5	8	220	Auto	4	4605	58
GMC Sonoma	Rear	17	Regular	20	4.3	6	190	Auto	4	3560	61
GMC Yukon	Rear	13	Regular	30	6.7	8	250	Auto	4	5335	52
Geo Metro	Front	29	Regular	10.6	1.3	4	70	Auto	3	2065	61
Geo Tracker	Rear	24	Regular	14.5	1.6	4	95	Auto	5	2780	52
Honda Accord	Front	21	Regular	17	2.7	6	170	Auto	4	3255	64
Honda Civic	Front	31	Regular	11.9	1.5	4	106	Auto	4	2440	62
Honda Odyssay	Front	21	Regular	17.2	2.2	4	140	Auto	4	3480	59
Honda Passport	Rear	16	Regular	21.9	3.2	6	190	Auto	4	4080	51
Hundai Accent	Front	28	Regular	11.9	1.5	4	92	Auto	4	2290	63
Hundai Elantra	Front	25	Regular	14.5	1.8	4	130	Auto	4	2725	63
Hundai Sonata	Front	21	Regular	17.2	3	6	142	Auto	4	3095	63
Infiniti 130	Front	23	Premium	18.5	3	6	190	Auto	4	3195	63
Isuzu Oasis	Front	21	Regular	17.2	2.2	4	140	Auto	4	3480	59
Isuzu Rodeo	Rear	16	Regular	21.9	3.2	6	190	Auto	4	4080	51
Isuzu Trooper	Rear	19	Regular	22.5	3.2	6	190	Auto	4	4430	51

(continued on next page)

TABLE 6.13

1997 Automobile Gasoline Mileage Data (continued)

Make and Model	Drive	Miles/ Gallon	Fuel Type	Tank Capacity	Engine Size	Number of Cylinders	Horse Power	Transmission Type	Speeds	Weight	Weight Distrbn
Jeep Grand Cherokee	Rear	15	Regular	23	5.2	8	220	Auto	4	4090	57
Land Rover Discovery	All	13	Premium	23.4	4	8	182	Auto	4	4535	48
Lexus ES300	Front	22	Premium	18.5	3	6	200	Auto	4	3390	62
Lincoln Continental	Front	18	Premium	17.8	4.5	8	260	Auto	4	3930	63
Mazda 626	Front	25	Regular	15.5	2	4	118	Auto	4	2860	63
Mazda B-Series	Rear	18	Regular	20	4	6	160	Auto	4	3680	60
Mazda MPV	Rear	16	Regular	18.8	3	6	155	Auto	4	4135	57
Mazda Millenia	Front	22	Premium	18	2.3	6	210	Auto	4	3415	63
Mazda Protégé	Front	26	Regular	13.2	1.8	4	122	Auto	4	2630	63
Mercedes Benz C Class	Rear	24	Premium	16.4	2.8	6	194	Auto	5	3320	55
Mercedes Benz E Class	Rear	22	Premium	21.1	3.2	6	217	Auto	5	3570	54
Mercury Cougar	Rear	20	Regular	18	4.6	8	205	Auto	4	3705	59
Mercury Grand Marquis	Rear	19	Regular	20	4.6	8	210	Auto	4	4010	55
Mercury Mystique	Front	23	Regular	14.5	2.5	6	170	Auto	4	3110	65
Mercury Sable	Front	22	Regular	16.1	3	6	145	Auto	4	3345	65
Mercury Tracer	Front	28	Regular	12.7	2	4	110	Manual	5	2585	64
Mercury Villager	Front	18	Regular	20	3	6	151	Auto	4	3900	59
Mitsubishi Gallant	Front	25	Regular	16.9	2.4	4	141	Auto	4	2970	62
Mitsubishi Mirage	Front	27	Regular	13.2	1.8	4	113	Auto	4	2525	63
Nissan 200 SX	Front	28	Regular	13.2	2	4	140	Manual	5	2580	63
Nissan Altima	Front	23	Regular	15.9	2.4	4	150	Auto	4	3050	63
Nissan Maxima	Front	24	Premium	18.5	3	6	190	Auto	4	3070	64
Nissan Pathfinder	Rear	19	Regular	21.1	3.3	6	168	Auto	4	4090	56
Nissan Quest	Front	19	Regular	20	3	6	151	Auto	4	3900	59
Nissan Sentra	Front	28	Regular	13.3	1.6	4	115	Auto	4	2500	63
Oldsmobile Achiva	Front	24	Regular	15.2	2.4	4	150	Auto	4	3025	64
Oldsmobile Aurora	Front	17	Premium	20	4	8	250	Auto	4	3995	63

(continued on next page)

TABLE 6.13
1997 Automobile Gasoline Mileage Data (*continued*)

Make and Model	Drive	Miles/Gallon	Fuel Type	Tank Capacity	Engine Size	Number of Cylinders	Horse Power	Transmission Type	Transmission Speeds	Weight	Weight Distbn
Plymouth Breeze	Front	23	Regular	16	2	4	132	Auto	4	3050	63
Plymouth Grand Voyager	Front	18	Regular	20	3.3	6	158	Auto	4	4035	57
Plymouth Neon	Front	30	Regular	12	2	4	132	Manual	5	2545	64
Plymouth Voyager	Front	19	Regular	20	3.3	6	158	Auto	4	3985	59
Pontiac Bonneville	Front	18	Regular	18	3.8	6	240	Auto	4	3665	64
Pontiac Grand Am	Front	24	Regular	15.2	2.4	4	150	Auto	4	3010	66
Pontiac Grand Prix	Front	21	Regular	18	3.8	6	195	Auto	4	3400	65
Pontiac Sunfire	Front	26	Regular	15.2	2.2	4	120	Auto	3	2890	65
Saab 900	Front	22	Regular	18	2.5	6	170	Auto	4	3145	63
Saturn	Front	29	Regular	12.2	1.9	4	100	Auto	4	2405	61
Subaru Legacy	All	24	Regular	15.9	2.2	4	137	Auto	4	2980	56
Suzuki Esteem	Front	29	Regular	13.5	1.6	4	98	Auto	4	2290	62
Suzuki Sidekick	Rear	23	Regular	18.5	1.8	4	120	Manual	5	2910	53
Suzuki Swift	Front	29	Regular	10.6	1.3	4	70	Auto	3	1845	61
Suzuki X90	Rear	27	Regular	11.1	1.6	4	95	Manual	5	2490	57
Toyota 4 Runner	Rear	22	Regular	18.5	3.4	6	183	Auto	4	3930	55
Toyota Avalon	Front	22	Regular	18.5	3	6	200	Auto	4	3320	62
Toyota Camry	Front	23	Regular	18.5	3	6	194	Auto	4	3240	63
Toyota Celica	Front	28	Regular	15.9	2.2	4	135	Manual	5	2720	63
Toyota Land Cruiser	All	14	Regular	25.1	4.5	6	212	Auto	4	5150	52
Toyota RAV 4	Front	25	Regular	15.3	2	4	120	Manual	5	2905	59
Toyota Supra	Rear	22	Premium	18.5	3	6	320	Manual	5	3555	54
Toyota Tacoma	Rear	21	Regular	15.1	3.4	6	190	Auto	4	3040	59
Toyota Tercel	Front	32	Regular	11.9	1.5	4	93	Auto	4	2165	63
Volkswagen Golf	Front	30	Regular	14.5	2	4	115	Manual	5	2570	63
Volkswagen Jetta	Front	23	Regular	14.5	2.8	6	172	Manual	5	2955	62
Volkswagen Passat	Front	20	Regular	18.5	2.8	6	172	Auto	4	3180	62

12. **Variables Affecting SAT Scores.** The 1997 SAT scores of 43 high schools in Westchester County, New York, are presented in Table 6.14. It is conjectured that the percentage of minority students and the percentage receiving free lunches explain the variation in scores. Data sets and conjectures such as these are often published despite their overtones of bias. Such conjectures need to be critically examined. (Source: *Tarrytown Daily News*, January 15, 1998)

   ■ Evaluate the conjectured effect of student demographics. Is the conjecture supported by the data?
   ■ Is there any evidence that other variables are affecting the SAT scores?
   ■ Is there any evidence that the educational programs of some schools are better than others, or is it all due to demographics?
   ■ What model best describes the relationship between SAT scores and the other variables?
   ■ With what process are these variables associated? Classify the variables as input, process, or output variables. Draw a schematic of the process.
   ■ Is there any evidence of atypical data points or multicollinearity in this data set?

13. **Gasoline Mileage of Latest Model Automobiles.** Obtain the data published most recently by *Consumer Reports* on the gasoline mileage of automobiles.

   ■ Develop a model to identify variables that explain the variation observed in the gasoline mileage measurements of the different autos.
   ■ Compare your results with those obtained for the 1997 model year cars in Exercise 11. How do the results compare? What similarities and differences do you see? What could have caused these differences?
   ■ Are the gasoline mileage data of good quality and worthy of analysis? How were the gasoline measurements obtained? How many cars of each model were tested? Are the data of high quality? If so, why? If not, why not?

14. **R-squared as a Measure of Model Fit.** Using the call center data in Table 6.6, fit the three-variable model (time to answer, VRU, and call volume) and then the two-variable model (obtained by deleting the call volume variable). Compare the R-squared values (not the adjusted R-squared values) of the two models. Note the *t*-ratio for the call volume variable in the three-variable model.

   ■ What does this comparison tell you about using R-squared as a measure of model fit?
   ■ How do the adjusted R-squared values of the two models compare?
   ■ Which statistic is a better measure of model fit and why?

15. **"Goodwill" Data.** The data presented in Chapter 5, Table 5.3, represent the "goodwill," total assets, and amortization period for the last 52 acquisitions of a major multinational conglomerate. Recall that goodwill is the amount of money over and above the fair market value of a company (typically measured by market capitalization—that is, stock price × number of shares outstanding) that is required to acquire the company. In general, one must pay more than the fair market value to acquire a company to provide an incentive for the current owners to sell. The amortization period is the time period over which the goodwill is written

TABLE 6.14
1997 SAT Scores, Westchester County, New York

School	SAT Score	% of Students Who Took Exam	% of Students Who Receive Free Lunch	% of Minority Students
Scarsdale	1223	96	0	7
Horace Greely	1201	95	1	4
Bronxville	1200	100	1	1
Edgemont	1194	99	0	6
Blind Brook	1154	95	1	2
Bryman Hills	1148	98	1	3
Hastings	1133	83	1	10
John Jay	1125	*	3	3
Ardsley	1122	94	8	6
Rye	1117	99	2	8
Mamaroneck	1106	94	3	13
Pleasantville	1105	94	0	8
Briarcliff	1099	100	3	10
Yorktown	1095	91	2	7
Irvington	1091	100	0	8
Croton Harmon	1090	93	0	9
Fox Lane	1078	91	8	20
North Salem	1060	*	2	4
Valhalla	1053	62	6	16
Harrison	1047	90	3	12
Rye Neck	1045	*	0	8
Pelham Memorial	1044	94	8	17
Ossining	1040	85	17	34
Lakeland	1033	77	4	9
Hendrick Hudson	1027	76	3	1
Somers	1026	91	4	5
Dobbs Ferry	1024	97	4	12
Westlake	1018	78	6	5
Walter Panas	1012	83	4	8
New Rochelle	1001	66	32	49
White Plains	987	65	26	54
Peekskill	985	60	33	48
Eastchester	971	100	0	3
Woodlands	962	87	29	62
Sleepy Hollow	959	76	34	63
Tuchhoe	953	79	7	23
Alexander Hamilton	940	90	21	70
Saunders Trades & Tech.	893	56	57	55
Groton	891	63	66	63
Port Chester	887	61	37	60
Mount Vernon	870	57	38	88
Roosevelt	841	44	62	64
Lincoln	826	54	62	61

*Missing data

off the books. Typically, we want to write off large amounts of goodwill over a long time period to avoid a major hit to our bottom line in any one year. Consider goodwill as the response variable (y), and total assets and amortization period as predictor (x) variables. (For reasons to be explained in Chapter 9, both goodwill and total assets will be measured in a transformed metric [log transformation].)

- Develop a model expressing $y = \log(\text{goodwill})$ as a function of $x_1 = \log(\text{total}$ assets) and $x_2 =$ amortization period.
- What model would you recommend be used to predict the amount of goodwill?
- How well does this model describe the data?
- What is the nature of the effects of the two predictor variables?
- Does the model need further development? If so, what additional work would you recommend?

Note: Finance theory might suggest that a more logical approach would be to treat amortization as the y and goodwill as an x, with total assets a predictor of goodwill. For certain financial applications, however, we might wish to group acquisitions by amortization period and would then be interested in estimating the magnitude of goodwill values we are likely to see in each amortization grouping. Utilizing total assets in this model might help us predict goodwill more accurately.

16. **Anscombe's Straight Lines.** Four groups of x and y data—one x and one y per group—are presented in Table 6.15 (Anscombe, 1973). Fit a straight-line model to each data set and compare the model coefficients and regression results. What are your conclusions regarding the relationship between x and y in these four data sets?

TABLE 6.15   Anscombe's Data

Group 1		Group 2		Group 3		Group 4	
x_1	y_1	x_2	y_2	x_3	y_3	x_4	y_4
10	8.04	10	9.14	10	7.46	8	6.58
8	6.95	8	8.14	8	6.77	8	5.76
13	7.58	13	8.74	13	12.74	8	7.71
9	8.81	9	8.77	9	7.11	8	8.84
11	8.33	11	9.26	11	7.81	8	8.47
14	9.96	14	8.1	14	8.84	8	7.04
6	7.24	6	6.13	6	6.08	8	5.25
4	4.26	4	3.1	4	5.39	19	12.5
12	10.64	12	9.13	12	8.15	8	5.56
7	4.82	7	7.26	7	6.42	8	7.91
5	5.68	5	4.74	5	5.73	8	6.89

17. **Predicting Process Performance.** The data in Table 6.16 show the defect rate and cost/unit for a process over a 22-month period. The most recent data show that the process is operating at around \$.36/unit. Assuming a relationship exists between defect rate and cost/unit, what do you predict is the lowest cost/unit at

TABLE 6.16   Process Performance Study

Year	Month	Defects	Cost/Unit
98	1	126,000	0.91
98	2	111,000	0.68
98	3	61,000	0.35
98	4	62,000	0.35
98	5	73,000	0.36
98	6	73,000	0.43
98	7	55,000	0.32
98	8	41,000	0.26
98	9	29,000	0.16
98	10	37,000	0.24
98	11	17,000	0.16
98	12	22,000	0.18
99	1	28,000	0.24
99	2	36,000	0.23
99	3	110,000	0.53
99	4	73,000	0.31
99	5	55,000	0.27
99	6	75,000	0.42
99	7	120,000	0.47
99	8	150,000	0.48
99	9	131,000	0.38
99	10	111,000	0.36

which this process is capable of operating? State the assumptions on which your prediction is based.

18. **Correlating Process Output Measurements.** The data in Table 6.17 are from a study correlating three measurements made on rubber samples: x_1 = Hardness, x_2 = Tensile Strength, and y = Abrasion Loss (Davies & Goldsmith, 1972, pp. 238–240).

   ■ Develop a model predicting y as a function of x_1 and x_2. Both x_1 and x_2 are easier and less costly to measure than y. Is the model based on x_1 and x_2 a good predictor of y? Should we use this model to evaluate our product in the future?

   ■ Analyze the residuals for the model and then comment of the appropriateness of using the model to predict the abrasion loss of the samples.

19. **Sulfate Ammonia Production.** The data in Table 6.18 are from a study whose objective was to develop a model to predict the flow rate (y) of an ammonia production process given the following: x_1 = moisture content, x_2 = length/breadth ratio for crystals, and x_3 = impurity (Davies & Goldsmith, 1972, pp. 250–256).

TABLE 6.17 Properties of Rubber Samples

Abrasion Loss	Hardness	Tensile Strength
372	45	162
206	55	233
175	61	231
154	66	231
136	71	231
112	71	237
55	81	224
45	86	219
221	53	203
166	60	189
164	64	210
113	68	210
82	79	196
32	81	180
228	56	200
196	68	173
128	75	188
97	83	161
64	88	119
249	59	161
219	71	151
186	80	165
155	82	151
114	89	128
341	51	161
340	59	146
283	65	148
267	74	144
215	81	134
148	86	127

y = Abrasion Loss in g/hp-hour
x_1 = Hardness in I.R.H. (Shore units)
x_2 = Tensile Strength in kg/sq. cm

- Develop a model predicting y from x_1, x_2, and x_3. Do a complete regression analysis of the data.
- What are the key drivers of this process? What is the nature and magnitude of the effects of the three variables?
- Is the linear model adequate?
- Does this model give good predictions of the flow rates at the various production conditions in the study?

TABLE 6.18
Sulphate of Ammonia Process

Sample Number	Moisture	Crystal Ratio	Impurity	Flow Rate	Sample Number	Moisture	Crystal Ratio	Impurity	Flow Rate
1	21	24	0	500	25	17	22	3	459
2	20	24	0	481	26	17	24	4	500
3	16	24	0	446	27	17	24	0	382
4	18	25	0	481	28	15	24	2	368
5	16	32	0	446	29	17	22	3	515
6	18	31	1	385	30	21	22	4	294
7	12	32	1	321	31	23	22	10	318
8	12	27	0	325	32	22	20	7	228
9	13	27	0	455	33	21	19	4	500
10	13	27	0	485	34	24	21	8	243
11	17	27	0	400	35	37	23	14	0
12	24	28	0	362	36	21	24	2	410
13	11	25	0	515	37	28	24	5	370
14	10	26	0	376	38	29	24	7	336
15	17	20	0	490	39	23	36	7	379
16	14	20	0	413	40	32	33	8	340
17	14	20	1	510	41	26	35	4	151
18	14	19	0	505	42	28	35	12	0
19	20	21	2	427	43	21	30	3	172
20	12	19	1	490	44	22	30	6	233
21	11	20	2	455	45	34	30	8	238
22	10	20	7	532	46	29	35	5	368
23	10	20	2	439	47	17	35	3	420
24	16	20	2	485	48	11	32	2	500

x_1 = Initial Moisture Content (units of 0.01%); x_3 = Percent Impurity (units of 0.01%).
x_2 = Length/Breadth Ratio for Crystals; y = Flow Rate (g/sec).

REFERENCES

Ackoff, R. L. (1978). *The art of problem solving.* New York: Wiley.

Anscombe, F. J. (1973). Graphics in statistical analysis. *American Statistician, 27*(1), 17–21.

Box, G. E. P. (1979). Robustness is the strategy of scientific model building. In R. L. Launer & G. M. Wilkinson (Eds.), *Robustness in statistics.* New York: Academic Press.

Box, G. E. P., & Jenkins, G. M. (1976). *Time series analysis: Forecasting and control* (rev. ed.). San Francisco: Holden Day.

Box, G. E. P., Hunter, W. G., & Hunter, J. S. (1978). *Statistics for experimenters.* New York: Wiley.

Cryer, J. D. (1986). *Time series analysis,* Pacific Grove, CA: Duxbury Press.

Davis, P. (1994, March). LP-based delinquent account strategy pays off in consumer credit business. *Siam News, 27*(3).

Draper, N. R., & Smith, H. (1998). *Applied regression analysis* (3rd ed.). New York: John Wiley and Sons.

Hillier, F. S., & Lieberman, G. J. (1995). *Introduction to operations research* (6th ed.). New York: McGraw-Hill.

McClave, J. T., & Benson, P. G. (1985). *Statistics for business and economics* (3rd ed.). San Francisco: Dellen-Macmillan.

Montgomery, D. C., & Peck, E. A. (1992). *Introduction to linear regression analysis.* New York: John Wiley and Sons.

Snee, R. D. (1973). Techniques for the analysis of mixture data. *Technometrics, 15*(3), 517–528.

Vandaele, W. (1983). *Applied time series and Box-Jenkins models.* New York: Academic Press.

7

Using Process Experimentation to Build Models

Experiment, make it your motto day and night.

Cole Porter

7.1 Overview

The focus of Chapter 6 was on building models using existing process data to understand how a process works and to better manage and improve it. Building models is an iterative process in which we move back and forth between hypotheses about how the process works and data that confirm or deny them. This strategy was illustrated in Chapter 2 in the statistical thinking model, and it is quite useful. However, this strategy is limited by the quality of the existing process data used to build the model.

In this chapter, we examine a second strategy, which is to proactively experiment with the process to obtain high-quality data. Using this approach, we can test our ideas about cause-and-effect relationships. The technique used to do this is the *statistical design of experiments.* Using this methodology, we define how the experiments should be conducted, what specific process changes should be made, what data should be collected, and how the data should be analyzed to build the model. The thought process is the same as that used in Chapter 6, but the data used to develop the model are collected by experimenting with the process rather than by observing the process as it operates normally.

Design of experiments was also shown in Chapter 4 to be a key tool in the process improvement strategy in the Understand Cause-and-Effect phase. Design of experiments is a very broadly applicable technique, and it can certainly be used outside this strategy. It is also a key tool in the Six Sigma methodology discussed in Chapter 4 and Appendix D.

We begin by discussing two classical approaches to experimentation—haphazard experimentation and one-factor-at-a-time experimentation—and the need for the statistical approach. Next we provide some examples illustrating how statistical design of experiments can improve business processes and the resulting outputs. This is followed by an introduction to the statistical approach and detailed discussions of a two-variable and a three-variable experiment to illustrate how to plan and conduct designed experiments and analyze the results to build models. A discussion of the use of regression analysis to analyze the results of designed experiments is also included. Issues associated with large experiments, such as blocking, randomization, and center points, are also addressed. The chapter concludes with a summary of key points, project update, and exercises. For in-depth discussions of statistical design of experiments, we recommend Box, Hunter, and Hunter (1978), Montgomery (1991) and Moen, Nolan, and Provost (1991).

7.2 Why Do We Need a Statistical Approach?

The first question that probably comes to mind is "Why do we need a statistical approach?" The statistical design of experiments will improve the quality of the results of our experiments while reducing both the time and the amount of resources required to do the experimentation. Large corporations have research and development budgets that amount to $1 billion and more, and statistical strategies of experimentation make the most effective use of these resources. Much of the recent dramatic improvement in science and engineering has come through designed experiments, particularly in agriculture, industry, and pharmaceuticals.

Haphazard Experimentation

The most popular strategy of experimentation is *haphazard experimentation,* and this strategy was used by Thomas Edison. He and his colleagues did thousands of haphazard experiments before finding a light bulb filament that worked. In this strategy you test your best guess of what will work based on what has been done before and what you think will work better. Edison liked to point out that he learned from every experiment, and more often than not he learned what would *not* work.

Haphazard experiments have some value, particularly in the hands of true geniuses like Thomas Edison, but doing thousands of experiments hardly seems like an efficient strategy. We are better served by a structured strategy that is useful in a wide variety of situations and can be used by experimenters with many different backgrounds. The one-factor-at-a-time approach is one such strategy.

One-Factor-at-a-Time Experimentation

The classical approach to experimentation is known as one-factor-at-a-time (OFAAT). In this approach each process variable (factor) is studied individually by varying the variable (x) and measuring the response (y) while holding all other variables constant. This is repeated for each variable in the study. The OFAAT approach is widely taught in high school and college science classes, but it has limited practical value due to the underlying assumptions of the approach.

The first assumption is that the relationship between the variable (x) and the response (y) is a complicated function with multiple peaks and valleys. Within the range of process operations, however, practitioners have found that relationships between process variables (x) and responses (y) rarely pass through more than one maximum or minimum. That is, there is typically only one peak and one valley.

The second assumption is that the process variables (x's) function independently of each other. If the variables do not interact, the effect of any one variable is not dependent on the level of any other variable. In reality, significant interactions are often found and are what make some processes effective. The OFAAT approach performs poorly in the presence of interacting process variables.

A third assumption of the OFAAT approach is that the experimental variation associated with the response (y) is small. Unfortunately, small experimental variation is the exception rather than the rule in most processes, thereby significantly reducing the effectiveness of the OFAAT approach.

A final limitation of the OFAAT approach is that it takes a long time to test all the variables when you test them one at a time. As Koselka (1996) points out, "there's a very low probability that you're going to hit the right one before everyone gets sick of it and quits" (p. 114). The statistical approach discussed in the next section does not have these limitations.

The Statistical Approach

The statistical approach to design of experiments increases our ability to find the vital few key drivers of a process, understand the process, and build useful process models. Statistical models are developed by quantitatively measuring the relationship between one or more process variables (x's) and the process responses (y's). This is more effective than testing one factor at a time and it minimizes the number of tests required in a given experiment. Sufficient repeat tests are included to quantify the amount of experimental variation and to provide the desired sensitivity to detect small effects. Interactions between process variables are identified and quantified, and checks are included to detect and describe complicated response functions. The result is an experimental process that requires less time to complete than the haphazard and OFAAT approaches and produces better results overall.

The typical success rate for statistically designed experiments is 90% or better, whereas the success rate for analysis of process data for an as is process is typically less than 50%. The statistical approach has a higher success rate because experiments are designed to produce the data needed to answer the questions being studied. When analyzing process data, the investigator is limited to those data that have been or can be collected without interfering with the ongoing process.

Here are some of the reasons for the high success rate of the statistical approach:

- Increases the chance of identifying key drivers
- Provides a systematic approach to testing
- Enables collection of quality data
- Evaluates a large number of variables
- Controls nuisance variables
- Produces quantitative estimates of effects
- Identifies interaction effects
- Measures experimental uncertainty
- Enables effective and efficient use of data

This method has been developed and refined over more than 70 years and provides a systematic approach to testing. Quality data are developed because the experiment is designed to collect data in sufficient amounts and quality to answer the questions being studied. The statistical approach significantly reduces the time to complete experimental programs because of the ability to test a large number of variables simultaneously in a small number of tests. It is not uncommon to study 7 process variables in 8 to 16 tests or 15 process variables in 16 to 28 tests.

A key aspect of the approach is the control of nuisance variables that affect process performance but are not part of the experiment. The methodology systematically leads one to identify these variables and take them into account in the experiment. The analysis procedure produces quantitative estimates of the effects of the variables so that we know not only the direction of the effects but also how much we can expect the response (y) to go up or down when the process variable (x) is changed. Interactions between the process variables are identified and quantified, and quantitative estimates of the experimental variation (uncertainty) are also produced. For example, we might find that increasing advertising 10% increases sales 15% with an uncertainty of ±3%. The practical value of a variable effect is greatly increased when we know the precision with which it has been estimated.

Finally, the statistical approach makes effective and efficient use of the data collected, and test results are typically used to calculate the effects of each of the variables being studied. Test results are often used multiple times in the analysis, a characteristic called "hidden replication." The statistical approach discussed in this chapter is an effective general strategy for experimentation and is recommended for experimenting with processes of all types.

7.3   Examples of Process Experiments

Let's take a closer look at a case study from Chapter 2 and two new case studies that illustrate how process experiments can inform us about improving the process.

The Effect of Advertising on Sales

The case study of the effect of advertising on sales (Ackoff, 1978) discussed in Chapter 2 is an outstanding example of the use of designed experiments in improving a process. In this case the process is the sales process, and the objective is to determine how the amount of advertising affects the output of the sales process. It is appropriate at this time to return to Chapter 2 and reread this case, focusing on how the following issues affected the experimentation:

- The use of a series of experiments to understand and improve a process
- The iteration between theory and data to guide the experimentation and the development of process knowledge
- The use of a variety of experimental designs
- Experimenting within practical constraints; such as keeping within available budgets and not interfering with normal business operations
- Concern over biasing the experiments through the selection of test areas

It is also appropriate to think about how you would identify and deal with these issues in your area of interest.

Product Development Case Study

A major food products corporation wanted to develop a hot chocolate drink that would obtain maximum consumer acceptance (Matthew Wiant, personal communication, 1988). Previous research with consumer behavior and preferences for hot chocolate drinks indicated that these variables affect hot chocolate enjoyment:

1. Chocolate level
2. Creaminess (milk/cream level)
3. Thickness (texture)
4. Temperature
5. Sweetness (tied to chocolate level)
6. Hunger level

Experimentation had demonstrated that there is an optimal ratio of cocoa (chocolate) to sugar, so these two variables were not varied separately in the study (see variable 5). Variables 1, 2, 3, and 5 deal with the formulation of the hot chocolate drink. Variables 4 and 6 are environmental factors, which were held constant in the test. It had been determined that 150°F to 170°F is the optimal drinking temperature for any hot beverage, and this temperature was held constant. Consumers' hunger level can affect overall enjoyment of any food product, so any data directly compared must be taken at the same time or at equivalent hunger levels. In addition, it was recognized that only five or six samples could be evaluated at any one time because consumers could not differentiate between samples when more than six were presented.

Initially, haphazard testing attempted to match or beat a competitor's product, but results were slow in coming and a match was not reached. Next, testing was done holding all but one variable constant (OFAAT), and good results were obtained on creaminess level. But the variables were not varied simultaneously. Experimentation with an ingredient to thicken hot beverages was found in initial testing to have a large affect on consumer enjoyment. Clearly, this was an ideal opportunity to use a designed experiment to determine the optimum levels of chocolate, creaminess, and thickness in a hot drink. This work should lead to the most preferred mixture of these ingredients from a consumer's standpoint while considering cost. Once this optimum is reached, other variables, such as type of cream, flavorings, and type of thickener, could be evaluated.

Test Design To find the optimum combinations of chocolate, cream, and thickness, a $2 \times 2 \times 2$ factorial design was used (Table 7.1). *Factorial design* means that each level of x_1 was run with each possible level of the other x's, and similarly for x_2 and x_3. As a starting point, a hot chocolate formula that had good flavor/texture (organoleptic) attributes was selected. The formulations tested were variations on this starting formula.

Formulations Various hot chocolate mixes were made with differing levels of cocoa and sugar, spray dried whole milk, and guar gum (a natural thickener). The total range of realistic levels for each ingredient was determined, and then the range was split into the following categories for each of the three variables.

Category Number	Corresponding Level of Chocolate, Cream, Thickness
0	Extremely low/none
1	Very low
2	Low
3	Medium low
4	Slightly low
5	Medium/moderate
6	Slightly high
7	Medium high
8	High
9	Very high
10	Extremely high/overpowering

Each hot chocolate drink was served in a 6-ounce cup at 170°F, and tasters used a subjective scale to grade them. The starting midpoint sample was the hot chocolate formulation the organization had been using, and it had these properties:

5 creaminess	medium/moderate
5 chocolatiness	medium/moderate
3 thickness	medium low

All of the samples had the same level of other ingredients, such as flavorings.

Execution A total of three experiments were run (see Table 7.1). Formula 9 was the center point (all three variables at their middle levels) in each experiment. The best formulation (highest flavor score) identified in Experiment 1 was used as the center point for Experiment 2. The best formulation from Experiment 2 was used as the center point for Experiment 3. The center point in Experiment 3 was found to be the best formulation, and any variation from it led to a decrease in consumer approval (i.e., lower flavor scores).

TABLE 7.1   Chocolate Drink Development Experiment

Experiment	Test	Cream	Chocolate	Thickness	Score
1	1	3	3	1	21
	2	7	3	1	55
	3	3	7	1	34
	4	7	7	1	67
	5	3	3	5	38
	6	7	3	5	68
	7	3	7	5	52
	8	7	7	5	78
	9	5	5	3	53
2	1	6	6	4	71
	2	8	6	4	91
	3	6	8	4	74
	4	8	8	4	83
	5	6	6	6	76
	6	8	6	6	85
	7	6	8	6	70
	8	8	8	6	79
	9	7	7	5	76
3	1	7	5	3	77
	2	9	5	3	84
	3	7	7	3	73
	4	9	7	3	85
	5	7	5	5	75
	6	9	5	5	84
	7	7	7	5	81
	8	9	7	5	89
	9	8	6	4	93

Note: In each experiment, Tests 1, 4, 6, 7, and 9 were run in the first block of tests and the remaining tests in the second block. Test 9 was the overall center point in each experiment.

The results of Experiment 1 indicated that consumers preferred the thickest, creamiest, and most chocolaty drink of the 9 tested. The analysis indicated that all three variables (cream, chocolate, thickness) had a positive effect on the flavor score of the drink, but cream appeared to have the largest effect. This was a major learning for the design team, who had thought that chocolate level would be the most important variable.

The results of Experiment 2 indicated that the formulation that was very creamy (8), only quite chocolaty (6), and not very thick (4) was the best of the 9 tested. Looking at the results, it is apparent that with these levels of the variables only creaminess is a major factor. An increase in chocolate and thickness from the middle level (center point) actually had a small negative effect on the acceptability of the drink.

In Experiment 3 the center point formulation was the best. This formulation was also found to be best in Experiment 2. Any variation from this formulation resulted in lower flavor ratings. It was conjectured that the best possible formulation (global optimum) had been reached, and more testing could be done to confirm this. In any event the product formulation group was delighted that three experiments testing only 27 formulations had produced a very good product. In this instance only *one* new formulation was of interest, but in other situations two or more formulations might be desired to satisfy the needs of different segments of the market.

It was originally thought that a drink consisting of a "9 chocolate, 5 cream, and 3 thickness" formulation would have the highest preference. This thinking overlooked the fact that consumers wanted a very creamy beverage and not a very chocolaty beverage. This was not intuitively obvious at the beginning of these experiments. These consumers perceived a quality hot chocolate drink as being very creamy.

Reducing Defects in Plastic Parts Case Study

In this case study the defect rate of molded plastic parts was running around 1.7% and needed to be reduced. The parts were being produced using two different processes (old, new) and two different sources of raw materials (A, B). Both processes could be operated over the same ranges of temperature (25°C–75°C) and pressure (15–45 psi). The operating conditions in use at the time were 50°C and 30 psi. Snee and Hare (1992) describe a series of experiments whose purpose was to determine how to eliminate or at least significantly reduce the percentage of defective parts produced by this process.

The effects of temperature and pressure were studied separately for each process and type of raw material, using three levels of temperature (25°C, 50°C, and 75°C) and three levels of pressure (15, 30, and 45 psi). The resulting combinations of temperature and pressure form a 3 × 3 factorial design (Figure 7.1). Four

FIGURE 7.1   Plastic Part Process Improvement 3 × 3 Factorial Design

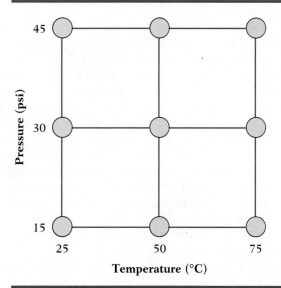

Temperature (°C)

TABLE 7.2   Plastic Part Improvement Experiment

Pressure (psi)	Raw Material A Temperature (°C)				Raw Material B Temperature (°C)		
	25	50	75		25	50	75
				Old Process			
15	0	0	0.7		0.1	0	0
30	0.1	0	4.0		0.6	0	12.7
45	0	2.0	6.1		0.3	0	23.7
				New Process			
15	0	0	0.4		0.1	0	0
30	0	0	0.6		0	0	2.4
45	0	13.9	2.4		0.4	0	19.7

such experiments were run—one for each of the four combinations of process type (old, new) and type of raw material (A, B). The response was percentage of defective parts in a sample of 700 parts.

The data from this first experiment are summarized in Table 7.2. We see immediately, without any statistical analysis, that there are no defective parts observed at 50°C when the pressure is at 15 or 30 psi. These findings applied for both processes and both types of raw material.

The next step was to verify these results by running a second set of 3 × 3 factorial designs in the region of a temperature–pressure combination that produced no defective parts. In the *verification test,* temperature was studied at 30°C, 45°C, and 60°C with pressure at 10, 15 and 20 psi (Figure 7.2). No defective parts were observed at any of the 9 combinations of temperature and pressure for either process and both types of raw material lots. These were exciting results as a defect-free process had not been seen before.

The initial and verification tests had been run in the laboratory. It was now time to run a *plant test* to see whether the manufacturing process would behave similarly. The plant test was run at the center point of the 3 × 3 verification test (45°C, 15 psi, see Figure 7.2), and no defects were observed in the plant test. A change in the standard operating procedures for the process was issued. In monitoring over subsequent months, the process defect rate dropped from its original 1.7% to 0.1%. This resulted in an annual savings of $500,000, to say nothing of the benefits, financial and otherwise, of increased customer satisfaction.

This case study illustrates three key points. First, the iterative nature of experimentation is evident. The team progressed from the initial test, to the verification test, and finally to the plant test. Second, it illustrates the power of the 3 × 3 factorial design and a key scientific principle: the need to establish the range of validity of the results by studying different processes and lots of raw material. Finally, it demonstrates that the factorial design does an effective job of sampling the experimental region. Hence, inspection of the data makes clear where the process should be operated. This is particularly true when only one response variable is of interest, as in this example (percentage of defective parts).

FIGURE 7.2 Plastic Part Process Development, Experimental Region

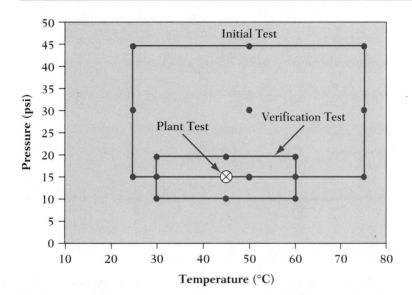

7.4   The Statistical Approach to Experimentation

We introduced the concept of an iterative approach to testing hypotheses and learning in Chapter 2. This produces a general framework for experimentation:

Hypothesis A $\Rightarrow$ Design A $\Rightarrow$ Data A $\Rightarrow$ Analysis A $\Rightarrow$ Hypothesis B

Hypothesis B $\Rightarrow$ Design B $\Rightarrow$ Data B $\Rightarrow$ Analysis B $\Rightarrow$ Hypothesis C, and so on,

in which we start with Hypothesis A for the process (e.g., variables x_1, x_2, and x_3 are likely to be important). Based on Hypothesis A, we design an experiment (Design A), run the experiment and collect the data (Data A), analyze the data (Analysis A), and interpret the results to revise Hypothesis A to create a new hypothesis (Hypothesis B). We may stop our experiments at this point or continue on and run experiments to evaluate Hypothesis B, Hypothesis C, and so on.

At each step along the way we must ask, "What design should we use to test the hypothesis?" The design depends on the experimental environment, and three major experimental environments flow naturally one from another: screening, characterization, and optimization (Table 7.3). These experiments are defined by the number of variables to be studied and the type of information desired (questions to be answered). In the screening environment we are in the early stages of the experimentation with potentially many variables to study (6–30) and we are interested in crude information such as "Which variables have the largest effect?"

TABLE 7.3   Comparison of Experimental Environments

Characteristics	Type of Experimental Environment		
	Screening	*Characterization*	*Optimization*
Number of factors	6–30	3–8	2–6
Desired information	Important factors	Understand how systems work	Prediction equation and optimization
Model form	Linear or main effects	Linear and interaction effects	Linear, interaction, and curvilinear effects

Source: Adapted from Pfeiffer, 1988.

In the characterization stage we are working with fewer variables (3–8), and these variables typically had the largest effects in the screening phase. For these variables we want to know which have the largest effects, the quantitative sizes of the effects, the important interaction effects, and whether there is any curvature in the individual variable response functions.

Finally, in the optimization stage we are working typically with 2 to 6 variables to create a prediction equation that will enable us to find maximum or minimum values of the response and to identify optimum operating conditions. In the statistical literature optimization studies are often referred to as response surface experiments and the associated designs as response surface designs (Box, Hunter, & Hunter, 1978; Myers & Montgomery, 1995). The prediction equations are developed using the regression analysis techniques discussed in Chapter 6.

The data in Table 7.4 are from a screening experiment studying the factors affecting a viscosity measurement system. The variation of this measurement was excessive and needed to be reduced if the measurement system was to be used to control a critical production process. Seven variables were studied, and five were found to have significant effects that, when controlled, significantly reduced the variation in the measurements. The design used here is called a *fractional factorial design* because it only uses a fraction of the experiments that would be in a factorial design. Plackett-Burman designs are another type of screening experiment (see Box, Hunter, & Hunter, 1978; Montgomery, 1991).

The plastic parts experiment summarized in Table 7.2 describes a characterization experiment whose objective was to quantify factor effects and their interactions. The chocolate drink experiment (see Table 7.1) has aspects of both the characterization and optimization stages. A center point was included to detect the presence of curvature in the response functions. In place of developing a prediction equation, three designs were run sequentially to optimize the quality of the drink.

TABLE 7.4

Screening Experiment for Viscosity Measurement System

Run	Test Sequence	x_1 Sample Prep	x_2 Moisture Measurement	x_3 Mixing Speed	x_4 Mixing Time	x_5 Healing Time	x_6 Spindle	x_7 Protective Lid	y Viscosity
1	5	1	1	1	2	2	2	1	2220
2	4	2	1	1	1	1	2	2	2460
3	18	1	2	1	1	2	1	2	2904
4	19, 20	2	2	1	2	1	1	1	2464, 2348
5	7	2	1	2	2	1	1	2	3216
6	11	2	2	2	1	2	1	1	3772
7	12	1	2	2	1	1	2	1	2420
8	6, 13	2	2	2	2	2	2	2	2340, 2380
9	9	1	2	2	1	1	1	2	3376
10	3	2	2	2	2	2	2	1	3196
11	2	1	1	2	2	1	2	1	2380
12	1, 16	2	1	2	1	2	2	2	2800, 2700
13	14	1	2	1	1	2	2	1	2320
14	10	2	2	1	2	1	2	2	2080
15	15	1	1	1	2	2	1	2	2548
16	8, 17	2	1	1	1	1	1	1	2796, 2788
17		1	2	1	2	1	1	1	2384
18		1	2	2	2	1	1	1	2976
19		1	2	2	2	1	2	1	2180
20		1	2	2	2	1	2	1	2300

Runs 17–20 were conducted at least one week after Runs 1–16. Factor level codes: 1 = low level, 2 = high level

TABLE 7.5   Ammonia Test Optimization

pH	Enzyme Concentration	Molarity	Sensitivity
7.25	160	0.04	182
7.25	90	0.04	146
7.25	125	0.05	166
7.25	160	0.06	176
7.25	90	0.06	140
7.45	125	0.04	181
7.45	160	0.05	177
7.45	125	0.05	187
7.45	90	0.05	180
7.45	125	0.06	181
7.65	160	0.04	169
7.65	90	0.04	178
7.65	125	0.05	179
7.65	160	0.06	168
7.65	90	0.06	179

The data in Table 7.5 are from an optimization experiment designed to maximize the sensitivity of a clinical ammonia test marketed by the DuPont Company. In this experiment each of the three variables is studied at three levels (low, middle, high) to enable the estimation of curvilinear response functions. A prediction equation was developed and response surface optimization techniques (see Box, Hunter, & Hunter, 1978; Myers & Montgomery, 1995) were used to find test method settings that maximized the sensitivity of the method.

Different experimental programs move through the screening, characterization, and optimization stages in different ways. Some go through all three stages sequentially. Some skip the screening and characterization stages and go immediately to the optimization stage. Others begin at the characterization stage and then move to the optimization stage. Still others use only one stage. We will focus on the characterization stage, which is the most frequently used stage. For a discussion of the other two stages and the fractional factorial, Plackett-Burman, and response surface designs used, we recommend Box, Hunter, and Hunter (1978), Moen, Nolan, and Provost (1991), Montgomery (1991), and Myers and Montgomery (1995).

Planning Test Programs

Careful attention must be paid to the setup of the design and the conduct of the associated experiments. Here are the key steps in planning and conducting the test program:

- Obtain a clear statement of the problem
- Collect background information

- Design the test program
  - Conference of all parties
  - Develop the design
  - Review the design and revise if necessary
- Plan and conduct the experimental work
- Analyze the data
- Report the results orally and in writing

The first step is to obtain a clear statement of the problem. This includes the problem to be solved, the questions that need to be answered (needed results of the experiment), the available resources, and the time frame for completion of the work. The next step is collection of background information, including work reported in the literature, work done in your own organization, and ideas proposed by colleagues and in published articles. We are now ready to design the test program. We assemble all of the interested parties, develop the design, and review and revise it as needed. Depending on the complexity of the process being studied, this step may be done in less than a day or take weeks or months. Review of the design should include all of the stakeholders—everyone who will be involved in the test program or who will be impacted by the test program or its results.

Once the design has been constructed, the next step is to plan and conduct the experimental work. Here careful attention must be paid to developing clear instructions for how the experiment is to be conducted and ensuring that the experiment is conducted as planned. Many experiments are ruined because the individuals participating were not instructed in how the test was to be conducted or chose not to follow the instructions properly.

Once the data have been collected, we move to the analysis step. We know how the analysis will proceed even before the data are collected because the design defines the analysis. Of course, if the experiment was not run as planned, for whatever reason, the analysis procedure will have to be altered to reflect the reality of the situation. It is appropriate to conduct interim analyses when experiments run over long periods of time or when it is critical to know the results very soon after the experiment has been completed. Careful attention to the results throughout the course of the experiment can also help to detect when the protocols are not being followed or unanticipated results are occurring requiring corrective action.

The final step is reporting and discussing the results. We recommend that this first be done orally. Once the discussions have been held and all the issues and implications have been identified, written documentation can be prepared. The discussion of the results with the various stakeholder groups typically defines what type of documentation is required and what further experimentation should be done.

Designing the Experiment

When designing the experiment, some key considerations need attention:

- Agree on the overall objective
- Identify output variables (y's) to be measured

- Identify input and process variables (x's) to be studied
- Select levels for input and process variables (x's) to be studied
- Available resources determine experiment size
  - Time schedule
  - Budget
- Determine the amount of replication needed
- Decide whether randomization is required
- Analysis is determined by the design

The first step is for the group to agree on the overall objectives of the experiment. This is a critical step because it is not uncommon for different people to have different perspectives or objectives. Agreement must be obtained to ensure success. The next step is to identify the responses (y's), or measures of process performance, that we want to improve. These responses are typically, but not always, well defined by the objective for the experiment. We are now ready to select the process variables (x's) to be studied. Focus on those believed to have the greatest effect on the process responses (y's). Brainstorming can be used to create this list. This list is typically long, and prioritization can be used to shorten the list to a workable number of variables.

For each of the process variables selected for the experiment, determine the levels of the variables to be studied (e.g., low and high). For most experiments two or three levels are sufficient, and rarely do we need to study more than five levels. In the case of quantitative variables, select equally spread levels (e.g., temperature at 50°, 60°, and 70°) unless there is a good reason not to. Qualitative variables can also be studied (e.g., male, female), and quantitative and qualitative variables can be studied in the same experiment.

The size of the experiment (number of tests) depends on the available resources (time and funds) and the amount of desired replication. When setting objectives, determine when results are needed and how much money can be spent on personnel and experimentation. A good strategy is not to use all of your resources in a single experiment. Things can go wrong, and it is always a good idea to confirm your results by running some additional experiments. The budget should be allocated to cover the initial experiments, any needed follow-up experiments, and a confirmation or verification run to verify that the recommended process will deliver the desired results. Box, Hunter, and Hunter (1978) recommend that no more than 20% of the budget be spent on the first experiment. This percentage will vary with the situation, and there is much more opportunity for follow-up experiments in a research and development environment than in manufacturing and business environments.

The number of tests is an important consideration. Increasing the number of tests increases the sensitivity of the experiment and enables you to detect smaller differences. Eight tests are a bare minimum, and 16 tests are better. Experiments with 20 to 30 tests or even as high as 50 tests are not uncommon. The appropriate number of tests depends on the situation and type of experimentation. Experiments involving human subjects usually require more tests and certainly more replication.

Replication means repeating all, or a portion, of the tests in the experiment. If the design called for 8 tests, 2 replicates for each test would result in an experiment of 16 total tests. Replication is very important as it increases the sensitivity of the experiment, enabling us to detect smaller differences. Replication also provides an estimate of experimental variation (uncertainty), a key benefit of the statistical approach.

A single observation ($n = 1$) for each test often works because of the hidden replication contained in designs, but duplicates ($n = 2$) at each test are sometimes needed. Larger sample sizes are used less often. Large effects can usually be identified with $n = 1$ or $n = 2$. Larger sample sizes can be used if data are inexpensive to obtain and analyze or greater sensitivity is needed. The reader is referred to Montgomery (1991) for a discussion of sample size calculation in using designed experiments.

The next consideration is how much randomization we should do. Randomization means running the tests in an experiment in a random order. Randomization provides an insurance policy against any unknown changes that may have happened during the conduct of the experiment. For example, in the hot chocolate drink experiment, it is possible that the taste panel may rate the first few drinks they taste very high and the last few low simply because they are getting sick of hot chocolate at the end of the experiment. If we ordered the experiments by tasting all the low chocolate level drinks first and the high chocolate level drinks last, we could be fooled by the results. We might think that the poor ratings for the high chocolate level drinks were due to people not liking chocolate, when in fact it was simply due to these drinks being tested late in the day. Randomization minimizes the possibility of being fooled in this way. Total randomization is recommended when possible. Partial randomization (only randomizing a portion of the experiment) can also be used when appropriate.

The final consideration is planning for the analysis of the results. Will the data be collected electronically or manually? If electronically, what computer software will be needed to collect the data? How will the analysis be done? Will any interim analyses be undertaken? As noted earlier, the analysis is dictated by the design, but planning the analysis ahead of time will greatly improve the effectiveness of the test program.

A key to successful experimentation is circulation of a short proposal for an experiment for comment and approval. In any environment (including student projects) this helps ensure that you have the political support and resources (people, time, equipment, money) to conduct the experiment. The comments received can help strengthen the experiment by identifying any gaps in your logic or thinking. The proposal should be as short as possible, one or two pages are ideal, and should include these elements (Bisgaard, 1999):

- A concise statement of the objectives of the experiment
- A list of factors and their levels
- A short description of how to measure the responses
- A table showing the design
- A layout of the data collection sheets

- A flowchart and a description of the experimental procedure
- A plan with a timetable and calendar for the experiment
- An outline sketch of how to analyze the data
- A budget for time, money, and other resources needed
- Anticipated problems and how to deal with them
- A list of team members and responsibilities

7.5 Two-Factor Experiments: A Case Study

A telemarketing firm wants to increase the percentage of sales calls that result in a sale. The management team agrees that better training of the sales representatives (reps) should increase the success rate. There is also debate on the value of a script for the sales reps to follow. Some people feel that the script restricts the flexibility of the sales reps in dealing with the customer's needs, whereas others feel that the script will provide clear direction on what the sales rep should say.

It is decided that both variables or factors will be tested. This results in a total of four possible test groups:

Test Group	Script	Training
1	No	No
2	Yes	No
3	No	Yes
4	Yes	Yes

Group 1 is the "control group" and represents business as usual: no script and no training. Group 2 receives the script. Group 3 receives training. Group 4 receives both the script and training.

A total of 20 sales reps were made available for the test and assigned at random to the four test groups, resulting in 5 sales reps per group. The response of interest is the percentage of the first 100 calls made by the sales reps that result in a sale.

In statistical terms we refer to this design as a two-level factorial design with two factors and five replicates per test group. "Two-level" means that each factor in the design is studied at two levels, and "factorial" means that all possible combinations of the two levels are studied.

Factor	Low Level	High Level
x_1 = Script	No	Yes
x_2 = Training	No	Yes

The response (y) is the percentage of successful sales calls.

The resulting data are summarized in Table 7.6, and the sales rate results vary from 8% to 46%. The question of interest is, "Is the observed variation in sales rate due to the use of script, training, both script and training, or neither variable?" One effective way to answer this question is to plot the data as shown in Figures 7.3 and 7.4, which display individual data and group averages (Table 7.7), respectively. These plots show us that the use of both the script and training increases

TABLE 7.6 Telemarketing Experiment, Effect of Script and Training on Percentage of Sales

Script	Training	
	No (−)	*Yes (+)*
No (−)	11	23
	13	21
	12	22
	10	19
	8	18
Yes (+)	15	42
	18	38
	16	39
	14	44
	13	46

FIGURE 7.3 Telemarketing Experiment, Effect of Script and Training

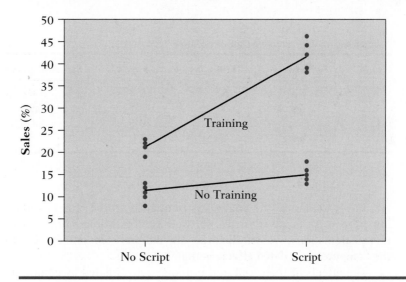

the sales rate and that the script and training have a synergistic effect on the sales rate. Training has a much larger effect when the script is used than when it is not used. We conclude from these plots of the data that both factors are important and have positive effects on the sales rate. The analyses summarized in the following paragraphs will confirm this conclusion.

FIGURE 7.4   Telemarketing Experiment, Interaction Between Script and Training

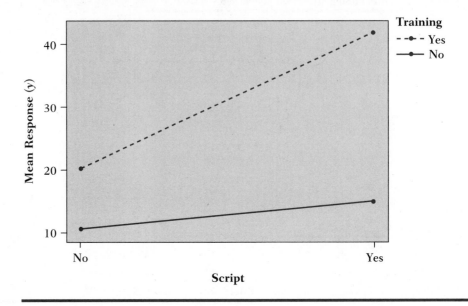

TABLE 7.7   Telemarketing Experiment Summary Data

Group	Script	Training	Average	Standard Deviation
A	No	No	10.8	1.92
B	Yes	No	15.2	1.92
C	No	Yes	20.6	2.07
D	Yes	Yes	41.8	3.35

The next step in the analysis is to quantify these effects and evaluate their importance. Although statistical software such as Minitab or JMP can be used for these calculations, we will illustrate the manual calculations here to show what the computer calculated effects actually represent.

The effects of the script and training are estimated by computing the difference in the average response at the low and high levels of the factors. In the following equations for the effects, A, B, C, and D are the averages of the four test groups (see Table 7.7).

Script Effect = (Average Response With Script) – (Average Response Without Script)
= (B + D)/2 – (A + C)/2
= (15.2 + 41.8)/2 – (10.8 + 20.6)/2
= 28.5 – 15.7 = 12.8

Training Effect = (Average With Training) − (Average Without Training)
= (C + D)/2 − (A + B)/2
= (20.6 + 41.8)/2 − (10.8 + 15.2)/2
= 31.2 − 13.0 = 18.2

Interaction Effect = (A + D)/2 − (B + C)/2
= (10.8 + 41.8)/2 − (15.2 + 20.6)/2
= 26.3 − 17.9 = 8.4

These results tell us that the use of the script increases the sales rate 12.8%. Training contributes 18.2% and the script–training interaction adds 8.4%.

In modeling terms we say that we have identified two key variables: "use of script" and "training." Both variables have positive effects. It appears that the effect of training (+18.2) is larger than the effect of using the script (+12.8). We should, however, avoid jumping to this conclusion. The two variables were found to interact, which indicates that the effect of each variable is dependent on the level of the other variable. Let's take a closer look at what this interaction represents.

Interaction Between Factors

We found a two-factor interaction in the telemarketing study, but what is an interaction? Has your doctor ever prescribed a medication that contained a label warning you not to consume alcohol while taking the medication? Why is this a problem? The medication won't hurt you; it's been clinically tested and shown to be both safe and effective. A couple of beers won't hurt you; you drank a few beers before and no big problems resulted. So how could doing both together hurt you? If the medication and beer worked independently, there probably wouldn't be a problem. The warning label suggests, however, that they don't work independently. They interact, and problems can (will?) occur if alcohol is consumed while taking the medication.

Process variables interact in the same way. An interaction between two factors, A and B, indicates that the effect of Factor A is dependent on the level of Factor B. It is equally valid to say that the effect of Factor B is dependent on the level of Factor A. In the telemarketing example, training increased the sales rate from 10.8% to 20.6% when no script was provided, an increase of 9.8%. However, when a script was provided, the sales rate went from 15.2% to 41.8%, an increase of 26.6%, which is almost three times the effect when no script was provided. This change in effect is called an interaction.

The interaction effect measures how much influence the level of Factor B has on the effect of Factor A. For example, if the average effect (or main effect) of Factor A is 20, and the interaction effect between Factor A and B is 5, the effect of Factor A will vary from its nominal (average) value of 20 by 5 points, depending on the level of Factor B. Therefore, if we only look at conditions where Factor B is at its high level, factor A would have an effect of 25. If we look at conditions where Factor B is at its low level, Factor A would have an effect of 15.

TABLE 7.8 Telemarketing Experiment, Comparison of Effects and Regression Coefficients

	Effect		Regression Coefficients		
Factor	*Value*	*SD*	*Value*	*SD*	*t-Ratio*
Script	12.8	1.07	6.4	.535	11.96
Training	18.2	1.07	9.1	.535	17.01
S × T Interaction	8.4	1.07	4.2	.535	7.85

Note that both the values of the effects and standard deviations (SD) of the effects are exactly twice the corresponding values of the regression coefficients.

In Table 7.8 we see that in the telemarketing study the main (overall average) effect of training is +18.2 units, and the training–script interaction effect is +8.4 units. This tells us that when the script variable is at its low level (no script) the effect of training is 18.2 − 8.4 = 9.8 units. When the script variable is at its high level (script used), the effect of training is 18.2 + 8.4 = 26.6 units. These effects are identical to the effects computed earlier using the averages shown in Table 7.7. The standard deviation of the effects is calculated from the replicated values in each group.

A key characteristic of factorial designs is that they can be used to detect interactions. Interactions can be synergistic or antagonistic. A *synergistic interaction* (interaction effect is positive) is one in which the two variables involved produce an effect that is larger than would be predicted if the effects of the two factors were additive. Similarly, an *antagonistic interaction* (interaction effect is negative) is one in which the effect of the two factors is smaller than would be predicted by the additive effects of the two factors.

The easiest way to interpret interactions is to construct a plot of the averages of the four groups. If the two lines are parallel, there is no interaction. If the lines are not parallel, the factors interact. Some examples of interaction are shown in Figure 7.5.

Regression Analysis of Two-Level Designs

In the model for the telemarketing data, we noted the positive effects of script and training and the interaction between the two variables. These effects can be combined into an analytical model using the regression analysis techniques discussed in Chapter 6. In fact, regression analysis is a general methodology for analyzing the results of two-level factorial designs, providing estimates of the main effects and interaction effects and tests of the statistical significance of these effects.

The regression analysis is accomplished by coding the levels of the factors (low = −1, high = +1), fitting a regression model, and evaluating the regression results. In the case of two factors, the regression model would be

$$y = b_0 + b_1 x_1 + b_2 x_2 + b_{12} x_1 x_2 + e$$

where y is the response, x_1 and x_2 are the two factors studied in the experiment, e is the error or residual, and b_0, b_1, b_2, and b_{12} are regression coefficients to be es-

FIGURE 7.5   Understanding Interaction Effects

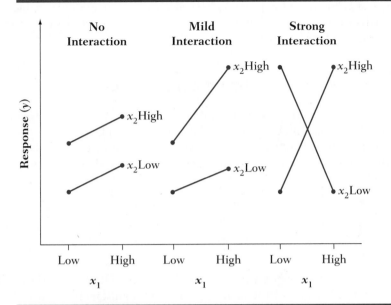

timated from the data. Note that with only a two-level design we cannot estimate squared terms.

The regression results are equivalent to the analysis just described. The constant term, b_0, is the overall average of the data. The regression coefficients are one-half the factor effects computed earlier. Remember that the effect measures the overall change in the response from the low level of a factor ($x = -1$) to the high level of the factor ($x = +1$), a distance of 2 units ($1 - (-1) = 2$). A regression coefficient measures the change in the response per unit change in x. Hence, for any factor:

Effect = 2 × (regression coefficient)

The results of a regression analysis of the telemarketing data are summarized in Table 7.8. The regression b_0 coefficient is the overall average value of 22.1.

The standard deviations (SD) of the effects and regression coefficients are related in a similar way.

SD (effect) = 2 × SD (regression coefficient)

The results of the regression analysis of the telemarketing data summarized in Table 7.8 confirm this relationship. All three effects are statistically significant, with t-ratios considerably larger than our guideline for statistical significance of 2.0.

The analysis of the factor effects and the regression analysis give identical results. The preferred method depends on the application. When working with qualitative variables, such as "script" and "training" in the telemarketing study, it makes more sense to calculate and interpret the factor effects. These factors are not continuous, and a "one-unit change in *x*" used in the interpretation of a regression coefficient makes no physical sense for qualitative variables. Regression analysis is much more useful when the factors are quantitative (continuous), such as time and temperature.

7.6   Three-Factor Experiments: A Case Study

A large supermarket chain wants to identify ways to increase sales of a new product. It is conjectured that where the product is displayed in the store is important and that the type of packaging has an effect on sales. There is also a concern that the size of the store is an important factor. These three factors were studied at each of two levels:

Factor	Low Level	High Level
Store size	Small	Large
Display type	Shelf	End of aisle
Package type	Paper	Plastic

The two types of displays selected for the experiment were "on the shelf" and "displayed at the end of the aisle," where it was anticipated that more customers would see the product. Traditionally, products of this type had been packaged in paper. It was felt that a colorful, see-through plastic package would enable the customer to see the product and would enhance sales.

Designing the Experiment

Three factors, each at two levels, results in $2 \times 2 \times 2 = 8$ combinations of factors to be tested. The resulting design is called a $2 \times 2 \times 2$ (or 2^3) factorial design. The resulting eight combinations are shown in the first three columns of Table 7.9.

Sixteen stores were selected for the study: 8 large and 8 small. This permitted each of the eight combinations of the three factors (store size, display type, package type) in the two-level factorial design to be tested in each of two stores. The total sales of the product for the 3-week test period in each store are summarized in Table 7.9. This represents the measured response. The variation between the total sales of the two stores associated with each combination of factors will be used as our measure of experimental variation. The average is the average of the total sales of the two stores associated with each of the eight combinations in the design.

The shelf displays were arranged in two different spots in the stores, and the aisle displays were constructed at the ends of two different aisles in the store. The values reported are the total of the two shelf or aisle locations in the store. Shelf displays and aisle displays were not tested in the same store.

Analysis of Results

The fifth column of Table 7.9 gives the average sales over the 3-week period for each of the eight combinations. Average sales varied from a low of 44.5 units to a

TABLE 7.9 Product Display Experiment

Store Size	Display Type	Package Type	Units Sold	Average	Standard Deviation
Small	Shelf	Paper	57, 46	51.5	7.8
Large	Shelf	Paper	43, 46	44.5	2.1
Small	Aisle end	Paper	50, 62	56.0	8.5
Large	Aisle end	Paper	77, 79	78.0	1.4
Small	Shelf	Plastic	49, 59	54.0	7.1
Large	Shelf	Plastic	51, 55	53.0	2.8
Small	Aisle end	Plastic	66, 69	67.5	2.1
Large	Aisle end	Plastic	95, 98	96.5	2.1

Each of the eight combinations of store size, display type, and package type was tested in two stores. The response is the total units sold in each store during a 3-week period. For example, in the case of the fourth combination (large, aisle end, paper), the total sales in the two stores for the 3-week period were 77 and 79 units, respectively.

high of 96.5 units sold. The question of interest is, "Which, if any, of the three factors in the design have an effect on the units sold?"

The first step in answering this question is to construct a three-dimensional plot of the factor space studied. This results in the cube shown in Figure 7.6. At the corners of the cube are the average weekly sales of the eight different combinations. The main (overall or average) effects can be studied by computing the changes in the response along each edge of the cube where one factor varies while the other two factors are held constant. This results in measuring the effect of each factor four times with the other factors held constant, even though all three

FIGURE 7.6 Product Display Equipment, Cube Plot (Data Means) for Units Sold

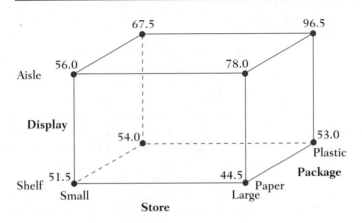

factors were varied in the experiment. For example, the effect of store size (right side versus left side) along each of the four edges is:

$$
\begin{aligned}
44.5 - 51.5 &= -7.0 \\
78.0 - 56.0 &= +22.0 \\
53.0 - 54.0 &= -1.0 \\
96.5 - 67.5 &= +29.0 \\
\hline
\text{Average} \quad &= \quad 10.8
\end{aligned}
$$

This results in an average effect of store size equal to 10.8 units sold. In other words, the larger store sold an average of 10.8 units more than the smaller store. It was thought that the larger store would have higher sales, but the actual size of the effect was unknown. This experiment provided an answer.

We can make efficient use of the data and use the same data to measure the effects of the other two variables by comparing the responses along the edges of the cube. The cube has four edges to measure the effect of each of the three factors (top to bottom, left to right, front to back).

Display Type (top to bottom)	Package Type (front to back)
56.0 − 51.5 = 4.5	54.0 − 51.5 = 2.5
78.0 − 44.5 = 33.5	53.0 − 44.5 = 8.5
67.5 − 54.0 = 13.5	67.5 − 56.0 = 11.5
96.5 − 53.0 = 43.5	96.5 − 78.0 = 18.5
Average = 23.8	Average = 10.2

We will also use these same eight data points to estimate interactions. This multiple use of data from factorial experiments is called hidden replication, which will be discussed further shortly.

Table 7.10 summarizes the main effects for the three factors. On average, the sales of end of aisle display is 23.8 units higher than the shelf display, and plastic packaging sales are 10.2 units higher than paper packaging. But that's not the whole story. Factorial experiments enable us to identify interactions, and in a three-factor experiment there are three possible two-factor interactions (x_1x_2, x_1x_3, x_2x_3) and one three-factor interaction ($x_1x_2x_3$). Although they are rarely important in real applications, a three-factor interaction would mean that the interaction of x_1 and x_2 is dependent on the level of x_3.

A general procedure for calculating the effects and interactions is summarized in Table 7.11. Because you will be using computer software to make these calculations, we will defer the computational procedure for now (see the box on page 310) and focus on the effects line of the table, which gives the effects and interactions. The first three values are the main effects for store size, display type, and package type of 10.8, 23.8, and 10.2, respectively, which are identical to those shown in Table 7.10. The remaining four values, 14.8, 3.2, 4.8, and 0.2, are the interaction effects for x_1x_2, x_1x_3, x_2x_3, and $x_1x_2x_3$, respectively. As with main effects, the larger the effect, in absolute value, the more important the variable. Here we see a large interaction between store size and display type of 14.8 units. The nature of this interaction is shown in Figure 7.7 (page 311).

TABLE 7.10   Product Display Factor Effects

Store Size	Display Type	Package Type
−7.0	4.5	2.5
+22.0	33.5	8.5
−1.0	13.5	11.5
+29.0	43.5	18.5
10.8*	23.8*	10.2*

*Average Effect

TABLE 7.11
Product Display Experiment, Calculation of Effects

Design Point	x_1 Store Size	x_2 Display Type	x_3 Package Type	x_1x_2	x_1x_3	x_2x_3	$x_1x_2x_3$	Average Sales
1	−	−	−	+	+	+	−	51.5
2	+	−	−	−	−	+	+	44.5
3	−	+	−	−	+	−	+	56
4	+	+	−	+	−	−	−	78
5	−	−	+	+	−	−	+	54
6	+	−	+	−	+	−	−	53
7	−	+	+	−	−	+	−	67.5
8	+	+	+	+	+	+	+	96.5
Sum +	272	298	271	280	257	260	251	
Sum −	229	203	230	221	244	241	250	
Avg +	68	74.5	67.7	70	64.2	65	62.7	
Avg −	57.2	50.7	57.5	55.2	61	60.2	62.5	
Effect	10.8	23.8	10.2	14.8	3.2	4.8	0.2	
t-Ratio	4.24	9.33	4.00	5.80	1.25	1.88	0.08	

There is very little difference between the shelf and end of aisle displays in small stores. However, the end of aisle displays have significantly higher sales in the large stores. This leads us to the conclusion that end of aisle displays for this product should be used in large stores and that either display can be used in small stores. There are no interactions with package type, and the plastic package can be used in all stores and both display types. This, of course, assumes that the price of the plastic package is cost-effective. In all cases, we have to make sure that the recommended changes have a favorable cost–benefit ratio when the needed changes are taken into account.

The importance of the x_1x_2 interaction and the small size of the interactions involving x_3 can be seen in the individual estimates of the effects summarized in

Calculation of Main Effects and Interactions

1. Write factor levels in standard order alternating the − and + signs in the first column individually, then in pairs in the second column, and four at a time in the fourth column. This results in the x_1, x_2, and x_3 columns in Table 7.11.

2. Create interaction columns by multiplying the corresponding elements of the associated main effect columns using the usual rules:

$$(-)(-) = +, \ (-)(+) = -, \ (+)(+) = +, \ (+)(-) = -$$

This results in x_1x_2, x_1x_3, x_2x_3, and $x_1x_2x_3$ columns in Table 7.11.

3. Add the average response as the last column of the table. This is the average sales column in Table 7.11.

4. For each effect column calculate five numbers: Sum+, Sum−, Average+, Average−, and effect where

   ■ Sum+ and Sum− are the sum of the responses corresponding to the plus signs and minus signs in the effect column.
   ■ Average+ and Average− are the average of the values that make up Sum+ and Sum−.
   ■ Effect = (Average+) minus (Average−).

   For example, the x_1 effect in Table 7.11 is computed as follows:

   $Sum-$ $= 51.5 + 56.0 + 54.0 + 67.5 = 229$
   $Sum+$ $= 44.5 + 78.0 + 53.0 + 96.5 = 272$
   $Avg+$ $= (Sum+)/4 = 272/4 = 68$
   $Avg-$ $= (Sum-)/4 = 229/4 = 57.25$
   $Effect$ $= 68 - 57.25 = 10.75$

5. Note: $(Sum+) + (Sum-)$ has the same value for all columns (effects).

Table 7.10. There is considerable variation among the values for the x_1 = store size effect (range is from −7 to +29) and x_2 = display type (ranges from 4.5 to 43.5) effects. These values vary because of the x_1x_2 interaction. There is a much smaller range in the x_3 = package type values (range from 2.5 to 18.5), as the interaction effects involving x_3 are much smaller.

In modeling terms we say that this experiment has identified three key variables: store size, display type, and package type. We also found that two factors (store size and display type) interact with each other and that the effect of package type is independent of the levels of the other two variables. All the other interactions were small. The numerical values of the effects tell us that all the effects are positive. The interaction effect is large, being greater than two of the main effects. These effects can be combined into an analytical model using regression analysis.

FIGURE 7.7   Product Display Equipment, Interaction Between Store Size and Display Location

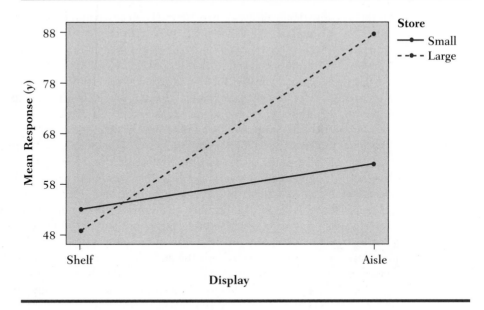

Importance of the Factor Effects

In the previous section we observed the factor effects and made decisions about which effects were large (important) and which were not. Before making any statement regarding the practical importance of an effect, we need to also assess the statistical significance of the effect.

This is done by performing a regression analysis of the data. The regression analysis calculates the main effects, interaction effects, and associated t-ratios that are used to test the statistical significance of the effects. The t-ratios for the seven effects (see Table 7.11) are 4.24, 9.33, 4.00, 5.80, 1.25, 1.88, and 0.08, respectively. Using our $t = 2.0$ guideline, the effects for x_1 = store size, x_2 = display type, x_3 = package type, and the store size–display type interaction are statistically significant. The other interaction effects are not. This is the same conclusion we reached by observing the relative magnitude of the effects.

Efficiency and Hidden Replication

Computing the effects using the method shown in Table 7.11 illustrates the efficiency of two-level factorial designs. Each response value is used to compute each effect; hence, each test result is used several times. No data are wasted. Effects are computed by looking at the difference between averages of $4n$ observations where n is the number of observations per group. This property of factorial designs is frequently referred to as hidden replication. In this example each effect is the difference between the averages of the sales of eight stores. For example, the store size effect is the difference between the average sales of the eight large stores minus the average sales of the eight small stores.

7.7 Larger Experiments

In theory, statistically designed experiments can accommodate any number of factors and any number of levels of the factors. However, as the number of factors or number of levels increases, the total size of the experiment increases, resulting in increased costs and increased work required. Most real applications of statistically designed experiments involve factors at two or three levels, although studying factors at four, five, six, or seven levels also occurs. It is also not uncommon in screening experiments to investigate as many as 15 to 20 factors.

In practice, the amount of experimentation we can do is often restricted because the total number of factor combinations to be investigated is an exponential function of the number of factors and number of levels per factor. For example, a two-level design with three variables has $2^3 = 8$ combinations, and a three-level design with three variables has $3^3 = 27$ combinations. Hence, when the number of factor levels is large, the number of factors studied tends to be small. Conversely, when the number of factors is large, the number of factor levels is small. For example, experiments involving more than six factors rarely have more than two levels per factor, perhaps with the addition of a center point, and those with more than two or three factors rarely have more than three levels.

Some examples may be helpful. In a study of the effect of advertising on sales, Ackoff (1978) used a 4×2 factorial design to study the effects of four types of advertising media and two levels of national TV advertising (yes or no) on sales. Each of the eight combinations was tested in 5 different areas as shown in Table 7.12. This results in a total of $8 \times 5 = 40$ test areas being involved in the experiment.

In another study discussed in Ackoff (1978), the effects of advertising, amount spent on sales effort, and amount spent on sales materials on the output variable "increase in sales" were evaluated. Each of the three variables were studied at three levels using a $3 \times 3 \times 3$ factorial design. Each of the 27 combinations was tested in one area involving a total of 27 different market areas (Table 7.13). One market area per combination is effective in this case because of the "hidden replication" present in factorial designs discussed earlier.

TABLE 7.12 Experiment to Study the Effect of Advertising Media on Sales

Media	National TV	
	No	Yes
Local TV	5*	5
Billboard	5	5
Radio	5	5
Newspaper	5	5

*Each combination was evaluated in 5 market areas, resulting in a total of 40 market areas.

TABLE 7.13 Effect of Advertising and Marketing Effort on Sales

Sales Effort	Sales Materials	Advertising Budget		
		Decreased 50%	*Standard*	*Increased 50%*
Decreased	Decreased	X	X	X
	Standard	X	X	X
	Increased	X	X	X
Standard	Decreased	X	X	X
	Standard	X	X	X
	Increased	X	X	X
Increased	Decreased	X	X	X
	Standard	X	X	X
	Increased	X	X	X

Sales Effort = amount spent on sales force.

Sales Materials = amount spent on displays, signs, and so on.

The 27 X's denote the 27 market areas involved in the study.

In another experiment, the advertising–sales response curve was studied using seven levels of advertising: −100%, −50%, 0%, +50%, +100%, +150%, +200% of standard budget. Each level of advertising was tested in six areas over a 12-month period. The number of test areas in the two extreme levels (−100% and +200%) was actually reduced. As noted earlier, as the number of levels studied increases, the number of factors studied decreases. In this experiment only one factor (level of advertising) was studied. Analyses of these larger, more complex experiments can be done using graphical methods, regression analysis, or analysis of variance (ANOVA), which is discussed in Chapters 8 and 9. See also Appendix E.

7.8 Blocking, Randomization, and Center Points

Three key techniques in design of experiments that require further explanation are blocking, randomization, and the use of center points. Let's look at blocking first.

Experiments are conducted in blocks, or groups of runs, to eliminate the effects of extraneous or nuisance variables such as time effects (hour of day, day of week, week of year, season of year) and location effects (different sites, different machines). By definition, the effect of the blocking variable is of little or no interest, but it must be taken into account in the design for the results of the experiment to be valid.

Let's use the telemarketing case study as an example (see Section 7.5). In this study 5 sales reps in each of four test groups were studied, for a total of 20 sales reps. Fortunately, the call center employed more than 400 sales reps, so finding 20 sales reps for the study was easy. Let's suppose that a number of smaller call centers were available for the study. In our experimental design we might choose 5 sales reps in each of the four groups, for a total of 20 sales reps in each call center. To increase the number of unique call centers studied, we could run the same study in four different call centers. By running the same study in each center, our results would not be biased if one call center (one block) were different from the others. However, if we ran the script plus training combination in one center, the no script and no training combination in another, and so on, our results could be biased by differences in the centers. The script plus training combination might produce good results not because this is a good combination of x's, but only because we happened to run this combination at a good call center. Clearly, it is better to design the study so that each of the four combinations is run at each center. In this case, the call centers are the blocking factor, and we would say that the experiment was conducted in four blocks. We are not particularly interested in studying the effect of call centers in this design (although we could); rather, we are protecting our results from bias in case there are differences.

Now let's consider how blocking was used in the chocolate drink study (see Section 7.3). Each of the three experiments in the chocolate drink study was run in two blocks: tests 1, 4, 6, 7, and 9 in Block 1 and tests 2, 3, 5, and 8 in Block 2. The purpose of the blocking in this instance was to guard against any uncontrollable shifts or trends that may occur during the course of the experiment. In particular, it was felt that tasters would become sick of drinking hot chocolate after four or five drinks and might rate the last few drinks poorly just because they were sick of hot chocolate. Therefore, the testing was split in half. The first block was done in the morning, and the second block was run in the afternoon. This gave tasters some time between tasting all nine drinks (including the center point). Two-level experiments are often blocked in this manner. (See Box, Hunter, & Hunter, 1978, or Montgomery, 1991, for details.)

How does blocking actually help us avoid bias? Careful examination of the specific points included in Blocks 1 and 2 in the hot chocolate experiment (see Table 7.1) reveals that Block 1 has half its combinations with low chocolate, half with high chocolate. Similarly, half have high cream, half have low cream. The same is true for thickener, and of course this holds for Block 2 as well. Using these combinations we can be sure that any change in overall ratings between morning (Block 1) and afternoon (Block 2) would not have an impact on our estimates of the main effects or two-factor interactions. The reason for this is that for each effect, say chocolate, any decrease in afternoon ratings would be decreasing two high chocolate combinations, and decreasing two low combinations by the same amount. When we calculate the chocolate effect, these decreases will cancel each other out, producing the same chocolate main effect as we originally had. This type of blocking helps to eliminate extraneous variation from our main effects and two-factor interactions.

Randomization is also used in designing experiments to reduce, if not remove, the effects of unknown extraneous or nuisance factors. If we know the factors that might cause extraneous variation, as in the hot chocolate experiment, we can remove their effects by blocking the experiment. To protect ourselves against other extraneous factors that are unknown, we use randomization. Randomization does not totally eliminate the extraneous variation, but it ensures that it is randomly allocated to all effects rather than significantly biasing our estimate of the effect of one factor in the design. There are two main types of randomization: complete and restricted. To completely randomize the telemarketing study, the 80 sales reps used in the experiment would have to be selected at random from the total call center population and assigned at random to one of the four groups. This was not how this experiment was conducted. Selection and assignment of sales reps was done at random within each location, but it was decided to have exactly 20 from each location. This was much more practical than moving sales reps to new locations, but it did put a restriction on the randomization. This study is, therefore, an example of restricted randomization. Complete randomization is the preferred method when practical. Restricted randomization is the next best method.

When designing experiments, randomize the sequence of testing as much as possible—the benefit is unbiased results. For example, suppose we ran all the low chocolate drinks followed by the high chocolate drinks, without any blocking or randomization. If the tasters were getting sick of drinking hot chocolate, the ratings for the high chocolate drinks might be lower. When analyzing the results, however, we would likely conclude that this difference was due to people preferring low levels of chocolate. Randomization helps avoid such biases

Center points are used regularly in response surface experiments where it is of interest to detect and quantify curved response functions. This is typically done by studying variables at three or more levels (e.g., low, medium, and high). The experiment in Table 7.5 is an example of such a design. In characterization experiments involving two-level experiments, we typically include one overall center point (a test with all the factors at their middle level) to see if curvature is present. The single overall center point will not tell us which variable or variables produced the curvature, however. Additional experiments will have to be completed to answer this question. The chocolate drink experiment used a single center point, with Run 9 in each of the three experiments the center point. These center points revealed that there were no optimal drink formulations hidden in the middle of experiments 1 and 2. This led the experimenters to explore regions outside the current design to improve taste ratings. In experiment 3, the center point revealed that an optimum had been obtained.

Blocking, randomization, and center points are standard practices in the statistical approach to experimentation and should be considered in any design. Blocking is the preferred method when we can anticipate the extraneous factors. Randomization can be used as insurance against those factors we cannot anticipate. Very often, both of these methods are incorporated in the experimental design.

7.9 Summary

The statistical approach to process experimentation enables us to identify key process drivers, understand the process, and build models that will help us better manage and improve our processes. Along the way, high-quality data on how the process performs are produced and the effects of changes in process inputs and operating conditions on process performance are identified and quantified. The cause-and-effect relationships between these process variables (x's) and process responses (y's) enable us to build models. These models are subsequently used to communicate the effects of the vital few key drivers, identify root causes and fix problems permanently, improve the process, control and manage the process, and predict how the process will perform in the future.

This chapter has focused on the concepts, fundamentals, and power of statistically designed experiments. Although we have concentrated on experiments with three or fewer variables, many experiments involve 5, 10, or 20 variables. Some involve the estimation of curvature effects as well as the estimation of linear and interaction effects. As you use the statistical approach to experimentation, keep these principles in mind to better understand processes and build models:

- Define a clear objective for the experiment.
- Create and test hypotheses that will help satisfy the objective.
- Be bold, but not reckless; study large numbers of variables and study the x's over a wide but realistic range.
- Use a sequential approach with realistic experiment sizes.
- Be patient; several experiments may be needed to solve some problems.
- Understand how the data will be analyzed before the experiment is run.
- Always plot the data.
- Iterate between hypothesis and data to confirm or deny hypotheses and build models.
- Good administration of the experimentation process is critical; be sure that the factor levels are set and that the data are collected as specified—avoid missed communications.
- Test any suspect combination of factor levels first and proceed with the rest of the design if no problems are encountered; consider redesigning the experiments if problems are found.
- Measure several responses (process outputs or y's) in addition to the responses of primary interest; the additional cost to do this is usually small.
- Randomize the runs in the experiment when you can, but don't let problems with randomization slow down your experimentation and improvement efforts.

7.10   Project Update

You may have collected and analyzed a variety of previously collected data for your project by now. At this time, consider how to design an experiment for your process that will enable you to evaluate your current hypotheses about it as well as develop new knowledge. You should have learned a lot of important information about the process already, which will enable you to enter the experimentation in the characterization phase, passing over the screening phase.

A simple "plain vanilla" design, which is also quite effective, is the two-level, three-variable design (2^3) design. This design is often run with a center point and perhaps some degree of replication. Of course, you may use whatever design is most appropriate for your project at this point in time. To run a design, you must be able to vary the x variables. If the x variables are totally outside your control, such as interest rates, it will be difficult to run a true designed experiment. If you are unable to design an experiment for your project, complete Exercise 1 to gain experience in designing an experiment.

EXERCISES

1. **Experimenting With Your Personal Interests.** From your own interests (work, home activities, hobbies) identify an activity (process) that needs improvement and could be studied through the use of a designed experiment. Design an experiment to improve this process. Be sure to include these elements in your discussion:

   - The experimental situation
   - The purpose of the study (e.g., problem to be solved)
   - The factors to be studied
   - The design to be used
   - The amount of replication required
   - How the data will be collected, analyzed, and reported

   Conduct the experiments you propose, collect and analyze the data, identify the improvements (changes) suggested by the experiments, and make these improvements. Did the improvements work? Did the process work better?

2. **Product Display Experiment.** Using the regression analysis techniques discussed in Chapter 6, estimate the effects of the three factors and their interactions in the *product display* experiment study discussed in Section 7.6. Be careful to code the factors: low level = -1, high level = $+1$. Estimate all main effects, two-factor interactions, and the one three-factor interaction. What is the relationship between the values and standard deviations of the values for the effects and the regression coefficients? Has the interpretation of the results changed?

3. **Effect of Advertising on Sales.** What are the key aspects of experimentation illustrated by the "effect of advertising in sales" case discussed in Chapter 2 and again in Section 7.7?

4. **Venture Guidance Appraisal Study.** The data in Table 7.14 are from a study to evaluate the effectiveness of a new option in an engineering capital project venture guidance appraisal (VGA) process used by a major chemical company. The VGA is a crude estimate of what an engineering project will cost. Projects are prioritized using the VGA (cost) and the project benefits. Projects that pass this screen are evaluated further and refined cost estimates prepared.

   The engineering team conjectured that the accuracy of the VGA approval could be increased by a front end loading (FEL) process that required getting the business team (the group requesting the project) involved early in the creation of the VGA. As VGAs were prepared, those that were candidates for the FEL option

TABLE 7.14   Effect of Front End Loading on the Accuracy of Venture
Guidance Appraisal Cost Estimate

Project	FEL Used	Difference (%)*
1	No	−31
2	No	−4
3	Yes	−4
4	Yes	−6
5	Yes	−1
6	No	−13
7	No	−2
8	No	21
9	No	−23
10	No	12
11	No	−13
12	No	−7
13	Yes	1
14	No	−11
15	No	30
16	No	9
17	Yes	0
18	No	−13
19	Yes	−12
20	No	−16
21	Yes	−1
22	No	−10
23	Yes	−16
24	No	−4
25	No	6
26	Yes	0
27	Yes	4
28	Yes	6
29	Yes	0
30	No	−9

*Percentage difference between VGA cost estimate and final cost estimate

were identified and the business team became involved. The response of interest
is the percentage difference between the VGA cost estimate and the final cost estimate obtained on those projects that the business team determined were appropriate for further refined cost estimates.

Analyze these data to determine whether the new FEL option has a significant
effect on the accuracy of the VGA estimate. Comment on the use of randomization in the study. Was it adequate? Could it have been done differently? Better?
What recommendation would you make to the engineering team regarding these
results?

5. **Reducing Invoice Errors.** Moen, Nolan, and Provost (1991) discuss an experiment designed to evaluate the effects of three variables on invoice errors. Invoice

TABLE 7.15 Invoice Error Experiment

Customer Size	Customer Location	Product Type	Number of Errors
−	−	−	15
+	−	−	18
−	+	−	6
+	+	−	2
−	−	+	19
+	−	+	23
−	+	+	16
+	+	+	21

Customer size: Small (−), Large (+)
Customer Location: Foreign (−), Domestic (+)
Product Type: Commodity (−), Specialty (+)

errors had been a major contributor to lengthening the time customers took to pay their invoices and increasing the accounts receivables for a major chemical company. It was conjectured that the errors might be due to the size of the customer (larger customers have more complex orders), the customer location (foreign orders are more complicated), and the type of product. A subset of the data is summarized in Table 7.15.

What type of design is this? What is the nature of the effects of the factors studied in this experiment? What strategy would you use to reduce invoice errors given the results of this experiment?

6. **Improving Bowling Performance.** A professional bowler is having trouble with his bowling performance and his income is suffering. He conjectures that the use of a wristband and a lighter ball will improve his score. How would you recommend he test these ideas? Be specific about the situation to be tested, how many tests you would recommend, and the order in which the tests should be done.

What measurement would you use: a frame (one set of pins), a game, several games? Describe the specific experiments you would conduct, including factor combinations, test order, number of tests, and measurements to be made.

7. **Weld Process Development.** Welding is the metallurgical process of joining two pieces of metal together. A construction supply company is developing a welding process based on a new proprietary alloy. The data in Table 7.16 are from an experiment designed to study the effects of two formulations of the alloy and arc current on the strength of the weld.

   ■ What is the role of alloy formulation and arc current in the welding process?
   ■ What kind of design is this?
   ■ What is the nature, quantitatively and qualitatively, of the effects of alloy type and arc current on weld strength?

8. **Height of Easter Lilies.** A horticulturist employed by a large florist is assigned responsibility for determining the effect of length of storage and conditioning time after storage on the height of Easter lilies at first bloom. Tall lilies are in great

TABLE 7.16 Weld Process Development Study

Weld	Alloy	Arc Current	Weld Strength
1	A1	140	2.5
2	A2	140	2.9
3	A2	200	0.9
4	A1	200	8.6
5	A2	200	2.0
6	A1	140	4.7
7	A2	140	3.5
8	A1	140	3.0
9	A1	200	8.3
10	A2	200	2.0
11	A1	200	9.4
12	A2	140	2.5

demand and sell at a premium price. Two storage periods (SP_1 and SP_2) and two conditioning times after storage (T_1 and T_2) were chosen for this experiment. The heights of three plants were measured at each of the four combinations of storage period and conditioning time (Table 7.17). It was felt that this amount of replication would give an adequate estimate of the plant height at each of the experimental conditions. All other variables were held constant, and the 12 plants were randomly distributed around the greenhouse to ensure that any unanticipated changes would not bias the results of the experiment.

- What kind of design is this?
- What is the nature of the effects of conditioning time and storage period?
- Under what conditions should Easter lilies be stored to maximize height at first bloom?

TABLE 7.17 Height of Easter Lilies at First Bloom

Conditioning Time	Storage Period	Height (inches)		
T1	SP1	28	26	30
T1	SP2	49	37	38
T2	SP1	31	35	31
T2	SP2	37	37	29

9. **Water Filtration System Development.** Reducing the amount of impurities in the filter material will improve the filtering properties of a certain filtering system. The data in Table 7.18 are from an experiment designed to determine the effects of process variables associated with the filter manufacturing process (blowing time and steam pressure) on the impurity level of the filter material (adapted from Draper & Smith, 1981).

TABLE 7.18 Impurity Level of a Water Filter Material

Steam Pressure	*Blowing Time*		
	1 Hour	*2 Hours*	*3 Hours*
10 pounds	45.2	40.0	35.9
	46.0	39.0	34.1
20 pounds	41.8	27.8	22.5
	20.6	19.0	17.7
30 pounds	23.5	44.6	42.7
	33.1	52.2	48.6

- What kind of design is this?
- What is the nature of the effects of blowing time and steam pressure?
- How would you recommend that the process be operated to reduce the amount of impurity in the filter material?

10. **Metal Etching Process Development.** A metal etching process used in the development of a communications device was the focus of an experiment at Bell Telephone Laboratories (Ott, 1961). In the process, tantalum sheets were weighed before and after an etching process. Their loss of weight was recorded in micrograms. The tantalum sheets were placed into a solution of nitric acid, hydrofluoric acid, and water as the etching commenced. Three factors were suspected of having an effect on the etch rate. The tantalum sheets, in a solution of 30°C or 0°C, were stirred in some cases and not in others. When taken from the solution, the sheets were rinsed and held for 3 or 12 minutes before weighing. The experiment was run as a factorial design with two replicate tests per combination (Table 7.19).

- What kind of design is this?
- What are the key drivers of this process?

TABLE 7.19 Metal Etching Process Experiment

Stirred	Solution Temperature (°C)	Cure Time (minutes)	Weight Loss (micrograms)	
Yes	0	3	851	528
		12	596	763
	30	3	2252	2978
		12	3339	3296
No	0	3	592	803
		12	794	664
	30	3	1261	1150
		12	1279	1169

- Is there anything unique about the performance of the process as the factors are varied?
- How would you operate the process to maximize the metal etching rate?

11. **Low-Carbon Steel Wire Manufacture.** A wire manufacturing company is developing a new product for industrial markets. An experiment was conducted to study the effect of electrolytic chromium plate as a source for the chromium impregnation of low-carbon steel wire. Eighteen treatments were considered, using all combinations of three diffusion temperatures (2200°F, 2350°F, 2500°F), three diffusion times (4, 8, and 12 hours), and with and without degassing. Each treatment was used on 4 wires, giving a total of 72 wires used. The response studied was average resistivity in microhms per cubic centimeter (a measure of electrical conductivity). The average resistivities are shown in Table 7.20.

- What kind of design is this?
- Using graphical methods, identify the key drivers of this process.
- Construct a graph that illustrates the nature of the effects of the key drivers and simultaneously summarizes the results of this experiment.

TABLE 7.20 Low-Carbon Steel Wire Manufacturing Test (Resistivity)

Temperature	2200		2350		2500	
Degassing	No	Yes	No	Yes	No	Yes
Time						
4 Hours	18.1	17.9	22.1	21.2	22.9	22.8
	18.9	18.0	20.2	20.4	24.0	22.3
	18.6	18.7	21.3	21.2	23.0	22.7
	19.1	19.0	22.6	21.2	23.0	23.3
8 Hours	19.2	19.2	23.2	22.7	25.5	26.9
	19.3	19.0	21.8	22.7	26.6	26.9
	20.7	20.4	22.9	22.5	25.9	26.3
	20.4	19.2	22.3	22.5	26.8	26.9
12 Hours	20.0	18.9	23.9	23.3	27.0	26.5
	20.2	20.1	23.6	23.5	26.2	26.8
	20.1	20.0	23.2	23.5	25.9	25.4
	20.5	20.8	23.7	22.9	26.9	27.2

Source: Anderson & Bancroft, 1952, p. 290.

12. **Microwave Popcorn Experiment.** The data in Table 7.21 are from an experiment studying the effects of five factors in the process of making microwave popcorn. This design is known as a half replicate of a 2^5 design because it uses half of the runs in a full 2^5 design. Four extra runs were also made. This design will permit you to estimate the five main effects and the 10 two-factor interactions $x_1x_2, x_1x_3, x_2x_3, \ldots, x_4x_5$.

The factors are self-explanatory except x_5 = elevate, where "no" means that the bag of popcorn was left flat on the floor of the microwave and "yes" means that

TABLE 7.21 Microwave Popcorn Experiment

Test	Run Order	A	B	C	D	E	Bullets	Taste
1	12	−1	−1	−1	−1	1	1.5	7.5
2	9	1	−1	−1	−1	−1	1.4	8.0
3	6	−1	1	−1	−1	−1	1.9	9.0
4	18	1	1	−1	−1	1	0.6	6.5
5	1	−1	−1	1	−1	−1	1.8	7.0
6	14	1	−1	1	−1	1	0.3	7.5
7	7	−1	1	1	−1	1	0.2	2.5
8	5	1	1	1	−1	−1	0.9	1.0
9	17	−1	−1	−1	1	−1	1.7	7.0
10	15	1	−1	−1	1	1	0.8	6.0
11	3	−1	1	−1	1	1	0.6	4.5
12	16	1	1	−1	1	−1	0.9	4.0
13	4	−1	−1	1	1	1	0.6	9.0
14	13	1	−1	1	1	−1	1.3	7.5
15	NA	−1	1	1	1	−1	Missing	—
16	NA	1	1	1	1	1	Missing	—
17	2	−1	−1	−1	1	−1	3.2	8.5
18	8	1	0	1	1	1	0.1	4.0
19	11	1	0	1	−1	−1	0.8	5.0
20	10	−1	0	1	1	−1	1.6	5.5

A = Price (generic, brand), B = Time (4 minutes, 6 minutes), C = Power (medium, high), D = Preheat (no, yes),
E = Elevate (no, yes).
Source: Anderson & Anderson, 1993.

the bag was put on a microwave-safe rack in the center of the chamber. The two responses were "bullets," which is the percentage of kernels not popped, and "taste," which was determined on a subjective 10-point scale with 10 = best and 1 = worst. Consumers often complain about unpopped kernels and perceive fewer bullets as higher quality popcorn.

Use regression analysis to analyze these data, estimating all the main effects and two-factor interactions. (Regression analysis is the tool of choice here because of the two missing combinations and the four extra combinations run.)

- Which factors have a significant effect?
- How do you recommend popcorn be made with this microwave?
- For the two combinations not run, what would you predict the taste of the popcorn would have been if these combinations had been run?
- Does this give you any insight as to why these two runs may not have been made?

13. **Displaying Results.** Create a graphical display that shows how the experimental regions of the three experiments in the chocolate drink case study fit together. What recommendations would you make to the development team regarding these results?

14. **Alternative Designs.** Review the product display case study and identify one or more alternate ways to design the experiment. Discuss the strengths and limitations of your proposed design and how it compares to the design discussed in this chapter.

15. **Blocking Experiments.** As in Exercise 1, describe an experiment in your field of interest or activity that would require blocking. Describe how the experiment would be conducted, the blocking strategy used, and how blocking would improve the results of the experiment. You may use the same experimental situation you used in Exercise 1.

16. **Blocking.** In the chocolate drink case study, each of the three experiments was run in two blocks. Block 1 consisted of tests 1, 4, 6, 7, and 9. Block 2 consisted of tests 3, 5, 7, and 8. Construct a diagram of these nine tests in the three-dimensional design space (cream, chocolate, thickness), and comment on the balance of the two blocks with respect to the three factors tested. Were the three factors treated equally in the two blocks? Would repeating Test 9 in Block 2 be of benefit? If so, how?

17. **Randomization.** For the experiment described in Exercise 1, discuss options for randomization, the technique and method for each option, and your recommended approach.

18. **Examples of Interactions.** Review the literature for your field of interest and/or your daily work, school, or hobbies and identify 5 to 10 examples of two-factor interactions. Draw interaction plots to illustrate each interaction.

19. **Comparing Strategies for Experiments.** Review the literature available on the Internet on the work of Thomas Edison, the famous inventor. Identify the strategies that he used to conduct experiments. Prepare a written or an oral report that includes at least two examples illustrating these strategies for experimentation.

20. **Wright Brothers' Strategy for Experimentation.** Review the literature available on the Internet on the work of the Wright brothers and identify the strategy they used to conduct their experiments. In what ways did they practice the principle of good experimentation? Prepare a written or oral report that includes two or more illustrative examples of the Wright brothers' approach to experimentation.

21. **OFAAT Experimental Strategy and Interaction Effects.** We noted that the OFAAT strategy does not work well when there are interactions among the factor effects. Construct a two-factor example that illustrates this point complete with quantitative values for the response (y) and the two predictor variables (x_1 and x_2).

22. **Internet Web Site Design.** You have been given the assignment of designing an Internet Web site to market a new product, and questions have been raised about how many graphics to use, the number of colors to use, and the best print density. You have been asked to collect data that will quantify the effects of these three variables on the total amount of product sold. The goal is to find the Web design that will maximize the amount of product sold through the Web site.

    - How would you design the experiment to assess the effects of the three variables on product sales and thereby define the design of the Web site? Be sure to discuss the use of randomization and replication.

- How would the experiment be carried out and administered? Over what time period would the experiment be conducted?
- How would you measure the response of each test?
- How would you analyze the data?

23. **Mortgage Approval Time Study.** A major financial services company wishes to better understand their mortgage approval process. In particular, the company is interested in learning about the effects of credit history (good versus bad), size of mortgage (<$50,000 versus >$500,000), and region of the United States on the amount of time it takes to get a mortgage approved. The database of mortgages approved in the last year is accessed, and a random sample of five approved mortgages is selected for each of the eight combinations of the three variables. The data are shown in Table 7.22.

   - Analyze the data to determine the nature and magnitude of the effects of the three variables on mortgage approval times. What are the key drivers of this process?
   - Comment on the sample of only five mortgages per combination. Under what circumstances would it have been appropriate to select a larger sample? Is a sample of five mortgages adequate to access the relative magnitudes of the effects of the variables? What sample size would you recommend? What could you learn from a larger sample size?
   - What other response might be of interest to measure and study? (*Hint:* If you were getting a mortgage or a loan, what are the two most important measures of the process you would have to go through?)
   - This is a different kind of experiment. A statistical design was used to collect data from an existing database rather than setting up each of the tests in a random order and measuring the response. What are the strengths and limitations of such a study? Are there any conditions in which such studies should not be used?

TABLE 7.22 **Mortgage Approval Time Study**

Credit History	Mortgage Size	Region	Approval Times (days)				
1	1	1	59	50	64	62	47
2	1	1	81	58	69	65	74
1	2	1	38	52	58	60	65
2	2	1	146	159	133	143	129
1	1	2	28	26	38	41	21
2	1	2	42	53	40	50	64
1	2	2	49	31	49	42	38
2	2	2	106	115	126	118	138

Credit History: Good (1), Bad (2)
Mortgage Size: < $50,000 (1), > $500,000 (2)
Region of United States: East (1), West (2)

REFERENCES

Ackoff, R. L. (1978). *The art of problem solving.* New York: Wiley.

Anderson, M. J., & Anderson, H. P. (1993, July–August). Applying DOE to microwave popcorn. *PI Quality,* 1–3.

Anderson, R. L., & Bancroft, T. A. (1952). *Statistical theory in research.* New York: McGraw-Hill.

Bisgaard, S. (1999). Proposals: A mechanism for achieving better experiments. *Quality Engineering, 11*(4), 645–649.

Box, G. E. P., Hunter, W. G., & Hunter, J. S. (1978). *Statistics for experimenters.* New York: Wiley (Wiley Interscience).

Draper, N. R., & Smith, H. (1981). *Applied regression analysis* (2nd ed.). New York: Wiley.

Koselka, R. (1996, March 11). The new mantra: MVT (multivariable testing). *Forbes,* 114–118.

Moen, R. D., Nolan, T. W., & Provost, L. P. (1991). *Improving quality through planned experimentation.* New York: McGraw-Hill.

Montgomery, D. C. (1991). *Design and analysis of experiments.* New York: Wiley.

Myers, R. H., & Montgomery, D. C. (1995). *Response surface methodology.* New York: Wiley (Wiley Interscience).

Ott, E. R. (1961, May). *Notes on interpretation of data.* New Brunswick, NJ: Rutgers Statistics Center.

Pfeiffer, C. G. (1988, May). Planning efficient and effective experiments. *Materials Engineering,* 35–39.

Snee, R. D., & Hare, L. B. (1992). The statistical approach to design of experiments. In D. C. Hoaglin & D. S. Moore (Eds.), *Perspectives on contemporary statistics* (pp. 71–91). Washington, DC: The Mathematical Association of America.

8

Applications of Statistical Inference Tools

The great tragedy of science—the slaying of a beautiful hypothesis by an ugly fact.

Thomas Huxley

8.1 Overview

In this chapter we return to statistical inference tools to fill in the missing piece in the process improvement strategy discussed in Chapter 4. The main learning objectives for this chapter are to develop understanding of the concepts of statistical inference and to develop functional capability to apply some of the more commonly applied inference tools. There are many potential types of confidence/ prediction intervals and hypothesis tests, and in this chapter we present some of the more commonly used types. Appendix F provides summary information for several more of these tools. Note also that the underlying theory of statistical inference, such as probability distributions, will be addressed in Chapter 9.

Statistical inference essentially means drawing conclusions about a population or process based on sample data. In other words, we are inferring about the population or process rather than doing 100% sampling. Conceptually, of course, it is impossible to do 100% sampling from a process, because we cannot yet sample future output of the process. We can do 100% sampling of previous output, but if we want to draw conclusions about future output, we still have to infer.

In most cases the conclusions will be obvious from simple plots of the data. In these situations we do not need to use formal statistical inference procedures. In some cases, however, more than one conclusion may be drawn from the plots. When the conclusions are not obvious, statistical inference provides an objective means for drawing appropriate conclusions. Performing the tests also documents our conclusions in a more formal and convincing manner than simply noting that a plot indicates a certain conclusion. This is particularly helpful in written reports, where the reader may not see every plot made.

327

Statistical inference includes many tools. When we draw conclusions about the process from the basic tools presented in Chapter 5, we are using inference. In this chapter we deal with formal (i.e., mathematical) inference and focus on two main types of statistical inference: confidence/prediction intervals and hypothesis testing. These methodologies enable us to make formal mathematical conclusions about the process based on sample data. For example, suppose we use a regression model to forecast next quarter's sales and get predicted sales of $22 million. Understanding statistical thinking, we realize that we will not have *exactly* $22 million in sales, there will be some variation. An important question we might want to ask is, "How far from $22 million is my actual sales likely to vary?" We can quantify our uncertainty in this forecast with a prediction interval. Based on the existing data and some important assumptions, we might be able to say that we are 95% sure that next quarter's sales will be between $20 million and $24 million. Similarly, if the regression equation includes advertising expenses as one of the independent variables in the equation, we might be interested in the slope, or coefficient, of this variable. The estimated coefficient is 1.2, indicating we observe an additional sales volume of $1.2 for every additional $1 of advertising expense. Again, we realize that this is an estimate based on this sample of data and that the true relationship between advertising and sales will be somewhat different from this. How different might it be? A 95% confidence interval for the true slope would quantify our uncertainty around this true relationship.

Confidence intervals and prediction intervals are similar concepts—intervals that quantify the degree of uncertainty in an estimate or prediction. Statisticians generally use the term *confidence interval* when referring to the uncertainty in an estimated statistic, such as a regression coefficient or an average. The term *prediction interval* is used when referring to uncertainty in forecasting a future observation, such as next month's sales.

Formal hypothesis tests determine mathematically if we have sufficient evidence from the data and our assumptions to draw definitive conclusions. For example, suppose a soft drink manufacturer is accused by a consumer advocate of providing less than the advertised 12 oz of soft drink per can. This is a very serious allegation, as the Department of Commerce considers this fraud, a criminal offense. If anyone at the soft drink manufacturer knowingly supplied less than the advertised amount, they are subject to criminal prosecution and potential incarceration. Suppose the commerce department samples cans off store shelves and finds an average of 11.8 oz per can. Does this prove that the company is cheating the public? Is it possible that the cans really do average 12 oz but by chance this particular sample of cans happened to be less than 12 oz? Of course, if the cans do average 12 oz, then we would expect half of the samples the commerce department might select to average less than 12 oz and half to average more than 12 oz. Hypothesis testing provides an answer as to whether we can conclude "beyond a reasonable doubt" that the company is producing cans that average less than 12 oz. It provides a formal means of testing our theories about the process.

As noted in previous chapters, the quality and quantity of data used to perform statistical inference are critical concerns. In Chapter 5 we discussed data quality and gave "rule of thumb" formulas for desired sample sizes. In this chapter we provide exact formulas to derive the sample size needed to obtain confidence intervals of a desired width. These formulas enable us to reduce the uncertainty in our inference applications to the level that we are willing to accept.

8.2 Examples of Statistical Inference Tools

We have already seen several examples of statistical inference tools. We formally test how confident we can be that individual coefficients in a regression model are not zero (a zero coefficient would imply that this x variable has no impact on the response, or y, variable). In addition, confidence intervals reflect the uncertainty in our estimates of these coefficients. In many cases, the main reason we developed the equation in the first place was to predict future observations, such as sales, stock price, interest rates, and so on. Of course, we would want to quantify the degree of uncertainty in these predictions via a prediction interval.

The data in Table 8.1 were taken from a consumer products company's marketing research efforts. The company was testing physical properties of various paper towel brands and asking customers to rate them on various attributes such as ability to absorb liquids. The purpose was to determine the degree to which internal measures, which could be done quickly and cheaply in a laboratory, would correlate with consumers' perceptions of the product. These data represent tests of water absorbency (in grams) for three brands. A preliminary question was whether these data indicated any real difference in water absorbency between brands.

Of course, we begin by questioning what specific process we are studying and how the data were collected to convince ourselves that there was no bias or other data quality issues. Remember that the process always provides the context for any data we analyze. Without a proper context, it is difficult to understand what the data are telling us. We shall assume for our purposes that there were no serious data quality issues.

Simple plots of the data revealed that Brand Y appeared worse than the others, but Brand X and Brand Z may be essentially the same. Figure 8.1 shows a raw data plot of these data, using dots for each data point. Because results for these two

TABLE 8.1 Water Absorbency Test (in grams)

Brand X	Brand Y	Brand Z
501	410	488
475	398	467
495	391	483
490	412	484
478	391	498
493	404	491
474	382	481
481	411	502
499	407	511
499	430	524

FIGURE 8.1   Dot Plots of Absorbency by Brand

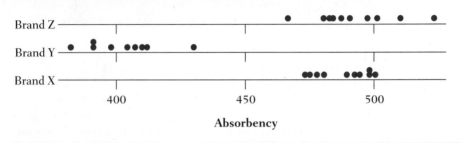

brands appear to be close, but perhaps not the same, some formal statistical inference may clarify the results. Clearly the averages of Brand X and Brand Z in this data set are different. The key question for statistical inference, however, is whether we can conclude from this sample data that the average of the entire population of Brand X is different from the average of the entire population of Brand Z.

In this case a 95% confidence interval for the difference between the average water absorbency of the two brands (in the entire population) would be appropriate. If calculated using the method shown in Appendix F, this interval is −17.5 to 8.7. This means that we are about 95% sure that, if we did a very large number of tests, the average difference in water absorbency between these brands would be between −17.5 and 8.7. A difference of −17.5 would mean that Brand Z is 17.5 higher, whereas 8.7 would mean Brand X is 8.7 higher. A negative number simply means that when we take the average of Brand X minus the average of Brand Z we get a negative number; that is, the average of Brand Z is greater. Notice that it is plausible that Brand X is higher, but it is also plausible that Brand Z is higher. Because zero is included in the interval, it is also plausible that the brands are actually the same. In this case the company decided that there was still too much uncertainty (the width of the confidence interval was too wide for their needs), and they collected more data (which will be discussed in more detail later in the chapter).

Another example of statistical inference involves a consulting firm. Retaining employees is a major issue in many organizations because of the significant costs involved in interviewing, hiring, and training new employees, not to mention the talent lost. In this particular case, the HR manager had noted in a staff meeting that he felt the organization was having trouble retaining female consultants. Other managers disagreed, stating that resignations appeared to be about the same for males as they had been for females. After a lengthy debate, someone suggested obtaining retention data for the past year (due to a major acquisition, previous data were not considered relevant), and these data are shown in Table 8.2. Note that this is all the data for employees this past year. The population of interest, however, would include retention in the future, hence it is appropriate to utilize statistical inference. Percentage resignations for females was 12/156 = .077, or 7.7%, and for males it was .031, or 3.1%. Some felt these data proved a difference, but others felt that it was just "noise" in the data and suggested that if additional data were collected for the following year the numbers might be reversed. After all, whenever

TABLE 8.2　Table of Resignations by Gender

	Male	*Female*	*Total*
Resigned	22	12	34
Remained	686	144	830
Total	708	156	864

you compare two numbers, there will always be a difference; that does not prove that female employees in general feel differently about this company than do males. How can we apply statistical inference to get a definitive answer?

There is a statistical hypothesis test to determine how confident we can be that there really is a difference in the *process*—that as a group female employees feel differently about staying with the company than males and will continue to resign more frequently in the future, assuming conditions stay the same. For discrete data (each employee either resigned or did not resign) the specific test is the χ^2 test (χ is the Greek letter "chi," pronounced "kigh," which rhymes with "high"). We therefore refer to this test as the chi-squared test. We can see how this test works using the data for those resigned versus those who did not resign, by gender (see Table 8.2).

Note that 34/864, or 3.9%, of all employees resigned. If our assumption is that males and females will resign at the same rate long term, we would expect this percentage of both males and females to resign in a given year. In this case this would be about 28 males and 6 females. We call these the "expected counts" in that they represent what we would expect theoretically based on our original assumption of no difference between males and females. Next we compare the actual data with the expected results.

Table 8.3 shows the chi-square analysis for these data, measuring the degree to which the actual results deviate from those we would theoretically expect, assuming no difference between males and females. Using the laws of probability, we can calculate the probability of getting this large a difference between expected and actual counts (quantified by the χ^2 value). This probability is called the *p*-value. In this case the *p*-value is .008, which means that less than 1% of the time we would observe a difference this large by chance if, in fact, males and females behave the same long term. The smaller this *p*-value is, the more difficult it is to

TABLE 8.3　χ^2 Analysis of Resignations

		Male	*Female*	*Total*
Resigned	Actual	22	12	34
	Expected	27.86	6.14	
Remained	Actual	686	144	830
	Expected	680.14	149.86	
Total		708	156	864

χ^2 = 7.109, df = 1, *p*-value = .008

What Does "df" Mean?   Most statistical software programs print out a statistic labeled df in statistical inference procedures, such as confidence intervals and hypothesis tests. This stands for degrees of freedom, which relates to sample size. Typically, the degrees of freedom are the sample size minus 1 or 2. In this case it is the number of rows minus 1 times the number of columns minus 1, which equals 1. The degrees of freedom are parameters of the statistical distributions used in statistical inference, such as the *t* distribution or the *F* distribution.

believe this result occurred by chance. If it did not occur by chance, then our original assumption must be false.

Generally, obtaining a *p*-value less than .05 is interpreted as having enough evidence to declare that our original assumption is false. This is not a magic number but just a rule of thumb. In practice our conclusions with a *p*-value of .051 and a *p*-value of .049 would be essentially the same—evidence that the assumption is false. In this case, the *p*-value of .008 means we have evidence that there is a real difference between males and females in this organization in terms of resignations. Note that this conclusion is not just referring to this data set but to the process in general, or to the entire population. We would be willing to predict that next year will again show females resigning at a greater pace if no action is taken. This is what we mean by inferring about the population or process from sample data. In this particular case the data and subsequent analysis did convince the organization that it had a problem retaining females, and it began an extensive study to identify the root causes, collecting additional variables, such as organizational level, position, years of service, and so on, to shed additional light on the root causes of resignations and the specific role of gender. In many situations we may perform a "high level" analysis to understand whether there are real differences, and then more detailed analyses to determine root causes.

8.3 The Process of Applying Statistical Inference

The examples presented in Section 8.2 were each unique but they all follow a similar pattern. The process of applying statistical inference, depicted in Figure 8.2, consists of four major steps:

1. Identify the specific process or population of interest.
2. Define the sampling frame, and collect sample data.
3. Calculate relevant sample statistics.
4. Draw appropriate inferences about the process using sample statistics in conjunction with subject matter knowledge.

It is very important to carefully define the specific population or process of interest. Is it all current customers, some important segment of customers, poten-

FIGURE 8.2   Statistical Inference Process

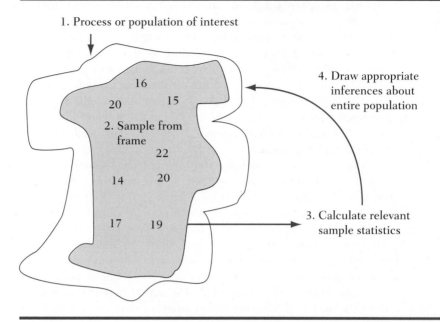

1. Process or population of interest

16
20   15
2. Sample from frame
22
14   20
17   19

4. Draw appropriate inferences about entire population

3. Calculate relevant sample statistics

tial customers, or some combination of the above? We also need to clarify the specific type of statistical inference we are interested in performing. For example, in hypothesis testing we need to specifically state the hypothesis we are testing. Once we have carefully articulated the hypothesis and population of interest, the next step is to sample from that specific population. We rarely have the opportunity to sample from exactly the population of interest due to practical constraints. For example, some people may refuse to answer our questions in a survey. In these cases we identify the "frame," or that subset of the population of interest, from which we actually sampled and are careful not to draw conclusions about segments of the population from which we could not sample. The critical issues of obtaining the right quality of data, as well as the right quantity of data, were discussed in Chapter 5.

In practice we often decide on the process or population of interest and even the type of inference desired after seeing the data. As noted in Chapter 2, data collected often result in more questions than answers. For example, why do we see such high sales growth in the East, but not in the West? Why does closing the books take so much longer for 5-week months than for 4-week months? Surprises in the data will lead us to new hypotheses that we may wish to formally test via inference, even though the data were not originally collected for this purpose. Carefully consider what specific process generated the data; this gives the data a context and helps identify any data quality issues.

Once we have the data, we can calculate the statistics of interest, such as average, standard deviation, percentage, or perhaps a regression coefficient. The calculated statistics provide what we need to perform confidence or prediction intervals, hypothesis tests, or any other formal statistical inference technique. We then perform some type of inference calculation to draw conclusions about the entire process or population of interest from a sample from the process—that is, to infer things about data that we did not sample. This often seems counterintuitive or impossible to students, but if done properly it can work quite well. For example, modern election polls in the United States typically sample only about 1000 voters, but these polls can make accurate predictions of election results for the entire country. The trick is to carefully quantify our uncertainty. In election polls, statements such as "This poll has an uncertainty of ±4%" are typically made.

A key conceptual question that needs to be answered at this point is how broadly the inference can be applied. To use the paper towels example, suppose the rolls of Brand Y were taken off a truck leaving the production facility in Buffalo, New York. Can we conclude that this facility makes inferior towels in general, or just today, or just on this one shipment? What about other facilities of the same company? Clearly, we can only draw conclusions about the population or process from which the data were actually sampled. We cannot draw any statistical conclusions about the other facilities, or even other trucks, because they were not in the frame from which we sampled. The actual frame from which we sampled is only this one truck, whereas the population of interest would most likely be all product produced by this company.

Practical considerations often prevent us from sampling from exactly the process of interest, and in these cases we must rely on subject matter knowledge rather than statistics to determine how broadly the conclusions can be inferred. In other words, use common sense and subject matter knowledge to decide if the difference between the frame and the process of interest is of practical importance. If it does not appear to be important, conclusions about the entire process or population of interest can be drawn, noting the limitations of our frame. Recall that the integration of data analysis with subject matter knowledge is one of the fundamental principles of statistical thinking.

Suppose a study is done to test a drug that may help prevent osteoporosis in elderly women. Because the pharmaceutical company that makes the drug is located in Florida, the sample of elderly female subjects in the study are all from Florida. Clearly the population of interest is all elderly women. But we have only sampled from elderly women in Florida; hence, this is our actual frame. If we consider the issue of nonresponse bias (the fact that randomly selected women may have declined to participate in the study), the frame would actually be even smaller. In considering the way the drug works researchers may conclude that there is nothing unique about women in Florida that would make the drug work differently with them. The logical decision at that point would be to publish the results and, if positive, claim that the drug does help prevent osteoporosis, noting the limitations in the sample. (Note that statistics does not allow us to make this conclusion, our subject matter knowledge does.)

What if the difference between the frame and the process or population of interest does appear to be important? For example, suppose we are doing a market research study of buyers' preferences in men's casual clothing. Because the company desired a study at minimal cost, the study was conducted via phone interviews of people in the New York City phone directory (those who were willing to respond) where the marketing research firm is located. The actual population of interest in this case may be all men who wear casual clothing in the United States, but the frame is only those in the New York City phone directory who were willing to respond. It is highly unlikely that people in New York City represent the preferences of people in more rural areas or in other regions of the country. It is very important to limit the inference to the actual frame itself or at least to that segment of the population that is similar to the frame. In this case, to be accurate, the market research firm would have to state that these results could be applied to men in the New York City phone book (ignoring the nonresponse bias for the time being). If the company were seriously interested in marketing outside this area, additional research would need to be done.

A lack of careful consideration of the actual frame versus the population or process of interest explains why so many published studies seem to contradict one another. For example, newspapers or news broadcasts may report a new study that indicates that some food or substance causes cancer. Very rarely is it actually mentioned that the study was conducted on rats that were given incredibly large doses of the substance in question. Later, another study concludes that this substance does not cause cancer. Again, only in the fine print do you read that this second study was an observational study done on humans, using levels of the substance normally found in typical usage. Neither study is wrong; the error occurs when the conclusions are inferred, or extrapolated, beyond the actual frame without careful thought.

We cannot sample from the future output of a process, but we generally want to infer about the future output. Therefore, the time dimension is particularly important when sampling from an ongoing process. If the process is reasonably stable, the process performance in the immediate future is likely to be similar to the present performance. However, the marketplace tends to be dynamic. In our sales versus advertising dollars study, for example, how do we know that the relationship between advertising and sales will remain stable in the future? Might new competitors, macro changes in the economy, or a gradual shift in consumer preferences significantly change the relationship? In all likelihood they will. We must therefore also consider the time dimension when determining frame and process of interest. If we are interested in next year's marketplace and are using a study conducted last year, we need to carefully consider this difference between the actual frame and the process of interest. As before, the practical importance of this difference must be determined with subject matter knowledge rather than statistics.

If the degree of uncertainty, measured by the width of the confidence interval, is too great for our needs, additional data must be obtained. This was the case in the paper towel example discussed previously. In general, the more data we have, the less the uncertainty, and therefore the narrower the confidence interval. Obviously, if we sampled the entire population, there would be no uncertainty in our estimates of average, standard deviation, and so forth—we would know the exact value. As we shall see, the width of the confidence interval is a function of sample size, with larger sample sizes producing narrower confidence intervals. As with statistical thinking in general, statistical inference is an iterative process. The results of our first analysis may suggest additional data are needed or other analyses to perform.

8.4 Statistical Confidence and Prediction Intervals

Confidence intervals quantify our uncertainty in sample estimates of "true" population values. For example, we may wish to know how long it takes to process loan applications. If determining the actual time has to be done manually, we may not wish to sample every single application. Suppose we sample 30 applications and calculate an average time of 4.3 hours. We know that the average processing time of all applications will not be exactly 4.3 hours, but how far might our sample estimate be from the "true" average time (the average time we would get if we did 100% sampling)? This uncertainty is quantified by a confidence interval. The "true" process or population average is generally referred to as μ (the Greek letter "mu").

A prediction interval is similar in that it quantifies the uncertainty in our sample estimates. It is different in that it quantifies uncertainty in predictions of a future observation. In the loan application example, we might wish to quantify the uncertainty in processing time of the next application we process. Note that in this case we are not concerned with the true average time but rather with one particular observation. In addition to the uncertainty associated with our sample estimate of the true average, we also have the additional uncertainty of this particular loan application. It could be an unusually long processing time or an unusually short processing time. Prediction intervals are generally wider than confidence intervals. A prediction interval can be calculated for the sample average of some future observations. Although this involves an average, it is still a prediction inter-

Greek Versus Latin Letters   Most statistics, such as the average, are sometimes written as Latin letters, such as $\bar{x}$, and sometimes as Greek letters, such as μ. Whenever we are referring to a statistic actually calculated from sample data, we use Latin letters, such as $\bar{x}$ or s. When we are referring to the "true" or long-term process or population values, such as the value we would get if we could do 100% sampling, we use Greek letters. This is standard statistical notation and avoids confusion over the two different (sample versus population) averages, standard deviations, proportions, and so on. Here are some examples:

	Sample	*Population or Process*
Average	$\bar{x}$	μ
Standard deviation	s	σ
Proportion	p	π

val because we are only interested in that one calculated average from one future sample, not the overall true average.

We can calculate confidence or prediction intervals for virtually any variable of interest, such as averages, variation, the difference between two averages, regression coefficients, proportion defective, and so on. Most statistical software programs will calculate these intervals, but let's review the more common intervals to illustrate how these intervals should be interpreted.

Confidence Interval for the Average

The formula for the confidence interval for a population average (μ) is illustrative of what influences our uncertainty. This formula for a 95% confidence interval for μ, the "true" process/population average, is:

$$\bar{x} \pm \frac{ts}{\sqrt{n}}$$

where $\bar{x}$ is the sample average
s is our sample estimate of σ, the population standard deviation
n is the sample size
t is the appropriate value from the t distribution for this sample size (see Chapter 9) and relates to 95% confidence. For large samples, say above 30, this will be approximately 1.96. (The theoretical justification of why 1.96 relates to 95% confidence will also be explained in Chapter 9.)

If we calculate this for the average of Brand X towels using the data in Table 8.1, we get:

$$488.5 \pm (2.26 \times 10.54)/\sqrt{10} = 488.5 \pm 7.5, \text{ or } 481.0 \text{ to } 496.0$$

We obtain the t-value of 2.26 from the table located in Appendix I, using a sample size of 10. (Statistical software such as Minitab or JMP will calculate these intervals for us, so there is rarely a need to use the printed tables.) To understand what this interval means, we have generated 20 random samples of size 10 from a normal distribution with an average of 488.5 and a standard deviation of 10.54. We have then calculated a 95% confidence interval for the average for each of the 20 samples. These are illustrated in Figure 8.3. Note that the true population average for each sample was 488.5. Because this is a 95% confidence interval, the true value of 488.5 should be within the confidence interval about $.95 \times 20 = 19$ times out of 20. Only sample 1 produced a confidence interval that did not include the population value of 488.5. If we repeat the experiment, we would get slightly different results; that is, we will not always get exactly 19 out of 20 capturing 488.5. The key point is that the population average will be included within the 95% confidence interval about 95% of the time—but not 100% of the time.

Theoretical Assumptions We assume that the process has a normal (Gaussian) distribution. This is not an important assumption because averages will tend to be normally distributed, even when the individual data points are not (see Chapter 9). If we know the actual population standard deviation, σ, we would

FIGURE 8.3 Twenty Confidence Intervals for the Average (population average = 488.5)

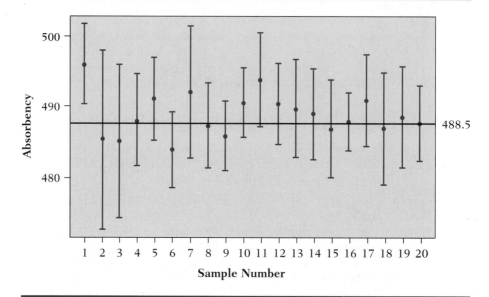

General Structure of Confidence Intervals   Almost all confidence intervals (with the exception of the confidence interval for a standard deviation) follow this general structure:

Estimate ± constant × standard deviation of estimate

The constant corresponds to the confidence level, with 1.96 being the appropriate large-sample value for 95% confidence. For smaller sample sizes we typically replace 1.96 with the appropriate t-value, which takes into account the uncertainty in estimating the population standard deviation. Other constants are used for other confidence levels. The standard deviation of the estimate (sometimes called the standard error) for a sample average is $s/\sqrt{n}$. The reason that the sample average (the "estimate" in this case) has a standard deviation of $s/\sqrt{n}$ will be explained in Chapter 9. Other estimates, such as a regression coefficient or proportion, will have different formulas for their standard deviation. Fortunately, statistical software such as Minitab or JMP will handle these calculations for us.

---

replace s with σ and use the constant of 1.96 rather than the t-value. This is because the t distribution documents our uncertainty in using s to estimate σ. Hence, if we know σ, we do not have this additional uncertainty. The t distribution will also be explained in greater detail in Chapter 9, and a t table is included in Appendix I. Perhaps the most important assumption is that the process from which the data have been collected is stable, both in terms of the average and the variation. This is a critical assumption. If the process is not stable, there is no such thing as a true average; the average today may be different from that of yesterday or tomorrow.

As the sample size goes up, the width of the interval (our uncertainty about μ) goes down. This is a critical point; the more data we have, the narrower the uncertainty, all other things being equal. Note also that the uncertainty, as measured by the width of the confidence interval, is not proportional to n but rather to $\sqrt{n}$. In other words, uncertainty decreases with the square root of sample size. We therefore need four times as much data to reduce the uncertainty in half.

Another key point is that the width of the confidence interval is dependent on s, our estimate of σ, the population/process standard deviation. The more variation there is in the process, the more uncertainty as to what the true population average is. If there were no variation in the process, each data point would equal the average. As previously discussed, variation clouds our view of the world.

Prediction Interval for One Observation

Suppose we wish to predict one observation in addition to the average discussed previously. What is the 95% prediction interval for this one value? The formula is:

$$\bar{x} \pm t\sqrt{s^2 + \frac{s^2}{n}}$$

where $\sqrt{}$ means the square root

$\bar{x}$ is the sample average

s is our estimate of σ, the population standard deviation (σ^2 is called the variance and is estimated by s^2, the sample variance)

n is the sample size

t is the appropriate value from the t distribution corresponding to 95% confidence and this sample size. For large samples (about 30 or more) this will be approximately 1.96.

If we select one more roll of Brand X paper towels and test water absorbency, we can be 95% sure that this one value will be between:

$$488.5 \pm 2.26\sqrt{10.54^2 + \frac{10.54^2}{10}} = 488.5 \pm 25, \text{ or } 463.5 \text{ to } 513.5$$

Note that this is wider than the confidence interval for the population average, which was 481 to 496. This is because the prediction interval includes not only the uncertainty associated with the population average but also the uncertainty associated with how far the additional roll of Brand X towels to be tested may vary from the population average. The s^2/n part of the equation relates to uncertainty in the sample average (how far it may deviate from the population average), and the first s^2 in the equation relates to the uncertainty in the additional roll of towels. We add the two variances (standard deviations squared) for the two uncertainties to get overall uncertainty, take the square root to get overall standard deviation, and then multiply by the appropriate *t*-value for 95% confidence. This illustrates a principle that will be explained in detail in Chapter 9 that variances generally add, but standard deviations do not.

Theoretical Assumptions   As with confidence intervals, we assume that the process has a normal (Gaussian) distribution. This is now an important assumption because we are not drawing conclusions about the average but about individual observations. If the underlying distribution was not approximately normal, we would need to use a formula specific to that distribution. If we have a large sample (>30) or know the population standard deviation σ, we replace s^2 with σ^2 and use 1.96 instead of the *t*-value. As noted previously, the *t* distribution takes into account uncertainty associated with estimating σ from *s*. If we know σ, or have a large enough sample size to estimate it accurately, we no longer need the *t* distribution. Perhaps the most important assumption is that the process from which the data have been collected is stable, both in terms of the average and the variation. This is a critical assumption. If the process is not stable, the next observation may come from a process very different from the one in which the original data were taken, resulting in erroneous predictions.

The prediction interval is similar to the confidence interval for the average and accounts for two sources of uncertainty. It accounts for the uncertainty in the location of the true process average μ (using the confidence interval formula) and considers uncertainty in what the next value will be, assuming we know the true average. Note that we add the two variances (the standard deviations squared) to get the total variance and then take the square root to convert this to a standard deviation. It is often the case in statistics that variances add, whereas standard deviations do not. (This phenomenon will be explained in Chapter 9.)

Confidence Interval for the Proportion

With discrete data, such as the proportion of females who resigned this year, we can also calculate the uncertainty in the overall population proportion. For example, in the resignation data presented previously, 12 out of 156 females resigned in a particular year, which is a proportion of .077. If we use this to estimate the overall process proportion (π) of female resignations (the proportion we might expect to observe over a long, stable time period), how far off might we be? In other words, how much uncertainty is there in this estimate? The specific formula used, based on a normal approximation to the binomial distribution, is given in Appendix F.

Theoretical Assumptions   This interval uses the normal approximation to the binomial distribution (see Chapter 9) and tends to work well with a reasonably large sample size and a reasonably large proportion. Specifically, $n \times p$ (sample size times proportion) and $n(1 - p)$ (sample size times 1 minus the proportion) should both be greater than 5. In the female resignation data, this would mean that we must have at least 5 females who resigned as well as at least 5 who did not. With a small sample size we can obtain an appropriate confidence interval using the binomial distribution directly. This is a little more work, but it can be done with modern statistical software such as Minitab or JMP. We are also assuming that all samples came from the same conceptual population or process. In other words, we are assuming that each employee has a similar probability of resigning. We would not want to combine positions that are known to be short assignments with ones that are considered permanent; these should be evaluated separately. As usual, we are assuming that the process was stable during the period the data were collected.

If we apply this formula to the female resignation data, we get a 95% confidence interval for the long-term annual female resignation proportion (π) of about .04 to .12. That is, over a very long time period the current process should produce an average annual female resignation rate anywhere from 4% to 12%. If this interval is wider than we would like, we can obtain additional data to reduce the width (reduce the uncertainty).

Confidence Interval for the Standard Deviation

Once we have estimated the true process standard deviation σ from a sample of data, we can quantify the uncertainty in this estimate by calculating a confidence interval. The formula for this confidence interval is not particularly illustrative, so we do not review it here. It can be found in Appendix F. Suppose we have a sample of 15 data points and observe a sample standard deviation, s, of 1.02. Using the formula in Appendix F, the 95% confidence interval for the true population standard deviation, σ, is .71 to 1.71. The upper and lower limits on the confidence interval are not symmetric about the sample standard deviation of 1.02 because the theoretical distribution of the standard deviation is skewed; that is, it is not symmetric about its own average. It is also true for the standard deviation, however, that the larger the sample size, the narrower the limits on the confidence interval.

How should we interpret this interval? If we sampled a very large number of observations from this process, we can be confident (about 95% confident) that the standard deviation we would observe would be somewhere between .71 and 1.71. It is likely to be close to 1.02, but it could be as low as .71 or as high as 1.71. It is very unlikely (but not impossible) that the true process standard deviation is outside these limits. If we calculate this confidence interval for the standard deviation of Brand X towel absorbency, we get a sample standard deviation of 10.54 and a 95% confidence interval for the long-term process standard deviation of about 6.7 to 22.4. This is a wider interval than for the average and illustrates the statistical principle that it is generally easier to estimate the average level than the variation.

To illustrate these points, we have taken the 20 samples randomly generated from a distribution with an average of 488.5 and standard deviation of 10.54 and calculated confidence intervals for the standard deviation. These are the same data for which we previously calculated confidence intervals for the average. Again, we would expect about 95%, or 19 of the 20 intervals, to include the population

FIGURE 8.4   Twenty Confidence Intervals for the Standard Deviation
(population standard deviation = 10.54)

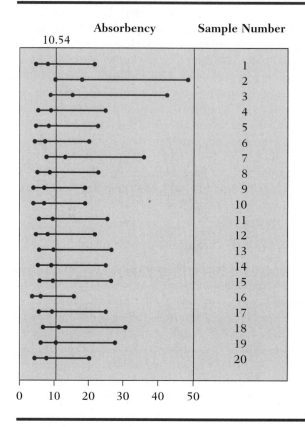

standard deviation of 10.54. These confidence intervals for the standard deviation are given in Figure 8.4. Note that in this case all 20 intervals include the population value of 10.54. We can also see that these intervals are not symmetric. In addition, by comparing Figures 8.3 and 8.4, we can see that for those samples with small sample standard deviations the width of the interval for the average is small (e.g., Sample 1). Similarly, when the sample standard deviation is large, the confidence interval for the average is wide (e.g., Sample 2). This is because the sample standard deviation is used in the calculation of the confidence interval for the average when we have to estimate σ from the sample data.

Theoretical Assumptions   The standard calculations that are done for this confidence interval assume a normal distribution, and in this case the assumption is reasonably important. However, it is also possible to calculate distribution-free or nonparametric confidence intervals, which do not make any distributional assumptions. Such intervals are not typically used unless needed because they tend to be wider. As with all the intervals, this interval assumes a stable process. This interval is appropriate for any sample size greater than 1, but with small samples it will tend to be very wide. Sample sizes of 30 or more are recommended. Recall that in general it is more difficult to estimate the variation than it is to estimate the average.

Confidence Interval for a Regression Coefficient

One aspect of regression analysis is estimating the impact of an x variable on a y variable, such as the impact of advertising on sales, the impact of price on market share, or the impact of call volume on waiting at a customer help center. In such cases we are often interested in estimating the specific coefficients in the model. That is, how much market share do I lose for every dollar increase in price? The regression model provides a best estimate of these values, but there is some uncertainty in these estimates. They will not be *exactly* correct (that's why we call them estimates), but a confidence interval for the coefficient will tell us how far from the true relationship we might be.

The formula for this 95% confidence interval follows the general format for confidence intervals presented previously:

$$b \pm t\text{SD}(b)$$

where b is the estimated regression coefficient (from the computer output), $\text{SD}(b)$ is the standard deviation of the estimated regression coefficient b (this is a measure of the uncertainty in our estimate of b and is provided in typical software used for regression analysis), and t is the appropriate value from the t distribution for 95% confidence.

Most software packages that do regression analysis will print out a table that lists each regression coefficient in the model as well as its standard deviation (sometimes referred to as the standard error). For large sample sizes the t distribution value for 95% confidence approaches 1.96, hence the confidence interval can then be estimated by adding plus or minus 1.96 standard deviations of the coefficient. This documents the uncertainty we have in estimating the true relationship. If this degree of uncertainty is unacceptable, we can obtain more data, which in general will reduce the width of the confidence interval. For example, Table 6.8 showed that, for the regression model on abandon rate at a call center, the variable "time to answer" had the following coefficient and coefficient standard deviation: $b = .86$, $\text{SD}(b) = .13$. Therefore, a 95% confidence interval for the true process coefficient (the coefficient we would get if we had all data that could ever be available) would be approximately: $.86 \pm 1.96 \times .13 = .86 \pm .25$, or .61 to 1.11. We can therefore be confident that, as time to answer goes up by 1, abandon rate goes up by between about .6 to 1.1.

Theoretical Assumptions   This formula also assumes a normal distribution, but this assumption is not critical because we are forming a confidence interval for the regression coefficient rather than for an individual value. A more critical assumption is that the form of the regression model we are using is correct. For example, if we are using a linear regression equation for data that displays curvature, we have violated this assumption. The diagnostic procedures outlined for evaluating regression models reviewed in Chapter 6 should be used to test the validity of this assumption. In addition, we are assuming that the process remains stable during the time period when data were collected other than the changes due to the *x* variables in the regression equation. As with several other intervals, if we have a small sample size, say, less than 30, we would use the appropriate *t*-value instead of 1.96 (see the table in Appendix I). (The actual sample size statistic that is important here is the sample size minus the number of parameters estimated in the regression model—a statistic known as residual degrees of freedom or degrees of freedom for error. This should be about 30 or more to use the constant value of 1.96.)

Prediction Interval for Future *y* Values Using a Regression Equation

It is also possible to develop a prediction interval for future values using a regression equation. For example, suppose we are using a regression equation to predict market share from selling price. The equation is:

$$ms = 1.0 - .01 \times \text{price (in dollars)}$$

This equation states that our market share will increase by .01 (where 1.0 is 100% of the market) for every $1 we decrease our selling price. We may therefore decide to lower our price by $5 with the intent of increasing market share by .05. But we will not get *exactly* a .05 increase in market share, and we may wish to develop a prediction interval for how much higher our market share will be next month. In this case we are concerned about the uncertainty associated with *one* future observation; hence, this is a prediction interval. To quantify the uncertainty in the *average* market share we will observe in the future at this price, we would construct a confidence interval. (Standard software packages for regression analysis provide these intervals.)

Suppose the 95% prediction interval for the increase in next month's sales was .01 to .09. That is, sales could increase next month by as low as .01 or as high as .09. This uncertainty is due to both the uncertainty in the regression equation and the normal variation in sales from month to month. If we calculated the confidence interval for the long-term average increase in monthly sales resulting from the price decrease, we would obtain a narrower interval because the only uncertainty would be in the regression equation. The normal month-to-month variation in sales would not be a factor in the long-term average. The confidence interval for the long-term average increase in sales might be .04 to .06, meaning we are confident that over the long term we will see an average increase in sales of close to .05 (no more than .06, no less than .04).

Theoretical Assumptions   The assumptions are similar to those for the confidence interval for a regression coefficient, but the issue of stability is even more important. For example, if sales are growing or declining due to issues other than price, these issues need to be included in the regression model or its predictions will be inaccurate. It is probably the exception rather than the rule for the relationship between price and market share to stay constant over long periods of time due to inflation, new competitors, evolving technology, and so on. For these reasons, it is risky to make financial or economic predictions far into the future.

Confidence Interval for the Difference Between Two Averages

In many cases the difference between two population averages rather than the averages themselves is of greatest importance. For example, in comparing the cycle time to approve mortgages between two branches of a bank we could develop two confidence intervals, one for each branch, and compare them. This approach results in two probability statements: we are 95% confident in the interval for Branch 1 and 95% confident in the interval for Branch 2. Unfortunately, the probability that both confidence intervals will be correct is only $.95 \times .95 = .903$. Therefore, it is a little more complicated to make conclusive statements.

A more logical approach is to define the variable of interest to be the difference between the two population average cycle times. For example, we might define the difference as the average of Branch 1 minus the average of Branch 2. From a practical point of view, we are concerned with whether the average difference is zero (no difference in the branches), positive (Branch 1 takes longer), or negative (Branch 2 takes longer). Of course, even if the branches are, in fact, identical, the value of the difference we observe from sample data will not be *exactly* zero due to random variation. The real question of interest is whether we can conclude from the sample data that there is a difference between the two branches' processes beyond this particular data set (i.e., if μ_1 is equal to μ_2). In other words, if we gathered additional data tomorrow, are we confident that we would observe the same difference we saw with this data set?

We can accomplish this objective by calculating a confidence interval for the difference in the population/process averages $(\mu_1 - \mu_2)$. The formula for this confidence interval depends on assumptions we make about the variation and is given in Appendix F. Suppose we calculate this interval at a 95% confidence level, and the interval turns out to be –2.1 to 4.3 days. That is, the average cycle time for Branch 1 could be as much as 4.3 days longer than that for Branch 2, or as much as 2.1 days *less* than Branch 2. It is also plausible that the true average difference is zero, because zero is included in the interval. We can pin down how large the difference might be, but we cannot conclude with certainty which, if either, of the branches takes longer.

Theoretical Assumptions   We assume that both processes have a normal (Gaussian) distribution. This, however, is not an important assumption because we are not evaluating single values. Typically, one assumes that the two variances are equal, but this is not mandatory with the appropriate software, such as Minitab or JMP. Perhaps the most important assumption is that the processes from which the data have been collected are stable in terms of both the average and the variation. This is a critical assumption. If the processes are not stable, there is no such thing as a true average. The average today may be different from yesterday or tomorrow.

As is generally the case, if we are dissatisfied with the width of the confidence interval, we can obtain additional data. As the number of data points increases, the width of the interval decreases. For example, if we obtained additional data on mortgage approval times and recalculated the confidence interval for the difference, it might now be .5 to 1.5 days. With these data we can conclude with confidence that Branch 1 takes longer than Branch 2 on average to approve mortgages. We estimate the average difference to be about 1 day, but it could be as little as .5 day, or as much as 1.5 days. Other examples where the difference between two averages would be of particular interest are comparing the average return of two different investment mutual funds or comparing the number of customer complaints received before and after a redesign of credit card statements mailed to customers.

Confidence Interval for the Difference Between Two Proportions

As with averages, we may be more interested in the difference between two proportions than we are with the individual proportions themselves. For example, in the resignation data we might be most interested in the difference between proportions of males and females resigning. This is done in an analogous way to averages. We create a new variable, difference, which is the difference between the two sample proportions. Even if the population proportions are the same, the sample data will rarely, if ever, be exactly the same. There will almost always be some difference in the sample data due to variation. The key question we want to answer is how different might the population proportions be? We again use the normal approximation to the binomial distribution (see Chapter 9 for details).

If we apply this formula to the resignation data, we get a 95% confidence interval for the long-term (population) difference between annual female and male resignation proportions $(\pi_1 - \pi_2)$ of about −0.09 to −0.002. In other words, it is plausible that the difference between male and female resignation rate (in this case, male proportion minus female proportion) over a very long time period is between −.09 and −.002. Note that zero is not included in this interval; hence, zero is not a plausible value. In other words, we are convinced beyond a reasonable doubt (using 95% confidence) that there really is a difference between male and female resignation rates. This is the same conclusion we reached through the formal hypothesis test discussed previously. This difference could be as large as 9% or as low as .2%. Our best estimate is about 4.6%, which is the difference between the female (.077) and the male (.031) resignation rates in the sample. Of course, this assumes that the process is stable during this long time period. If this interval is wider than we would like, we can obtain additional data to reduce the uncertainty.

Theoretical Assumptions This interval uses the normal approximation to the binomial distribution. The assumptions are similar to the assumptions used for confidence intervals for individual proportions. We again need reasonably large sample sizes—$n \times p$ (sample size times proportion) and $n(1 - p)$ (sample size times 1 minus the proportion) should both be greater than 5, for both proportions. If we have a small sample size, we can still obtain an appropriate confidence interval using the binomial distribution directly. This is a little more work but can be done with modern statistical software. We are also assuming that all samples came from the same conceptual population or process. In other words, we are assuming that each employee has a similar probability of resigning. We cannot combine short-term positions with permanent positions; these should be evaluated separately. As usual, we assume that the process was stable during the period the data were collected.

8.5 Statistical Hypothesis Tests

When comparing cycle times to approve mortgages, zero was a critical number for the confidence interval. If zero was included in the confidence interval, we could not conclude with any confidence that the branches were different relative to cycle times. However, if zero was not included in the interval, we could be confident that there really was a general difference between the branches. In some cases we are more interested in making a yes-no statement than in estimating the specific size of the difference. This is particularly true if we are comparing several things, such as 7 branches of the same bank. We could make 21 difference comparisons of one branch versus another, for example, but it would be preferable to simply ask if we can conclude with confidence that there is a real difference between the 7 branches. When we are only asking whether we can conclude there is a difference, we generally use statistical hypothesis tests as opposed to confidence intervals. When comparing only two things, however, calculating the confidence interval provides the limits of our uncertainty as well as determining whether there is a real difference.

Recall also that we are looking at sample data to draw conclusions about the entire population or process. When comparing data from two branches, the sample data will always differ due to variation in the process. The key question, however, is whether this proves that the branches themselves are really different. We need the tools of statistical inference to deal with the variation in these data.

The Hypothesis Testing Process

To explain the logic used in hypothesis testing, let's consider the hypothetical example of a retail exercise equipment business. This business sells exercise equipment, such as stair-steppers and treadmills, to the general public via the Internet, through a physical catalog, and in a large store with displays of all the equipment. The new marketing manager is dissatisfied with the number of service contracts the business sells on major pieces of exercise equipment. These contracts typically have higher profit margins than the equipment itself, and the marketing manager claims that if the prices of these contracts are reduced even slightly, say by 5%, she could increase the percentage of customers who buy them from the current 50% to perhaps as high as 75%. The sales manager does not believe this but agrees to allow the marketing manager to decrease the price of major equipment service contracts on equipment sold in the store (not in the catalog or on Internet sales) by 5% for one week as a special promotion. The business typically sells about 20 major pieces of equipment per day in the store, so they would expect about 100 sales during the week.

At the end of the week, after 100 sales, 50 service contracts have been sold. What would you conclude? There is no evidence that the rate of service contracts sold had increased from the historical 50%. But suppose 100 service contracts had been sold that week. Would you be convinced? Most people would be convinced now because the chances of selling 100 service contracts in 100 sales (assuming each sale had only a 50–50 chance of selling a service contract) is very, very small. The exact probability is .5 to the 100th power. The probability of this happening by chance is so unlikely that we conclude that the rate of selling service contracts at the reduced pricing must not be 50–50.

It is easy to draw conclusions using common sense when the results are so dramatic, but suppose we sold 60 service contracts? 70? 80? 90? At what point would you say you have convincing evidence that the rate of selling service contracts is now more than 50%? Obviously, there is variation in service contract sales, so about 50% of the time we would expect to sell more than 50 contracts and about 50% of the time less than 50 in 100 sales. Therefore, selling more than 50 service contracts is not necessarily convincing evidence. We have posed this or a similar question to numerous students over the years, and although responses vary considerably (anywhere from 51 to 100), students typically answer that they have convincing evidence if we sold more than about 70 or 75 service contracts in 100 sales. This is a common sense, or gut feeling, answer, but we can use statistical thinking, along with the laws of probability, to develop a more scientific answer.

We could consider selling 100 pieces of major exercise equipment as an experiment, with the number of service contracts sold shown as one data point. Alternatively, we could consider each sale to be an experiment, in which case we would observe 100 contract or no contract data points. If we make the assumption that the rate of selling service contracts is 50% (i.e., that we have a 50–50 chance of selling a contract for each sale), then we can develop a 95% prediction interval for the number of contracts we will sell using the laws of probability. In this case we can use the binomial distribution that will be more formally introduced in Chapter 9. This prediction interval turns out to be approximately 40 to 60; if the rate of selling service contracts is still 50%, we are approximately 95% sure we will sell between 40 and 60 contracts in 100 equipment sales. Therefore, if we observe any more than 60 contracts sold, we can say that we have convincing evidence that the rate of selling service contracts is not 50–50 but greater. Alternatively, we could calculate a 95% confidence interval for the proportion of contracts sold and see if 50% was included; this would produce the same answer.

This prediction interval is considerably tighter (narrower) than most people would anticipate using common sense. This is why we use statistical methods and probability to formally test hypotheses.

What was the hypothesis being tested here, and what was the process used to test it? When students are asked at what point they would be convinced that the store could really sell service contracts more than 50% of the time, they mentally go through an informal hypothesis testing process. If pressed to describe their thought processes, they will often state that they began with the original hypothesis to be tested; in this case, the assumption that the rate of selling contracts is 50–50. They then identify an experiment that will provide data or information that could be used to evaluate the hypothesis. (If no pieces of equipment were ever sold, we could debate forever whether the reduction in price resulted in more contract sales without resolution.) When the data obtained have been evaluated, they determine whether or not it was consistent with the original hypothesis.

If 50 contracts were sold in 100 equipment sales, this would be consistent with the assumption of a 50–50 chance of selling a contract. However, if 100 contracts were sold in 100 equipment sales, this would be inconsistent with that assumption, and we would be convinced that the rate of selling service contracts had increased from 50%. The major problem with informal hypothesis testing is that we have no scientific basis on which to make a decision. Is 60 contracts enough evidence? Or 70? Or 80? With statistical hypothesis testing, we will use the laws of probability to decide whether or not to reject the original hypothesis.

Another important point to be emphasized here is that we cannot prove that the original hypothesis is true. That is, even if we sell 50 contracts, that would not prove that the rate of service contract sales was exactly 50% over the long term. It is possible that the rate of contract sales is now 52% over the long term, but due to variation in the process we happened to sell exactly 50 contracts in this experiment. We can prove the hypothesis false, however, at least beyond a reasonable doubt. For example, based on the 95% prediction interval we identified, if we sell more than 60 contracts (or less than 40), we could say that we are confident that the rate of contract sales is no longer 50%. Only 5% of the time would we observe a result outside these values if the rate is 50%.

To summarize, the steps in the hypothesis testing process are as follows: specify the hypothesis; obtain data relevant to the hypothesis; evaluate the degree to which the data are consistent with the hypothesis; if inconsistent, reject the hypothesis (we have evidence it is not true); if consistent, do not reject the hypothesis. These are essentially the steps of the scientific method and are illustrated in Figure 8.5. The only real difference is that we use the laws of probability to decide if the data are consistent with the hypothesis or not.

In this case, the original hypothesis is that the rate of service contract sales is still 50% over the long term. We obtain data by observing 100 sales, or attempts to sell a service contract. We would then compare the observed number of contract sales with the expected range of 40–60, assuming the hypothesis is true. If the observed number of contract sales are outside this range, we have convincing evidence (at the 95% confidence level) that the contract sales rate is not 50–50. On the other hand, if the observed number of contract sales is between 40 and 60, this is consistent with what we expected, hence we have no reason to doubt the validity of the original hypothesis.

FIGURE 8.5

The Hypothesis Testing Process

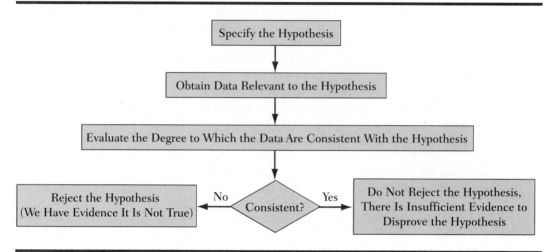

Connection to Confidence Intervals

We can perform a standard hypothesis test by calculating a confidence interval; if the hypothesized value is within the interval, we cannot reject the hypothesis. If it is outside the interval, we reject the hypothesis. For example, if we sell 100 pieces of equipment and 51 contracts, the 95% confidence interval for the long-term proportion of contract sales would be about 41 to 61. This includes 50; hence, 50% contract sales is a plausible value. In other words, we have no evidence to disprove the hypothesis that the long-term proportion of contract sales is 50%. If we performed a statistical test on the hypothesis that the long-term proportion is 50%, we would fail to reject this hypothesis. This would be true for any value that is included in the 95% confidence interval from 41 to 61. If we test the hypothesis that the long-term proportion of contract sales is .65 (at the 95% confidence level), we would reject this hypothesis because it is outside the confidence interval.

Note that there are many values that we could not reject in this case—further evidence that failure to reject a hypothesis does not prove it to be true. This is illustrated in Figure 8.6, which shows two scenarios: one with a confidence interval for the proportion that includes .50, and one that does not. In the first case .50 is not included, therefore it is not a plausible value. We have evidence that the long-term proportion is not .50. Any value within the confidence interval is plausible, but not .50. In the second case, .50 is one of many plausible values, and we have no evidence to conclude that the long-term proportion is not .50. These intervals illustrate why we can never prove the null hypothesis. The true long-term value could be any of the infinite number of values included in the confidence interval.

FIGURE 8.6 Two Confidence Intervals for the Proportion

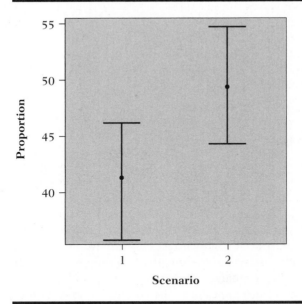

Stating the Hypotheses

Some common terminology and standards have been developed and accepted for hypothesis testing. When formally stating the hypotheses, we will always have two: the null hypothesis and the alternative hypothesis. The *null hypothesis* is so named because in pharmaceutical and agricultural research it typically states that there is no effect of a drug or fertilizer. The null hypothesis will always carry the equals sign, either literally or figuratively. In the service contract case discussed previously, the null hypothesis is that the proportion of sales resulting in a service contract sale is .50, or in mathematical terms, $\pi = .5$, where π represents the long-term probability of selling a contract. The *alternative hypothesis* is the alternative of the null, in this case, $\pi \neq .5$. Note that the alternative includes the possibilities that the percentage of service contracts sold may be greater than 50 as well as less than 50. This is called a two-sided alternative because there are two possible ways it could be true. Performing this type of test is analogous to calculating a confidence interval and determining if the null hypothesis value is in the interval.

In many cases we are trying to prove the alternative hypothesis, such as in testing a drug (the drug company always wants the drug to work). This makes sense because we can prove the null hypothesis false—hence, the alternative is true—but we can never prove that the null hypothesis is true.

An example of a one-sided alternative would be when the commerce department investigates claims that a soft drink manufacturer is deceiving the public. For example, if they claim that their cans contain 12 ounces, it is legally considered fraud to provide less than 12 oz in the cans. The government does not care if you provide more than 12 oz, however; this is perfectly legal. Therefore, if the commerce department is investigating claims of consumer fraud in such a case, their null hypothesis will be that the average amount of soft drink in the cans is 12.0 oz *or more* (μ 12.0). The alternative hypothesis would then be that the average is less than 12.0 oz. This is called a *one-sided alternative*. If the null hypothesis is rejected, they have proven beyond a reasonable doubt that the company is cheating the public and is guilty of fraud.

Obtaining the Data

Theoretically, data required to test the hypothesis should be obtained after stating the hypothesis. In many cases, however, we do not decide to formally test a hypothesis until after the data are collected. In such cases, analysis of the data is what led us to consider this hypothesis. As with other statistical procedures, the quality and quantity of the data are both important. Data quality issues can lead to bias in the data, resulting in incorrect conclusions. For example, faulty survey methodology could lead us to conclude that a certain political candidate will win an election when that candidate ultimately turns out to lose the election.

Data quantity is also important. The confidence interval for an average is a function of sample size; the larger the sample size, the narrower the interval. That is, uncertainty decreases as data quantity increases. This is generally true of all the confidence intervals, not just the average. Therefore, the more data, the more precisely we can estimate the average, assuming we do not have data quality issues. Recall also that if we perform a two-sided hypothesis test, any null hypothesis value that is included within the interval will result in a "cannot reject null hypothesis" conclusion, whereas any value outside the interval will result in a "reject null hypothesis" conclusion.

Evaluating the Consistency Between the Data and the Null Hypothesis

Formal probability techniques are used to examine the data to see whether it supports the null hypothesis or offers evidence that it is in fact false. Typically, we calculate how probable it would be to observe results this unusual if the null hypothesis is true. For example, suppose we sell 100 service contracts in 100 equipment sales. The probability of "observing results this unusual" is called the *p*-value, for probability value. This result is so unlikely (assuming the null hypothesis is true) that we conclude that the null hypothesis must be false and accept the alternative hypothesis. Typically, we reject the null hypothesis if the *p*-value is less than .05, which in the case of a two-sided alternative is equivalent to saying the null hypothesis value is outside of the 95% confidence interval. Recall that values outside the confidence interval are not plausible at the 95% confidence level. Increasing the confidence level to 99% would result in a wider confidence interval with fewer values outside of it. Those values outside of the 99% confidence interval would have *p*-values of less than .01. The lower the *p*-value, the more unlikely the result, assuming the null hypothesis is true.

To produce the appropriate *p*-value, we generally calculate a test statistic and compare it to its theoretical distribution, assuming the null hypothesis. When testing the equality of two averages, we use a *z* or *t* statistic. When testing the equality of several averages, we use an *F* statistic. When comparing several proportions, we use a χ^2 test. These statistics are calculated because their theoretical distributions are known; hence, we can calculate the probabilities of producing the observed results assuming the null hypothesis is true. Although the technical details of these statistics and their distributions can be complex, as long as we understand the concepts involved the computer can do all the calculations for us.

In terms of drawing conclusions from the observed data, consider the confidence interval for the long-term percentage of service contracts sold. Selling 51 contracts in 100 equipment sales would result in a 95% confidence interval for the long-term percentage of about 41% to 61%. If the null hypothesis were that the percentage of contracts sold equals 41%, we would fail to reject the null hypothesis. Based on our data, 41 is included in our confidence interval and therefore is a plausible value. We would also fail to reject the null hypothesis if it stated that the long-term percentage of contracts sold were any value between 41 and 61. A null hypothesis stating any of these values would not be rejected. However, it is not plausible to sell 51 contracts if our long-term rate of selling contracts is 70%. In this case we would reject the null hypothesis.

Why is this point important? By increasing the sample size we narrow the range of plausible values, and therefore increase our chances of rejecting the null hypothesis. The ability to reject the null hypothesis is a function of both the degree to which the hypothesis is true and the sample size. To illustrate this point, suppose we sold 1000 pieces of equipment (instead of 100) and 510 service contracts, or 51%. The 95% confidence interval for the long-term rate of selling service contracts using these data would be 46.5% to 55.5%, compared to the interval of 41% to 61% obtained with 51 contracts for 100 sales. Note that 57% contracts sold is no longer a plausible value. The null hypothesis stating that the long-term rate of selling service contracts (π) equals 57% would be rejected (at the 95% confidence level). This hypothesis could not be rejected when we sold 51% of the contracts in 100 sales, but it can be rejected with 51% service contracts in 1000 sales. The only difference is sample size.

Conversely, suppose we sold two pieces of equipment. Even if we sell service contracts on both, this does not prove beyond a reasonable doubt that the long-term rate of service contracts sold is greater than 50%. We would expect to sell two contracts in two sales about 25% of the time with a 50–50 rate of selling contracts. The marketing manager's original claim of being able to sell service contracts more than 50% of the time is impossible to prove with this sample size, regardless of the results. As our sample size increases, our ability to detect that the null hypothesis is false increases. The statistical term for this phenomenon is power.

The *power* of the test is its ability (probability) to detect that the null hypothesis is false. This ability is obviously affected by "how false" the null hypothesis is. For example, suppose we sell 100 pieces of equipment, and the actual long-term rate of selling service contracts is actually now 52%. If we test the null hypothesis that the rate of selling service contracts (π) equals 50%, the power to reject this null hypothesis is quite low because 52% is very close to 50%, and we are not likely to sell significantly more than 50 contracts. In other words, we have a very small chance of rejecting the null hypothesis. We would need to sell 61 or more contracts in the 100 sales to reject the null hypothesis, and selling this many contracts from a process that produces 52% over the long term is very unlikely. Conversely, rejecting the null hypothesis of 50% would be easy if the true long-term rate of service contracts sold were actually 80. The test now has very good power to detect that the true contract rate is not 50%. If we want to improve our power to detect a long-term contract rate of 52% versus the original 50%, we need a much larger sample size.

Failure to reject the null hypothesis does not prove that the null hypothesis is true. Failure to reject the null hypothesis could be due to an insufficient sample size. By calculating a confidence interval, we can see what the actual range of plausible values is. If we feel it is too wide to be useful for practical purposes, we can increase the sample size. It is also possible to make sample size calculations ahead of time and determine how large a sample is needed to generate a confidence interval of the desired width. We illustrate this in Section 8.9.

Our general recommendation is to use confidence intervals in lieu of formal hypothesis tests. We can still employ the desired tests by observing if null hypothesis values are inside or outside the confidence interval. We cannot reject values within the interval, but we do have convincing evidence at the 95% confidence level to reject values outside the interval. It should be noted that for some hypothesis tests there is no practical useful confidence interval to calculate. For example, when testing whether four outsourced data processing firms are producing the same level of errors, we are interested in comparing four different error levels, so there would be several confidence intervals comparing two of the error levels.

Rejecting or Failing to Reject

If we observe a *p*-value less than .05, or if the null hypothesis value is not included within our 95% confidence interval, we can reject the null hypothesis. We can state that we have convincing evidence that the null hypothesis is false and that the alternative is true. In doing pharmaceutical research, the null hypothesis might state that the effect of the drug is equal to the effect of a placebo. Being able to reject this hypothesis means that we have proven that the drug actually works. When rejecting the null hypothesis that two bank branches have the same average cycle time for approving mortgage applications, we can state that the branches are different. Remember that we are making a statement about the entire population or process of interest, not just about the sample data.

The conclusion is very different if we cannot reject the null hypothesis. In this case, we cannot state a definitive conclusion. It is possible that the null hypothesis is true, but it is also possible that the null hypothesis is false. For example, in testing the null hypothesis that the long-term rate of selling service contracts is 50%, suppose our confidence interval calculated from our data goes from 40 to 60. In this case, the null hypothesis value of 50 is certainly plausible, so we cannot reject. However, it is also plausible that the long-term rate of selling service contracts is 55, or 47, or 58. It could be anywhere from 40 to 60, so we cannot make a definitive statement about whether the null hypothesis is true or not.

Another key point to keep in mind is that using formal hypothesis testing or confidence intervals does not guarantee we will always make the right decision. For example, it is theoretically possible to sell 100 service contracts in 100 sales when the long-term rate is only 50%, although it is extremely unlikely. If we actually sold 100 contracts, we would no doubt reject the null hypothesis and state that the rate of selling contracts was not 50% but significantly greater. When we select a confidence level, such as 95%, we are accepting a probability of making a mistake of .05. That is, 5% of the time that the null hypothesis is actually true we will make a mistake and reject it. Similarly, 5% of the time our 95% confidence interval will not contain the actual population value we are estimating. Therefore, when we make statements about proving a null hypothesis false, we should actually say we have proven this beyond a reasonable doubt, or with 95% confidence.

We will now review some of the more common hypothesis tests used in business. For each of these, statistical software packages calculate p-values based on the relevant statistical distributions. These p-values represent the probability of producing the observed data assuming that the null hypothesis is true. Generally, we reject the null hypothesis if this p-value is less than .05 (for 95% confidence). The lower the p-value, the more confident we are that the null hypothesis is false.

8.6   Tests for Continuous Data

Test for One Average

This test is sometimes referred to as a one-sample t test and is used to compare a sample average to a hypothesized value. One example would be the commerce department testing to see if a soft drink company is providing an average of 12 ounces of drink per can as advertised.

Typical hypotheses (one-sided):

Null: Population average (μ) equals or exceeds 12 (in this case)
Alternative: Population average (μ) is less than 12 (typically the alternative
       would be "does not equal")

Distribution used: t distribution. For large sample sizes (>30) the t distribution is almost identical to the z distribution, hence this might be called a z test. (See Chapter 9 for more on this topic.)

Theoretical Assumptions   This test assumes a normal distribution for the population or process. This is not a critical assumption because averages tend to have a normal distribution, even if the raw data do not. This test assumes the sample data were collected during a stable time period wherein the average and the variation were constant. If we test the null hypothesis that the average absorbency of Brand X paper towels is 480 (see Table 8.1), we would get the following results:

Test of $\mu = 480.00$ versus μ not $= 480.00$

Variable	n	$\bar{x}$	s	t	p
Brand X	10	488.50	10.54	2.55	.03

The p-value of .03 indicates that it would be highly unusual to obtain sample results like this if the population average were actually 480. Because the p-value is less than .05, we would conclude that we have convincing evidence (at the 95% confidence level) that the original hypothesis is false (i.e., the population average is not 480).

Test for Comparing Two Averages

Also called a two-sample t test, this test compares the averages of two processes or populations. For example, we might want to compare average time to pay bills before and after a new accounts payable process was implemented. We could calculate a confidence interval for the difference in the averages (i.e., Average 1 minus Average 2), and see if zero was included in this interval, as an alternative.

Typical hypotheses (two-sided):

Null: Average 1 equals Average 2 ($\mu_1 = \mu_2$)
Alternative: Average 1 does not equal Average 2 ($\mu_1 \quad \mu_2$)

The averages here refer to the process/population averages, not the sample averages. Recall that we are trying to draw conclusions about the population based on sample data.

Distribution used: t distribution. For large sample sizes (>30), the t distribution is almost identical to the z distribution, hence this might be called a two-sample z test. (See Chapter 9 for more on this topic.)

Theoretical Assumptions   Same as test for one average (for both samples), plus one version of this test assumes that the two populations have equal variation. Most statistical software, such as Minitab or JMP, offers an option that does not make this assumption.

If we perform this test on the difference between Brand X absorbency and Brand Z absorbency, we obtain the following results:

t test of μ Brand X $= \mu$ Brand Z (versus not =): $t = -0.71$ $p = 0.48$ df $= 18$

	n	$\bar{x}$	s
Brand X	10	488.5	10.5
Brand Z	10	492.9	16.4

The p-value is much greater than .05, so we conclude that there is no evidence to disprove the original hypothesis that the population averages are the same. This is consistent with the conclusion from our confidence interval calculated earlier, which it should be.

If there is a one-to-one correspondence between the data points from the two samples, such as before and after weights for a diet plan (each person has their own before weight and after weight, hence these two data points are naturally matched), we would normally calculate "pounds lost," or the difference in the paired or matched data points, and then do a test on one average, typically testing the null hypothesis that the population average pounds lost is zero. This is called a paired *t* test for obvious reasons.

Test for Comparing Several Averages

Also referred to as analysis of variance, or ANOVA, this test is used to compare several averages. For example, we might want to compare average inventory turns (a measure of cycle time) for five warehouses. Note the misnomer that "analysis of variance" is used to test averages. The actual test statistic is an *F* statistic, calculated by taking the ratio of two variances, one of which is calculated from the sample averages. One can also use the technique of analysis of means (ANOM), which essentially creates a control chart of the sample averages with statistically based limits.

Typical hypotheses:

Null: All averages are equal ($\mu_1 = \mu_2 = \cdots = \mu_n$)
Alternative: At least one average is different

If we reject the null hypothesis, we are saying that there are *some* differences in the averages, not that all are different from each other. We would have to look at the individual averages in this case to understand which appear to be different.

Distribution used: *F* distribution. (See Chapter 9 for more on this topic.)

Theoretical Assumptions   Same as test for one average (for each sample) but also assumes that each population has equal variation. This later assumption is important. If it does not appear to be true, this test should not be used. Several tests comparing two averages can be done, or a special set of tests, called nonparametric tests, which make minimal assumptions, can be used. These tests are not as powerful as the *F* test, hence they are not recommended unless we have evidence that the variation differs between populations.

To test the hypothesis that all three paper towel brands listed in Table 8.1 have the same long-term average absorbency, we could perform an ANOVA, producing the following output:

Analysis of Variance

Source	DF	SS	MS	F	p
Factor	2	50673	25336	133.82	0.000
Error	27	5112	189		
Total	29	55785			

The key data point here is the *p*-value, which is 0 (rounded off to 3 decimal points). Because it is so small, much less than .05, we can be quite confident that all three paper towels do not have the same long-term average absorbency. This is consistent with simple plots of the data, which revealed that Brand Y had a significantly lower average (see Figure 8.1). We may still wish to know which of the three brands appears to be different, which is why graphing our data is still necessary.

Test for Comparing Two Variances (Standard Deviations)

Also called the F test, this method compares two population variances (σ^2) or standard deviations (σ). For example, we might wish to compare before and after variation for a project to reduce variation in cycle time of delivering customer orders. There is a test to compare one variance to a hypothesized value, but this is not commonly used. For this situation we could simply calculate a confidence interval for the variance and see if the hypothesized value fell within the interval.

Typical hypotheses (two-sided):

Null: Variance 1 equals Variance 2 ($\sigma^2_1 = \sigma^2_2$)
Alternative: Variance 1 does not equal Variance 2 ($\sigma^2_1 \quad \sigma^2_2$)

Distribution used: F distribution. (See Chapter 9 for more on this topic.)

Theoretical Assumptions Assumes that both populations have a normal distribution and that the data were collected while the process was stable. If one or both of the populations do not appear to be approximately normally distributed, an alternative test, called Levene's test, can be used (see below).

If we compare the variances of Brands X and Z, we obtain the following:

Brand X standard deviation: 10.5
Brand Z standard deviation: 16.4
F statistic: 2.42
p-value: .20

Even though there is a large difference in the sample standard deviations, 10.5 versus 16.4, the p-value is .20, which is greater than .05. Therefore, it is plausible that both samples come from populations with the same population standard deviation. We have failed to prove that they are different. Common sense would suggest that the population standard deviations are different, but these data fail to prove this point beyond a reasonable doubt. In such situations it is often helpful to obtain additional data to resolve the issue. Recall that as we gather additional data our uncertainty decreases. For example, if these two sample standard deviations were based on sample sizes of 19, the p-value would be .03, providing conclusive evidence of a difference in variation between the two brands.

Test for Comparing Several Variances (Standard Deviations)

Sometimes called homogeneity of variance, which literally means "same variance," this test could determine whether five different investment options have the same degree of volatility (variation).

Typical hypotheses:

Null: Variances are all equal ($\sigma^2_1 = \sigma^2_2 = \cdots = \sigma^2_n$)
Alternative: At least one variance is different

Distribution used: Several different tests are used. The recommended test, Levene's test, is based on a specialized statistic developed by Levene (1960).

Theoretical Assumptions   Levene's test does not make assumptions on the distribution of the populations. It does require that the data are continuous (we would generally not perform a test on variation for discrete data anyway) and that the sample data used in the test are from a stable process/population.

If we test the null hypothesis that all three brands have the same variation using Levene's test, we obtain the following results:

Brand X standard deviation: 10.5
Brand Y standard deviation: 13.7
Brand Z standard deviation: 16.4
Levene's test statistic: .399
p-value: .68

Because the p-value is much larger than .05, we fail to find evidence that the population variation among brands is different. Note that the p-value for this three-way test is much larger than it was comparing only Brand X and Brand Z. This is because the third brand, Y, has a standard deviation between those of Brands X and Z. In other words, Brand Y produced variation similar to that of Brands X and Z. If the standard deviation of Brand Y had been much greater than that of Brand Z, or much less than that of Brand X, we would expect the p-value to decrease, possibly to less than .05.

8.7   Test for Discrete Data

Test for Comparing Two or More Proportions

This test, sometimes called the chi-squared test because this is the statistical distribution used, can compare two or more proportions. This could be used to test whether five different branches of a financial institution are approving the same percentage of mortgage applications. For only two proportions, we could use the z test for proportions, which is included in most statistical software, such as Minitab or JMP. This z test uses a normal distribution approximation to the binomial and was used in the confidence interval discussion for the difference between two proportions.

Typical hypotheses:

Null: All population proportions are equal $(\pi_1 = \pi_2 = \cdots = \pi_n)$
Alternative: At least one proportion is different

Distribution used: Chi-squared. This distribution is often written as χ^2.

Theoretical Assumptions   The calculation of the chi-squared statistic uses an approximation. This approximation works well when we have a sufficient sample size in each possible outcome—for example, for mortgages approved as well as mortgages not approved for each branch of the financial institution. In general we should have five or more data points for each possible outcome. Theoretically, the number of expected observations, assuming that the null hypothesis is true, is more important than the actual number of observations. This expected value can be calculated by statistical software. Of course, we also assume that the process/population proportions under study are stable over time.

Recall the chi-squared test presented earlier in this chapter for male versus female resignation rates. The results were a χ^2 value of 7.109 with a p-value of .008, indicating that we do have convincing evidence to conclude that males and females are resigning at different rates. We can expect similar results in the future unless changes in the process are made.

The chi-squared test can also be used when there are more than two possible outcomes—for example, when we are looking at the proportion of registered Democrats, Republicans, and Independents. This would simply result in a three-column table, and standard statistical software can handle this table equally well.

8.8   Test for Regression Analysis

Test on a Regression Coefficient

When we discussed regression analysis in Chapter 6, we noted that we are typically interested in evaluating which of the process variables (x's) is really related to the output (y). That is, which will really help us predict the output more accurately? One of the methods for doing this, the t-ratio in the regression computer output, is actually a test statistic for hypothesis testing. The null hypothesis being tested is that the population regression coefficient (the coefficient we would get if we had all possible data one could ever collect) is equal to zero. A coefficient of zero indicates that y is unaffected by changes in this variable.

Typical hypotheses (two-sided):

Null: The regression coefficient is equal to zero (population coefficient)
Alternative: The regression coefficient is not equal to zero

A rule of thumb for the t-ratio is that a value larger than 2 (or smaller than –2) is generally considered sufficient evidence to reject the null hypothesis (see Chapter 6). This is because for large sample sizes a t-value of 1.96 (approximately 2) corresponds to 95% confidence. This depends on sample size, but the regression output typically provides an exact p-value so we can be more precise in this decision.

Distribution used: t distribution (same as z for large sample sizes). See Chapter 9 for more on this topic.

Theoretical Assumptions   A number of assumptions are associated with regression analysis. The most important assumptions are that the model accurately depicts the relationship between the process variables and the output (i.e., we have not suggested a linear model when in fact the relationship has significant curvature) and that the variation in the output is constant over the entire range of process variables studied.

Recall the regression model on abandon rate at a call center from Table 6.5. The variable "time to answer" had the following coefficient and coefficient standard deviation: $b = .93$, $SD(b) = .17$. This produces a t-ratio of $.93/.17 = 5.6$, with a p-value of 0.00 (zero when rounded to 2 decimal places). Clearly, the variable "time to answer" does have a relationship to abandon rate and, assuming that our analysis satisfied the key regression assumptions, we can conclude that increasing time to answer does increase abandon rate. This intuitively obvious conclusion has now been confirmed with data.

Note that the regression output from many popular software packages also calculates an overall F-ratio using ANOVA. This tests the null hypothesis that *all coefficients* are equal to zero. If the p-value is less than .05, we can be confident that at least one coefficient is not zero, but the ANOVA analysis does not tell us which one.

8.9 Sample Size Formulas

In Chapter 5 we provided rule of thumb sample sizes to help determine the amount of data needed for reasonable estimates of the population average, standard deviation, or proportion. These give reasonable sample size values but not exact values. Earlier in this chapter we explained that in many cases we begin by taking rule of thumb sample sizes and then calculate confidence intervals for the parameters of interest. If these are not sufficiently precise to satisfy our needs, we can then use the exact sample size formulas to calculate how much more data we need to obtain a confidence interval of the desired width. Here is the general process for obtaining an appropriate sample size:

1. Use rule of thumb sample sizes to start.
2. Estimate μ, σ, or other parameters of interest from these data.
3. Evaluate the uncertainty in the sample data by calculating confidence intervals for the parameters of interest.
4. If this degree of uncertainty is acceptable, stop.
5. If this degree of uncertainty is not acceptable, use sample estimates of μ, σ, and so forth in the exact formulas to determine the total sample size required.
6. Subtract the rule of thumb sample previously collected from the total required and obtain this number of additional data points.

Exact formulas use the confidence interval formulas, assume they provide an interval of the desired width, and solve for n, the sample size. Note that the exact formulas make assumptions about the standard deviation or proportion. To the extent these assumptions are correct, we will obtain confidence intervals of exactly the desired width. If the data collected deviate from these assumptions, however, the width of the resulting confidence interval may deviate from the width originally desired. In other words, the actual width will match the desired width to the degree that the actual data match the assumptions.

Sampling from an Infinite Population

In many cases we find ourselves sampling from a very large population, infinite for practical purposes. This is particularly true when sampling from an ongoing process to draw conclusions about the process itself (i.e., including future outputs) rather than just about the data set we have currently obtained. In rare cases we may be only concerned about 50 medical imaging devices we currently have in inventory. If these 50 devices are the entire population of interest, we use modified formulas for finite populations. Of course, in reality all populations are finite (even the number of atoms in the universe is finite), but most are large enough to be considered infinite for statistical purposes. As a rule of thumb, we consider the population to be infinite if we are sampling less than 20% of it. If we are going to sample more than 20% of the population of interest, we use the formulas for finite populations discussed in the next section.

To obtain a 95% confidence interval of width "W" for the population average (μ), we need a sample size of at least:

$$n = \frac{4(1.96\sigma)^2}{W^2}$$

where σ is the population standard deviation, and W is the desired width of the confidence interval.

For a 99% confidence interval of width W, we would use 2.58 instead of 1.96. For other confidence levels, we would use the z-value (normal distribution with zero mean and standard deviation of 1) that corresponds to the desired confidence level. (A z table is provided in the back of this book in Appendix J.) We do not use the t distribution here, because it depends on n, which is what we are trying to determine.

This is the total sample size needed, so if we already have collected 20 data points and this formula suggests we need 50, we need only an additional 30 data points. Of course, it is important that the process under study has remained stable in between these two samples, or it would not make sense to combine them. Note also that if the sample standard deviation from the complete 50 data points differs from the estimate we used in the formula based on 20 points, the final confidence interval we compute may not be exactly the width desired.

Another noteworthy point is that σ, the true population standard deviation, is used in this formula. In practice we never really know this value but estimate it from the data by calculating s. In many sampling situations we do not have high-quality data currently, which is why we consider gathering additional data. We therefore often start with the rule of thumb sample sizes and use the sample standard deviation s to estimate σ. This is the general process for determining sample size, and the rule of thumb sample sizes ensure that we have a good estimate of σ to use in this formula.

Recall the absorbency evaluation of the paper towel brands discussed earlier. We calculated a 95% confidence interval for the average absorbency of Brand X based on 10 data points. This interval was from 481.0 to 496.0, or of width 15, based on a sample standard deviation of 10.5. To obtain a 95% confidence interval of width 5, we would need a total sample size of:

$$n = 4(1.96 \times 10.5)^2/(5^2) = 67.8$$

We would round up this total to 68, because we will need at least this large of a sample. We already have 10 data points, so we need to obtain 58 additional data points to complete this sample.

To obtain a 95% confidence interval of width "W" for the population proportion (π), we need a sample size of at least:

$$n = \frac{4\left(1.96^2\right)\pi\left(1-\pi\right)}{W^2}$$

where π is the population proportion of interest, and W is the desired width of the confidence interval.

Note that the formula for the required sample size to estimate π depends on π. This is a "catch-22" in that we need to know the population proportion to calculate the sample size required to estimate the population proportion to the desired degree of precision. To deal with this problem, either use the rule of thumb sample size initially and estimate π from these data or use the worst-case sample size. ("Worst case" means the value of π that would require the largest sample size.) Mathematically, the value of π that maximizes this sample size calculation is .5. Therefore, if we use .5 as our estimated value of π, we will get a conservative sample size that will give us at least the desired degree of precision regardless of the actual value of π.

In the female resignation example, our confidence interval for the population proportion (π) was about .04 to .12. To reduce the width of this interval from .08 to .05, we would need a total sample size of:

$$n = [4(1.96^2) \times .08(.92)]/.05^2 = 452.3$$

We would round this number to 453. We already have 156 data points, so we need to collect 297 additional data points. We have used .08 as our estimate of π in this formula, which was calculated from the original data. There are no more females in this organization to go measure, hence, we would have to wait until we had annual data with at least this sample size.

To obtain a 95% confidence interval of width "W" for the population standard deviation (σ), we need a sample size of at least:

$$n = 1.0 + .5\left[\frac{1.96}{\log_e\left(1+\dfrac{W}{2\sigma}\right)}\right]^2$$

where $\log_e$ stands for logarithm base e. This is a mathematical function that is performed by most calculators.

This formula depends on the actual value of σ, which is what we are trying to estimate in the first place. We can either enter our best estimate of σ into the equation or state W in terms of σ. For example, suppose we decide we want a confidence interval whose width is .4σ. That is, we want to be able to estimate σ within ±20% (.8σ to 1.2σ). In this case W/(2σ) becomes .4σ/2σ, in which case the σ's cancel, and this becomes .4/2 = .2, which no longer depends on the value of σ. Using this in the previous formula, the value of *n* needed for estimating σ within ±20% becomes 59 regardless of the value of σ. This formula is an approximation developed by Nelson (1982).

For Brand X absorbency in the paper towel example, we obtained a 95% confidence interval for the population standard deviation (σ) of 6.7 to 22.4, with a sample estimate of 10.5. This confidence interval has a width of 22.4 − 6.7 = 15.7. To reduce this to a width of 5, we would need a total sample size of:

$$n = 1.0 + .5 \left[\frac{1.96}{\log_e \left(1 + \dfrac{5}{2 \times 10.5} \right)} \right]^2 = 43.1$$

We would round up this result to 44. We already have 10 data points, so we would need 34 additional data points to obtain a confidence interval with a width of approximately 5.

Sampling from Finite Populations

Finite populations are those with a limited number of units. For example, if we are interested in only 50 medical devices in a warehouse or 200 employees involved in a class-action lawsuit, we are dealing with finite samples. In these situations we are not trying to draw conclusions about *all* medical devices produced or *all* employees, nor are we trying to draw conclusions about the process itself. We are interested only in these particular devices or employees.

These formulas use a finite population correction factor (FPCF):

$$\text{FPCF} = \sqrt{(N - n)/N}$$

where N is the finite population size. For a fixed sample size, *n*, this ratio gets closer and closer to 1 as N increases. As a rule of thumb, we would consider the population finite if it is of fixed size and we are planning to sample more than 20% of it.

In the formula for the confidence interval for an average, we would replace $\dfrac{1.96\sigma}{\sqrt{n}}$ with $\dfrac{1.96\sigma}{\sqrt{n}} \sqrt{\dfrac{N-n}{N}}$. Solving this equation for *n* to obtain a sample size formula for 95% confidence, we get:

$$n = \frac{(1.96\sigma)^2 N}{N \left(\dfrac{W}{2} \right)^2 + (1.96\sigma)^2}$$

If we wanted to measure the weight of the paper towel rolls as well as their absorbency and if our population of interest is not all potential rolls of towels but only the 100 rolls shipped to a specific customer on a specific date, the population of interest would be finite and consist of 100 rolls. We had previously measured

10 rolls and had obtained an average of .2 lb with a standard deviation of .03 lb. To estimate the average weight of this shipment of rolls with a confidence interval width of .02 lb, the required sample size is:

$$n = \frac{(1.96 \times .03)^2 \, 100}{100\left(\dfrac{.02}{2}\right)^2 + (1.96 \times .03)^2} = 25.7$$

We would round up this result to 26. We already have 10 data points, so we would need to obtain an additional 16 data points.

For estimating proportions, the formula becomes

$$n = \frac{1.96^2 \, \pi(1-\pi)N}{N\left(\dfrac{W}{2}\right)^2 + 1.96^2 \, \pi(1-\pi)}$$

In the female resignation case, suppose the 156 females were part of a larger organization that included 500 females. To estimate the proportion of females out of 500 who actually resigned this year with a confidence interval width of .04, we would need a sample size of:

$$n = \frac{1.96^2 \times .08 \times .92 \times 500}{500\left(\dfrac{.04}{2}\right)^2 + 1.96^2 \times .08 \times .92} = 292.8$$

We would round this result up to 293. We already have 156 data points, so we would need an additional 137 data points in our sample.

In this case the problem was fundamentally restated. We are now no longer interested in estimating the process (long term) proportion but only the females who actually resigned this year. This is a significant shift of focus and is not typically what is of greatest interest.

The key drivers of sample size are these:

- *Standard deviation.* The larger the variation, the more data are required.
- *Confidence level.* The greater the confidence level desired, the more data are required.
- *W.* The more precise we want our estimate to be, the more data are required.
- *Population.* We actually need fewer data with a finite population because we have already sampled some of the data, leaving even less data about which we have to draw inferences.

Sample Sizes for Hypothesis Tests

We may desire a certain level of confidence (95%) to reject the null hypothesis, and there are sample size formulas for this type of hypothesis testing. These formulas are more complicated because the probability of rejection depends on exactly how the null hypothesis is false. For example, when testing the equality of five means, four might be the same and one different or each mean might be different. The size of the difference between the means also will have an impact on this probability. Each possible scenario will result in a different probability of rejecting the null hypothesis. These formulas are too complex to include here, but Montgomery (1991) is a good source for further inquiry on this topic.

8.10 Summary

1. The fundamental concept of statistical inference is to draw conclusions, or inferences, about an entire population or process on the basis of sample data. Predicting election results based on sample polls is an everyday application.
2. Laws of probability enable us to document the uncertainty based on the sample size, but subject matter knowledge must be used to determine how widely the results of the sample may be extrapolated to the entire population or process of interest.
3. Confidence intervals enable us to document the uncertainty in our sample estimates of the average, standard deviation, proportion, regression coefficients, and so on. Any value in these intervals is plausible for the true population/process value of one of these statistics, but values outside the interval are not plausible at the appropriate confidence level, typically 95%.
4. Formal hypothesis tests can be performed to test the plausibility of hypotheses about population statistics.
5. To reduce the uncertainty in our confidence intervals, we can use sample size formulas to determine how large our sample size must be to achieve a confidence interval of the desired width.
6. Statistical inference is most valuable when plots of the data suggest certain relationships but do not provide obvious conclusions.

8.11 Project Update

At this point in your project you have no doubt collected a significant amount of data and perhaps have applied regression analysis or design of experiments. Look over some of the graphical analyses you have performed and determine if any require formal statistical inference to resolve uncertainties. You may wish to calculate confidence intervals for averages, standard deviations, or other important statistics, or perform statistical hypothesis tests. You may wish to reanalyze some of your existing data, or you may decide to obtain additional data and apply formal statistical inference procedures.

If you have performed a regression analysis, reconsider any *t*-ratios calculated through the analysis given your enhanced understanding of what these ratios are testing. If any questions remain to be answered, use the sample size formulas to calculate how much additional data would be required to resolve these issues.

In addition to data quantity issues, consider your data quality. To what conceptual population do you feel the results can be applied? To what population or process can we infer from the sample data? For example, if you gathered data on your golf game, do you feel these results would apply to others or only to you? If the results apply to others, do they apply to all others or only to some that met certain conditions, such as similar skill level, left versus right handed, male versus female, similar age, and so forth? Can you draw general conclusions from your analyses? Why or why not? Do you now see the need to collect additional data?

EXERCISES

1. A friend notes that she has discussed the upcoming presidential election with 15 friends, and 10 said that they are voting for candidate Jones. She is therefore convinced that candidate Jones will win the election. Based on what you now know about statistical inference, is this a logical conclusion? Why or why not? How would you explain your conclusion to her without using any statistical jargon?

2. Compare and contrast hypothesis tests with confidence intervals. When would one or the other be preferred? Why?

3. For what types of questions would you want to calculate a prediction interval versus a confidence interval? Give three specific examples of situations where each would be preferred over the other.

4. Election polls typically sample about 1000 people. If they make the assumption that the race between two candidates will be close and that their sample is reasonably random from the desired population of voters, how precisely will this enable them to estimate the proportion voting for each candidate? What sample size would they need to estimate the proportion voting for each candidate within ±1%? What data quality issues in such polling might invalidate their assumptions?

5. Suppose there is a 40% nonresponse rate in election polling. What data quantity and quality issues would this raise? What would you suggest be done to address these issues?

6. Find an article in a paper or electronic publication where the data from a sample are extrapolated to an entire process or population. Determine whether the sample was collected from the entire population of interest or from only a subset of it. In other words, did the frame equal the population of interest? Write a brief essay explaining why the inference was or was not inappropriate, noting the conclusions that can actually be drawn from these data.

7. How large a sample size is required to estimate the population standard deviation to within ±5%? ±30%? ±50%?

8. Table 8.4 provides cost overruns (actual costs minus estimated costs) on similar-sized engineering projects completed this year. Calculate the following:

   ■ A 95% confidence interval for the average overrun this process would deliver over a long time period, assuming it is stable.
   ■ A 99% confidence interval for the same average. Explain why this is a different width than the 95% interval.

TABLE 8.4 Engineering Overruns (in $000s)

−16	−480	−222
−404	−77	−76
−397	221	−427
113	93	−268
−791	107	−230
−165	−272	287
−118	133	−285
9	276	−344
−651	36	−394
383	−298	161

Note: Negative numbers represent projects that came in under budget.

- A 95% prediction interval for the overrun on the next project.
- A 95% confidence interval for the long-term standard deviation of overruns.
- The company president claims that these data prove that the engineering organization is consciously "high-balling" cost estimates—that is, deliberately overestimating actual costs—to meet their budget. Do you agree or disagree with this claim? Why?
- If we consider the three columns as three separate types of projects, is there convincing evidence to conclude that the long-term average overruns of the three types of projects are different? How about the standard deviations?

9. Explain why it is helpful in practice to use rule of thumb sample sizes in addition to the exact formulas.

10. Three analysts looked at insurance claims and categorized them as covered, not covered, or requiring more information to make a determination. Given the following data, can we conclude that the analysts are equally "tough" in their analyses, or is there evidence that some are more inclined to pay claims than others? What assumptions about the sampling and allocation of claims to analysts do you need to make to answer this question?

Analyst	Number Covered	Number Not Covered	Number Requiring More Information
1	331	22	2
2	287	45	19
3	420	27	24

11. Suppose the number of equipment sales and service contracts sold over the last 6 months for treadmills and exercise bikes was as follows:

	Treadmill	Exercise Bike
Total sold	185	123
Service contracts	67	55

A service contract can only be sold on a new piece of equipment. Of the 185 treadmills sold, 67 included a service contract and 118 did not. Construct a 95%

confidence interval for the difference between the proportion of service contracts sold on treadmills versus exercise bikes. Is there a difference between these two pieces of equipment? If so, what is it? Could you have arrived at this conclusion without creating a confidence interval?

12. Table 8.5 lists the order to remittance (OTR) cycle time in days for software system installations in various countries by country code. Do these data support the

TABLE 8.5 Country Code and OTR Cycle Time for Software System Installation

Country Code	Cycle Time	Country Code	Cycle Time
1	20	5	29
1	24	6	40
1	46	7	157
1	26	8	19
14	38	5	24
1	15	1	81
1	15	7	53
17	23	7	26
1	31	1	28
1	31	1	34
6	64	1	34
5	29	7	50
5	44	1	52
1	32	1	19
1	15	1	44
7	11	14	150
7	14	7	29
1	89	17	25
17	41	6	79
7	41	17	13
1	36	6	32
8	43	7	61
17	21	8	42
8	28	8	46
7	18	7	88
8	47	14	24
6	26	7	7
6	47	1	33
5	9	5	129
7	42	17	41
5	5	17	43
6	27	14	42
6	27	14	42
1	33	7	53
7	44	7	53
1	21	7	48
1	22	5	21
1	50	1	19

hypothesis that the OTR tends to be the same in every country? Can you detect differences in variation? In addition, calculate the following and explain how these results should be interpreted:

- Calculate a 95% confidence interval for the average OTR of Country 1.
- How large a sample size would be required to reduce this confidence interval by half?
- Calculate a 95% confidence interval for the difference between the average OTRs of Country 1 and Country 6. Can you conclude that these two countries have different average OTRs?
- Calculate a 99% prediction interval for the OTR for the next installation in Country 1.
- Calculate a 95% confidence interval for the standard deviation of OTRs in Country 1.

13. Table 8.6 lists data on appliances sold as inclusions, meaning the appliances were included in the price of a new home. The appliance company is trying to grow this sales channel. People buying homes for hundreds of thousands of dollars are unlikely to quibble over paying $10 or $20 more for an appliance than the lowest

TABLE 8.6   Appliance Inclusion Data

Builder	Price ($)	Profit ($)	Builder	Price ($)	Profit ($)
1	229.46	485.47	6	239.83	495.8
1	185.98	84.93	6	212.94	626.22
1	228.25	1020.71	6	185.07	−95.52
1	191.25	278.4	6	173.51	407.55
1	213.56	370.5	6	222.54	265.82
2	214.35	305.92	7	196.8	278.42
2	180.57	160.05	7	187.65	17.18
2	185.72	180.63	7	185.46	48.67
2	185.36	216.74	7	194.59	116.8
2	238.38	897.62	7	214.8	541.57
3	193.29	82.86	8	193.05	123.29
3	176.15	211.48	8	169.15	−173.43
3	202.05	94.96	8	199.3	394.21
3	155.8	39.02	8	196.26	386.25
3	212.49	178.74	8	217.9	658.09
4	184.29	−28.67	9	228.57	568.04
4	193.91	97.21	9	198.06	50.11
4	210.76	328.02	9	207	380.13
4	170.15	3.85	9	181.06	84.67
4	214.84	548.68	9	230.95	931.77
5	188.59	186.61	10	238.8	461.67
5	152.77	−66.57	10	211.82	494.83
5	172.89	17.12	10	203.44	414.71
5	191.44	35.7	10	164.42	10.46
5	190.99	180.44	10	217.19	259.93

price they could find on the Internet or through a discount appliance store. The manufacturer therefore makes a higher margin, the home builders can sell the fact that they are simplifying the overall process of buying a new home, and the buyer has fewer details to worry about. Selling the appliances as inclusions can be a win-win situation for all parties. These data were collected in one particular sales region in the Midwest last month and include data from 10 different builders, the selling price of the home (in thousands), and the manufacturer's profit from the included appliances. Note that profit can be negative, indicating a loss.

- What other data might you want to collect to determine what had the biggest impact on profitability?
- Is there evidence that the builder has an impact on profitability? Why or why not?
- Construct a 95% prediction interval for the profitability from the next sale in this region.
- Construct a 99% confidence interval for the difference in the average profit from Builder 1 versus Builder 10. Is this result consistent with your answer to the question on builder profitability? Why or why not?
- Perform a regression analysis on profit using home selling price as your *x* variable. Construct an approximate 95% confidence interval for the coefficient for price. Does the home selling price have an impact on profit from the included appliances? Why or why not?
- Can you use these results to draw general conclusions about selling appliances as inclusions? What cautions would you suggest to people about how widely these results might apply?

14. Explain how the hypothesis testing process is similar to a criminal trial in the U.S. legal system.

15. This excerpt is from the *Schenectady (NY) Daily Gazette*, April 23, 2000.

Ever-Changing Medical Claims Confound Doctors

Daniel Q. Haney

This time it is fiber, which contrary to the collective wisdom of the brightest minds in medicine apparently does not ward off colon cancer after all. The specifics change, but the pattern is the same: Over and over, the conventional medical wisdom collapses under the weight of new evidence. Remember when salt was evil? When eggs were the soul of dietary wickedness? When estrogen seemed like an iron shield against heart disease? Now it is pretty clear that salt is not an important cause of high blood pressure. Most people probably can eat an egg for breakfast without triggering a heart attack. And estrogen? No one really knows how that will turn out, but there is doubt about the long-accepted assumption that it keeps the heart working smoothly after menopause.

So how does this happen? Why do health rules fall apart after they are chiseled in stone? And how do they get to be rules in the first place? Many health professionals say it comes down to the willingness of all involved— the scientists, the news media, and the public—to draw firm conclusions from a stew of often poorly conducted, contradictory, and incomplete observations. "One of the problems is that strong recommendations have often been made on very weak data," says Dr. Walter Willett of the Harvard School of Public Health. "It may have been the best guess at the moment. But often the recommendations are repeated so many times that people forget they were rough guesses in the first place and come to think they are hard facts."

This is not to say everything is wrong or likely to be overturned tomorrow. For instance, scientists feel absolutely certain that smoking is bad. . . . Willett says one reason for the today-it's-good-for-you, tomorrow-it's-not phenomenon is the well-intended "missionary zeal" of scientists who believe their own work and happily repeat the seemingly solid bottom line without going into the complexity and uncertainty of the whole business. Another essential player in this process, scientists like to point out, is the news media. "You get the two-sentence synopsis that turns a complicated issue into a black or white, a yes or no," says Lynn L. Moore, an epidemiologist at Boston University School of Medicine. . . .

Dr. Thomas Pearson of the University of Rochester notes that some scientists seem bent on encouraging the boldest headlines for their research, and get plenty of help from the reporters who interview them. . . . The fact is, science is a messy process. No single study, no matter how large or careful, is likely to settle an important health question. Sorting out the influence of genes, food, pollutants, living habits, and all the rest requires drawing together information from many different scientific approaches. These include experiments in lab dishes, tests on inbred rats, observations of large groups of people, and human experiments.

Data from all of these kinds of science went into the rise and fall of the idea that fiber prevents colon cancer. The theory began in the 1970s. Scientists noticed that poor people in rural Africa get much less colon cancer than do better-off Westerners. Of course, the differences between these two populations are too numerous to count, but an obvious one was the Africans' higher consumption of fiber. Over time, many lines of evidence seemed to support the theory. For instance, it was shown that people who immigrate to places where colon cancer is common take on a higher risk as they adopt the eating habits of their new home. In the lab, experiments showed that animals fed cancer-causing toxins seem to be protected by high-fiber diets. . . .

Finally, two large federally financed studies put the theory to the test by putting people on low-fat, high-fiber diets. The meticulously run experiments found no evidence this lowers the risk of polyps, which are the first stage of colon cancer. Willett's team came to the same conclusion by an entirely different method. . . . "When we published it last year, it was heresy to say that data don't support a major benefit of fiber in reducing colon cancer. . . . Now we know that if there is a benefit, it's not very large, because it is not just one study showing this."

Medical studies and research results are examples of the application of the statistical inference process. Answer these questions about the key points in this article:

- How can competent and well-meaning medical researchers come up with conflicting conclusions from their studies?
- Is this evidence of badly designed medical studies?
- In what sense are the media contributing to the problem?
- How could the results of medical studies be more appropriately communicated to the public?

REFERENCES

Levene, H. (1960). *Contributions to probability and statistics.* Stanford, CA: Stanford University Press.

Montgomery, D. C. (1991). *Design and analysis of experiments.* New York: Wiley.

Nelson, W. (1982). *Applied life data analysis.* New York: Wiley.

9
The Underlying Theory of Statistical Inference

Figure as far as you can, then add judgment.

Anonymous

9.1 Overview

The process improvement strategy uses statistical inference, a general methodology that includes several individual tools (such as confidence intervals and hypothesis tests). The theory of statistical inference is used to determine the appropriate formulas for confidence intervals or the hypothesis tests, such as t tests or F tests, and it includes the mathematical basis for these formulas.

This chapter will explain some of the key theoretical concepts that underlie statistical inference and explain the concepts of the normal and other probability distributions, sampling distributions, linear combinations, and transformations. This chapter will clarify the methods discussed earlier, so that they will not seem to be conducted in a mysterious "black box" into which one cannot see. Understanding these concepts will prove important in future statistics courses or training.

Why study statistical inference when your goal is to run a start-up? Of course, you do not need to be a mechanical engineer or mechanic to drive a car. Once you learn to use the brake pedal, ignition, steering wheel, and so on, you can drive. You may not see a need to understand the theories behind the combustion engine, distribution of power, or the complexities of the electrical system—you just want the practical skills needed to drive the car. However, if you hear a strange sound from under the hood, a little theoretical knowledge about the car's functions—is it an imminent explosion or simply a loose fan belt?—is very helpful. So in reality, some basic level of theoretical understanding is advantageous, even if our primary concern is just to drive the car. Similarly, the theory presented in this chapter will help

374

you better understand the tools of statistical inference and prepare you for future, more advanced, statistics courses. (Potential future directions for your statistical education will be discussed in Chapter 10.)

The main learning objectives for this chapter are therefore:

- To be able to apply appropriate statistical theory, such as the central limit theorem, in practice.
- To be able to explain the statistical theory underlying the basic tools of inference. The rationale for this learning objective is to prepare for future course work or training in statistical methods.
- To apply basic statistical theory to avoid potential pitfalls, such as applying an inappropriate tool, in applications.

Although we use the term *theory*, the material presented here is definitely relevant to practice. For example, the central limit theorem (which will be discussed in Section 9.6) has important ramifications for statistical thinking applications. To understand it, however, one needs to understand probability distributions and sampling distributions, which are both somewhat theoretical concepts and not required to apply the basic tools discussed in Chapters 4 and 5. The discussion of such theoretical topics is therefore appropriate now: This chapter places the "capstone" over the material presented previously.

9.2 Applications of the Theory

Previous chapters have presented applications of several statistical inference methods, including hypothesis tests, confidence intervals, regression analysis, and design of experiments. Formulas were often given for the application of these methods with very little explanation. For example, when discussing confidence intervals in Chapter 8 to estimate the average number of fluid ounces per can of soft drink, we presented the following formula:

$$\bar{x} \pm \frac{ts}{\sqrt{n}}$$

Why is this the correct formula? How do you know it will work? A basic understanding of statistical theory is required to determine how to attack a problem for which none of the standard formulas apply. The saying "If all you have is a hammer, everything looks like a nail" applies here—if you have only memorized formulas, you will not know which one to use when facing new types of problems.

Another example is the formula for the confidence interval for the difference between two averages. This method is useful in situations requiring linear combinations (discussed in Section 9.7)—for example, to estimate the difference in cycle times between two offices evaluating creditworthiness of potential customers. If we break total cycle time to determine creditworthiness into five discrete steps, the method of linear combinations will show us how variation in each of the five steps' cycle time relates to variation in overall cycle time. Similarly, in manufacturing, the interference fit between two components (that is, the relationship between the dimensions) is often more important than the dimensions of the components themselves. Examples include the dimensions of medication bottle lids versus the dimensions of the bottles themselves, or the dimensions of compact discs (CDs) versus the dimensions of the CD drive. In these cases, the variable that best shows whether the components fit is the relationship of the two dimensions, not the individual components' dimensions themselves. We analyze this relationship using the method of linear combinations.

Suppose we must analyze the cycle time to approve a small-business loan application. The process steps are

1. Receive application
2. Enter into computer
3. Initiate credit screening
4. Decision by loan officer

The average cycle time is 16 working hours, with a standard deviation of 4 hours. We want to reduce the total cycle time to an average of 4 hours, with a standard deviation of 1 hour. Our improvement team will split up into four subteams, each of which will focus on one of the subprocess steps. What average cycle time and standard deviation for each step would produce an overall average of 4 and overall standard deviation of 1? The theory of linear combinations tells us that the overall average (μ) will be the sum of the four subprocess averages, and the overall standard deviation (σ) will be the square root of the sum of squared standard deviations, as follows:

$$\mu = \mu_1 + \mu_2 + \mu_3 + \mu_4$$

$$\sigma = \sqrt{\sigma_1^2 + \sigma_2^2 + \sigma_3^2 + \sigma_4^2}$$

We can now use these formulas to tell us how much variation we can tolerate in each subprocess step and still achieve our overall variation objective. For example, standard deviations of 2.0 for each of the steps will produce the current overall standard deviation of 4.0, since $4.0 = \sqrt{(2^2 + 2^2 + 2^2 + 2^2)}$. Clearly, we will have to obtain lower standard deviations for the individual steps.

A probability distribution tells us how likely any particular outcome is. The terms "bell curve," "normal distribution," and "Gaussian distribution" are all names for one probability distribution. Many other probability distributions exist, however, and apply to a wide variety of practical situations. For example, to determine how likely you are to win a hand of blackjack when holding a hand of 19, you need to understand the probability distribution of hands in blackjack. This is how casinos set financial returns on various games, such as poker, blackjack, roulette, craps, or slot machines. They study and understand the probability distributions of these games to ensure that the house will make a profit.

In fact, all of the inference methods discussed previously have the same underlying theoretical basis: the theory of statistical inference, also referred to as mathematical statistics. Section 9.3 reviews the overall process for applying this theory; Sections 9.4–9.8 discuss some useful concepts in more detail.

9.3   The Theoretical Framework of Statistical Inference

Although there are many types of statistical inference tools, the theory underlying them all is the same. It is based on the field of probability. A basic understanding of probability helps apply statistical inference tools more thoughtfully. Chapter 8 explored an overall model for statistical inference. This chapter will present more technical detail about how to draw conclusions about the population of interest from the sample data, using probability theory.

Figure 9.1 shows an overall framework for the underlying theory of statistical inference and explains in greater detail how to draw quantitative conclusions about the population or process of interest. This framework should, therefore, be considered a subprocess of the statistical inference model presented in Chapter 8.

First, define the objective(s) and reasonable assumptions. Next, identify the key variable of interest and convert it into a generic format that is not dependent on the specifics of the current problem. This generic format makes it easier to derive the probability distribution, which is then used to make the calculations required for statistical inference.

Following is a closer look at each of these steps:

Step 1: Identify the objectives of the analysis and define necessary and reasonable assumptions to discern how to proceed. For example, in the loan cycle time example in Section 9.2, the objective is to understand what cycle time distributions (average and variation) are required in the four subprocesses, in order to reduce the total cycle time to an average of 4 hours, with a standard deviation of 1 hour. It is reasonable to assume that the total cycle time is the sum of the individual subprocess cycle times.

FIGURE 9.1   The Theory of Statistical Inference

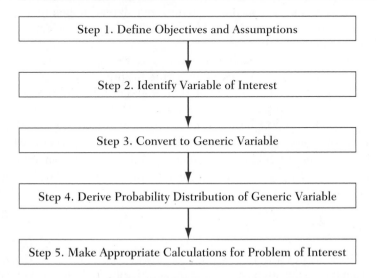

Step 2: Identify the variable of interest. The variable must be something quantitative—that is, something that can be counted or measured. If the variable is not quantifiable, it cannot be mathematically analyzed. It could be a discrete variable and be assigned values of 0 ("no") or 1 ("yes"). The variable of interest could be directly measured—such as dollars, temperature, time, and so on—or something calculated—such as profit, inventory turns, or price-to-earnings ratio. The variable of interest may be an average, a standard deviation, or other parameter. To compare two different bank branches, the variable of interest may be the difference in their average cycle times for transferring funds. The variable of interest may actually be a linear combination of other things. In virtually all cases, it is a population parameter that is estimated from a statistic calculated from a sample. The cycle time example contains two population parameters of possible interest: the average and also the standard deviation.

Step 3: Convert this variable into a "generic," or standardized, variable to the greatest degree possible. Standardization allows categorization of numerous problems, which are all unique, into a few generic "buckets" that have standard solutions. This helps avoid having to come up with a unique solution to every problem. For example, recall the *t* test for comparing two averages from Chapter 8. The following example was given in Section 8.6:

Test of $\mu = 480.00$ versus μ not $= 480.00$

Variable	n	$\bar{x}$	s	t	p
Brand X	10	488.50	10.54	2.55	.03

Note that the sample average of 488.5 and standard deviation of 10.54, based on 10 observations, are converted into a *t*-ratio. The original variable of interest was the population average, estimated from the sample average. However, every problem involving a sample average will be different, because the average itself, the standard deviation, and also the sample size will tend to be different. This would require us to come up with a unique solution to every problem, which would be exhausting and expensive.

A more logical approach is to standardize the sample average to create a generic variable. In this case, we subtract the hypothesized mean of the sample average, μ (i.e., if we took averages of 10 over and over again, what would be the average value of these sample averages? see Section 9.6 on sampling distributions for a more detailed explanation), and then divide by the standard deviation of the sample averages, $s/\sqrt{n}$. This produces a generic variable, which we call *t*. Performing these calculations, we obtain a *t* value of 2.55. In mathematical terms, for any sample average $\bar{x}$, based on *n* observations, with assumed population average μ, we have:

$$t = \frac{\bar{x} - \mu}{s/\sqrt{n}}$$

This *t* variable has a well-known probability distribution, for which we can make any calculations of interest. This approach is valuable because we can calculate this same *t* variable for any average and standard deviation, in any units of measurements, and get exactly the same t variable, with exactly the same distribution. We can therefore solve virtually any problem involving testing a single average using this same t variable; it doesn't matter what the sample average and standard deviation are. If we worked with the sample average itself, each problem would result in a different distribution and therefore a different solution.

To further emphasize the value of converting to a generic variable in Step 3, consider the χ^2 distribution, which can be used to test whether there is a real difference between several proportions. Suppose we are comparing the proportion of inaccurate price quotes from 10 salespeople. Each proportion has a different probability distribution, depending on sample size and magnitude of the proportion. A direct approach would require a unique solution for every such problem. A much simpler approach is to calculate a χ^2 value, which has a well-known distribution. This distribution is the same, regardless of sample size and magnitude of the proportions, hence we now have a generic solution to all such problems comparing two or more proportions. This is why exotic-sounding variables—such as F, t, or χ^2—are calculated. In the cycle time example we do not need a generic variable to estimate the average and standard deviation, but we would need to calculate t and χ^2 or F values to calculate confidence intervals or perform hypothesis tests.

Step 4: Now that we have a generic variable, which is not problem-specific, we must derive its probability distribution. This key step is often much more difficult than one might think. Fortunately, the mathematical work required to derive the probability distributions has been done for the majority of standard inference problems (such as those given in Chapter 8). For an individual measurement, we can often estimate the probability distribution from a histogram of sample data. However, suppose the variable of interest is the standard deviation. What is the probability distribution of a standard deviation calculated from sample data? If we obtained more data and calculated a sample standard deviation from the new data, would it exactly equal the sample standard deviation from the old data? Of course not, because of variation in the samples. Probability theory is the tool used to derive these sampling distributions for averages, standard deviations, proportions, and so on. Using probability theory, we can derive the average and standard deviation of total cycle time in the loan processing example as a function of the individual subprocess step averages and standard deviations. This derivation produced the formulas presented on page 376. For a more in-depth discussion of probability theory, see Walpole and Myers (1993).

Step 5: Make the appropriate calculations to estimate a regression coefficient, perform a hypothesis test or confidence interval, and so on. Modern statistical software, such as Minitab or JMP, typically makes this easy. Very little, if anything, needs to be calculated by the statistical thinking practitioner. For example, in the discussion of hypothesis testing in Chapter 8, a key question we would ask if a person sold 70 service contracts in 100 sales is "How unlikely is it to sell 70 or more service contracts in 100 sales, if we assume that the probability of selling a service contract is 50% on each sale?" Our software will readily answer such questions.

In the t test, we compared a sample average to a hypothesized value of 480 and calculated a t value of 2.55. How should this number be interpreted? The

FIGURE 9.2   *t* Distribution (*n* = 10) With Reference Line at 2.55

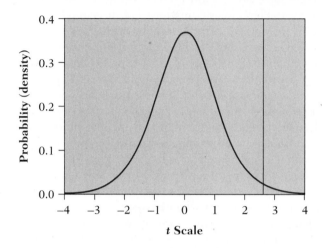

equivalent question is: "How unlikely is it to obtain a *t* value of 2.55 or more if the population average really is 0 (i.e., if the average of the original variable is 480)?" Figure 9.2 shows what the *t* distribution looks like in this situation and where 2.55 falls. Note that it is extremely unlikely to obtain a *t* value of 2.55 or more if the population average is 480. The probability of obtaining a *t* value greater than 2.55 (or less than –2.55) is only .03. Note that any sample average less than 480 would be converted to a negative t value, because we subtract 480 from the sample average, prior to dividing by $s/\sqrt{n}$. In other words, we would observe a result this unusual only 3% of the time, if the null hypothesis were true. Given that our observed result is so unusual, we conclude that the null hypothesis must not be true—that is, the population average must be greater than 480.

As another example of making the calculations for Step 5, consider the loan process cycle time example again. Recall that our improvement objective is to achieve an overall average cycle time of 4 hours, with a standard deviation of 1.0. We can use the formulas for average and standard deviation of total cycle time to back-calculate what averages and standard deviations we would need in the individual steps to achieve our objectives for total cycle time. For example, to achieve an average total cycle time of 4 hours, we would need each of the four subprocesses to average 1 hour. Note, however, that this is not the only possible solution! Another option is to let the first step average 2.5 hours, and the other three average .5 hour. Given that this adds to 4.0, it still satisfies our objectives.

To achieve the desired standard deviation of 1, we need the sum of the squared standard deviations to be 1.0—that is, we need $\sigma = \sqrt{\sigma_1^2 + \sigma_2^2 + \sigma_3^2 + \sigma_4^2}$ to equal 1.0. One solution to this equation is to have $\sigma_1^2 = \sigma_2^2 = \sigma_3^2 = \sigma_4^2 = .25$, which means that each standard deviation would have to be .5 hour. Again, this is one solution that works, but not the only one. This solves the problem we had, which was to determine what averages and standard deviations of the subprocesses would achieve our objectives ($\mu = 4.0$, $\sigma = 1.0$) for total cycle time.

This interpretation completes the cycle of applying the theory of statistical inference depicted in Figure 9.1. We apply these conclusions to the entire population of interest, not just to the sample data. Typically, we continue our improvement efforts with additional cycles through the inference process, each building upon those previously completed, resulting in continuous improvement.

By discussing population averages and standard deviations, we assume that we have stable processes for which population averages and standard deviations exist. In many real-world applications, the process drifts or is subject to periodic special causes—the process is dynamic rather than static. In such cases, it makes little sense to refer to a population average, because the average is changing over time. Before applying statistical inference tools based on the assumption of a static population, be sure to verify that the process really is stable. For unstable processes, we may need to isolate time periods during which the process was stable and apply formal inference tools, if such analysis helps us improve the process.

9.4   Types of Data

A detailed understanding of different types of data is necessary because a key aspect of applying the underlying theory of statistical inference is understanding probability distributions of the variables of interest. Understanding data types can also help us select the most appropriate data during the data collection phases of our statistical thinking applications. There are many types of probability distributions, which relate to the different types of data. The main types of data, with examples, are given in Table 9.1.

Nominal Data

The most basic type of data is called "nominal" data, or data measured on a nominal scale. Nominal data are those which fall into two or more categories, such as male and female, or Democrat, Republican, and Independent. For nominal data,

TABLE 9.1

Examples of Data Types

	Nominal	*Ordinal*	←——————— *Scale* ———————→ *Integer*	*Interval*	*Ratio*
Characteristics	Discrete categories	Ordered discrete categories	Counted integers	Continuous, no absolute zero	Continuous, absolute zero
Examples	Political affiliation	1st, 2nd, or 3rd place	Number of defects	Temperature (°F or °C)	Time, money, weight

there is no continuous measurement, just a designation of which category a person or item falls into. Neither is there a "natural" ordering to the categories. For example, being male is neither better nor worse than being female (despite occasional claims to the contrary). The simplest situation is where everything is classified into one of two categories, such as male or female, sale or no sale, or defective or not defective. (Recall from Chapter 3 that data on accuracy defects are one of the most common measurements we make in business applications.) This type of data may therefore be called "defectives data," because we may define each item as either defective or not defective. The binomial distribution is often applied to nominal data. This is the underlying distribution applied when using the control charts called "*p* charts," which were discussed in Chapter 5. The *p* chart is typically used with nominal data, because the binomial distribution is often applied. For more than two categories, we can extend the binomial distribution to the "multinomial" distribution.

Ordinal Data

In some cases, the discrete categories do have some logical ordering. An example would be gold, silver, and bronze medals in the Olympics. A gold medal is better than a silver medal, which is better than a bronze. This type of data is called ordinal data, or data on an ordinal scale, because there is an order. Another example would be data from a survey where the response options were strongly agree, agree, neutral, disagree, and strongly disagree. These are still categories rather than continuous measurements. However, in some cases, the categories are assigned a number range (1–5 for these survey data, for example), so that continuous probability distributions may be applied. More exact methods have been developed to analyze such data, usually referred to as categorical data analysis. These methods are beyond the scope of this book. See Agresti (1990) for more detail on categorical data analysis.

Integer Data

If we are counting something, such as the number of defects or injuries, we produce an integer scale, or integer data. The key difference between integer data and nominal or ordinal data is that an integer scale consists of actual numbers, not just categories. For example, we may count the number of people who select "agree" on a survey, but the actual scale for each person still consists of five categories. These data are therefore on an ordinal scale. If the variable of interest is the number of times a financial forecast has to be revised, the scale consists of the values 0, 1, 2, 3, and so on. This is an integer scale.

Integer data are not on a true continuous scale, however. For example, we may count three missing items in a delivery to a warehouse. These data are not quite continuous, however, because we cannot count 3.2756 missing items. All the data are whole numbers, or integers. Sometimes this type of data is also referred to as "defects data," because the data type may be applied to count the number of defects in an item. The Poisson distribution, the underlying distribution used in control charts called *c* charts, is often applied to integer data.

Note that not all integer data conform to the Poisson distribution, nor do all nominal data conform to the binomial. Other distributions may more accurately portray a given data set. Sometimes we may consider the integer data continuous, for practical purposes. For example, when analyzing financial data, we express dollars (or at least pennies) as integers. These numbers tend to be so large, however, they behave like continuous data (discussed below), and we usually treat financial data in any currency as continuous. (See Appendix G for more detail on other probability distributions.)

Continuous Data

If we are able to measure something, as opposed to count something, we have continuous data. When we say "measure," we mean that we could measure items to as fine a resolution as desired, if we had the appropriate measurement device. For example, we may choose to measure time in years, months, days, hours, or minutes; if we have a high-resolution clock, we can measure to hundredths or even thousandths of a second. Other continuous measurements include length, height, weight, and distance. The normal distribution (also known as the "Gaussian distribution" or "bell curve") is most often applied to continuous data. However, the normal distribution does not apply to all continuous data. Other continuous distributions include the chi-squared, exponential, and Weibull distributions.

Interval and Ratio Scales Continuous data can be further broken down into interval and ratio scales, which have the property that any interval, say an interval of 1.0, means the same thing wherever it is applied in the entire range of the data. For example, in looking at cycle time data, the difference between 4 seconds and 5 seconds (1.0 second) is the same as the difference between 100 seconds and 101 seconds (1.0 second also). As a counterexample, assume that Olympic medals are assigned values as follows: gold is 1, silver is 2, and bronze is 3. These values are on an ordinal scale, because 1 is better than 2, which is better than 3. On the other hand, they are not on an interval scale, because we cannot say that the difference between a gold and silver medal (1.0 using this coding) is the same as the difference between silver and bronze (also 1.0 using this coding). Most athletes would agree that the difference between winning a gold versus silver medal is huge, whereas the difference between silver and bronze is much smaller.

The key distinction between interval and ratio scales is that ratio scales have an absolute zero; interval scales do not. This relates to whether ratios of the values have practical meaning. For example, time has an absolute zero, hence it has a ratio scale. Ten minutes is twice as long as 5 minutes. Temperature in degrees Fahrenheit does not have an absolute zero: "Zero degrees" does not represent a total lack of heat. Therefore, 40° is not twice as warm as 20°, and temperature Fahrenheit has an interval scale, but not a ratio scale. Generally speaking, one should be very careful taking ratios of data if the data are not on a ratio scale. Distance, time, money, and weight are other examples of ratio scales.

The types of data form a spectrum, where nominal data are the most basic, and ratio is the most informative, scale. This is also illustrated in Table 9.1. The discussion of sample size illustrated that, in general, we need more data when we have a nominal versus interval or ratio scale. Also, the more informative the scale (the farther we are from nominal), the more options we have for data analysis. For these reasons, we prefer to capture continuous data, either interval or ratio, whenever possible.

9.5   Probability Distributions

Probability distributions are the building blocks of statistical inference. The key step in the process of applying statistical inference is identifying the probability distribution of the variable of interest, which is often a generic variable. We will briefly discuss the common continuous and discrete probability distributions you are likely to encounter in your project or subsequent statistics courses. These are listed in Table 9.2, along with some potential applications for each. Appendix G contains additional information about continuous and discrete probability distributions. For a more detailed treatment, see Walpole and Myers (1993).

A probability distribution is an equation that quantifies the probability of observing any particular value, or set of values, for the variable in question. For example, the probability distribution for cycle time of approving credit applications will enable us to calculate the probability of any application taking less than 4 hours, more than 15 hours, somewhere between 5 and 8 hours, and so on. These calculations are required to derive the limits of confidence intervals or p values in hypothesis tests. In fact, p values are just probabilities of observing a result this unusual, assuming the null hypothesis is true. We therefore need to quantify the probability distribution if we are to apply statistical inference. We plug into the equation a potential outcome, such as selling 5 service contracts in 10 sales, and the probability distribution calculates the probability of observing this result. As we shall see below, these distributions are best understood graphically, by plotting each possible outcome versus its probability of occurring.

Another reason to study probability distributions is that understanding a given distribution frequently provides insight into how to improve the process. For example, if data follow a Poisson distribution, the standard deviation will be equal to the square root of the average—that is, if the average is 4, the standard deviation

TABLE 9.2   Some Probability Distributions

Distribution	Potential Examples
Binomial	Number of service contracts sold in 100 appliance sales
Poisson	Number of hits to a Web site
Normal	Daily changes to a stock price
Exponential	Time to failure of a bearing
Standard normal	Normal variable minus μ, divided by σ
Chi-squared	Distribution of a sample variance
t	Same as standard normal, but uses s for σ
F	Distribution of a ratio of chi-squares (variances)

must be 2. Therefore, if our objective is to reduce the variation, we can only do this by reducing the average. A practical example shows how *not* to improve a process: The U.S. commercial airline industry began to grow significantly in the 1950s. Prior to this time the airline industry focused on military applications. Faced with socially unacceptable safety performance of their aircraft, many of the airlines began replacing critical parts on the aircraft more frequently. Unfortunately, this did not improve their safety/reliability performance in many cases, puzzling the airlines (see Chapter 12 of Moubray, 1992). Further research found that if time to failure of a part follows an exponential distribution, periodic replacement of the part has no impact on reliability. In fact, if time to failure follows other distributions with "bathtub" reliability curves (as do many electrical components), periodic replacement can actually make things worse. This is because some components have a high failure probability when they are first put into use and then become more reliable. Better understanding of these probability distributions would have avoided an expensive lesson.

To evaluate the underlying probability distribution, we typically must assume that the process is producing a stable output that will conform to a static distribution. In many, perhaps most real applications, this will not be true because of natural dynamic behavior of the process or perhaps the existence of special causes in the data. In these situations, we must either first stabilize the process or stratify the data into static subgroups to achieve a single, static probability distribution.

Note that probability distributions are theoretically derived, not estimated from data. A histogram may give us an estimate of the probability distribution, but it is not the population distribution itself. No real data will conform exactly to these theoretical distributions, but many real data sets will approximate them—because all models are wrong, but some are useful.

Discrete Distributions

Binomial Distribution The binomial distribution is used to document the probability of observing any number of "successes" in n observations of a nominal variable, with only two possible outcomes. For example, suppose we sell 100 appliances and count the number of service contracts we sell with the appliance. Using the above nomenclature, we would say we will have 100 observations (i.e., $n = 100$), and we will define selling a service contract as a "success." What is the probability of selling exactly 50 contracts? Exactly 70 contracts? More than 70 contracts? The binomial probability distribution will answer these questions for us.

Because the binomial formula (shown on page 387) is not very intuitive, we should look at graphs of the binomial distribution for various situations to understand it. Figure 9.3 shows the distribution of number of contract sales for 1, 10, and 100 appliance sales, respectively. The distribution of contracts in 1 sale is not particularly interesting, but shows that each of the two possible outcomes is equally likely (a probability of .5) in this case. (Of course, the probability of one outcome does not have to be .5 for the binomial distribution to apply.) The distributions of the number of contract sales in 10 and 100 sales are more interesting: They show that the number of possible outcomes increases as sample size

FIGURE 9.3   Distribution of Service Contracts Sold in 1, 10, and 100 Appliance Sales, Assuming a 50–50 Chance of Selling a Contract on Each Appliance

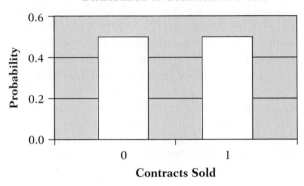

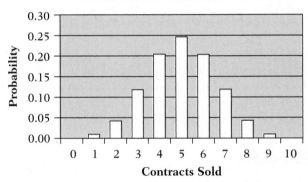

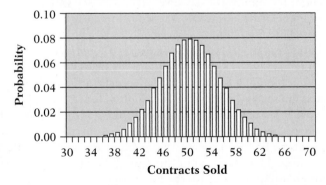

increases and that the distribution starts to look more like a bell shape, or normal distribution, as the sample size increases.

This second point, known as the central limit theorem, is a very important phenomenon in statistics. This theorem states that, as sample size increases, the distribution of averages gets closer and closer to a normal distribution. As Figure 9.3 shows, the central limit theorem also applies to sums, such as the number of contracts sold in 100 appliance sales, because the sum is just the average multiplied by the sample size—that is, the sum is just the average multiplied by a constant. Therefore, for reasonable sample sizes we can analyze binomial averages or sums using the normal distribution. This is a much simpler approach when we wish to estimate probabilities for a wide range of possible values, as we do with confidence intervals. If we had to calculate these probabilities using the binomial distribution, we would have to calculate each individual probability and add them up. The normal distribution provides a much more efficient approach.

The mathematical formula for the binomial distribution is:

$$P(x) = \frac{n!}{(n-x)!\,x!} \pi^x (1-\pi)^{n-x}$$

where:

P(x) means "the probability of observing exactly x successes"
n is the number of observations, or "trials"
x is the number of "successes" for which we are calculating the probability
π is the probability of a success
$n!$ means "n factorial," which is defined as $n \times (n-1) \times (n-2) \times \cdots \times 2 \times 1$
π^x means π raised to the power of x (if x is 2, this means π squared)
The average of the binomial distribution is $n \times \pi$ (i.e., the number of trials times the probability of success). For 100 appliances sales and a probability of selling a service contract of .5, $100 \times .5 = 50$. The average of this distribution is therefore 50, as can be seen in Figure 9.3.

The standard deviation of the binomial distribution is $\sqrt{n\pi(1-\pi)}$. For the same 100 appliance sales, this would be $\sqrt{100 \times .5 \times .5} = 5$.

Note that the variation in a binomial variable is dependent on the average. If the average changes, so will the variation. This is an important limitation when using percent-accurate or percent-defective as response variables in models, because both will show nonconstant variation in the residuals. This violates one of our key assumptions of regression analysis. However, the appropriate transformation, typically a square root transformation, can alleviate this problem. See the discussion on transformations in Section 9.8.

A key theoretical assumption is that each "trial" is independent—that is, each trial has a probability of success of π, regardless of what happened in previous trials. For example, selling a service contract to one customer does not affect the probability of selling one to the next customer.

Computer software, such as Minitab or JMP, will perform these calculations for the binomial and other probability distributions. The only reason we present

What Does the Standard Deviation of a Probability Distribution Mean?
Recall that the standard deviation is a measure of variation: The larger
the standard deviation, the more variation in the process. For many
probability distributions, most of the observed values will fall within 1
standard deviation of the average. In the appliance service contract
example, for every 100 appliances sold, we will likely sell between 45
and 55 contracts (the average—50—plus or minus 1 standard deviation
of 5). The *population* standard deviation (*not* the sample standard
deviation) for a given probability distribution is mathematically defined
as follows: The population variance σ^2 is defined as the expected value
of the quantity $(x - \mu)^2$, where x is a randomly selected value from the
probability distribution, and μ is the average of the distribution (population average). By "expected value" we mean the long-term average, or
the average we would get if we were to sample from this distribution
over a very long period of time and calculate this quantity over and over
again. This quantity, $(x - \mu)^2$, is the squared deviation from the average.
For example, this quantity for 70 contracts is $(70 - 50)^2 = 400$. The
standard deviation, σ, is then the square root of the variance: $\sqrt{\sigma^2}$. The
standard deviation could therefore be considered the "typical" deviation
from the average.

the formula on page 387 is to show that such a formula exists (in the computer's
memory if not in ours) for each distribution. For example, if we wish to calculate
the probability of selling exactly 50 contracts in 100 appliance sales, we would
input the following: $x = 50$, $n = 100$, and $\pi = .5$ (we have a 50–50 chance of selling a service contract on each appliance). The probability of selling exactly 50
contracts turns out to be about .08. This means that, for every 100 appliance
sales, we can expect to sell exactly 50 contracts 8% of the time. Although selling
exactly 50 contracts is not very likely, it is a more likely result than any other value.

Poisson Distribution   The French mathematician Poisson derived this distribution by looking at what happened to the binomial distribution when the
sample size n became very large, approaching infinity, but the number of observed
"successes" or "occurrences" remained constant. Obviously, this would result in
the probability of an occurrence decreasing toward zero. Poisson was looking at
the number of soldiers killed by horse kicks in the Prussian army (Grant, 1952).
Of course, one might try to simply count how many soldiers had been killed by
being kicked, but getting an exact count at any point in time was very difficult, especially given a large army that was geographically dispersed. Because he could
not count n but knew that it was very large, Poisson took the approach of determining mathematically what happened when n got very large in the binomial formula, but the average number of deaths remained constant. This led to the discovery of the Poisson distribution. Note that having the number of deaths remain

constant while *n* got very large implied that the probability of a death to any particular soldier was very small.

This background gives us insight into those situations where the Poisson distribution is likely to apply. It typically applies to situations where we are counting discrete occurrences (such as defects) and the probability of occurrence is low, but the opportunities for the occurrences are essentially infinite. For example, we might be looking at the number of people who access a particular Web site each day. How many people could access it each day? Theoretically, it could be the total number of people who have access to the Internet—a sample that is very large and difficult to accurately estimate. For practical purposes, it is infinite. However, only a very small portion of people with access to the Internet will access this particular Web site in a given day.

A theoretical assumption is that each possible occurrence is equally likely, just as in binomial data. This will never be exactly satisfied; in the Internet example, affluent people are much more likely than the general population to hit Web sites providing stock quotes. Fortunately, this assumption does not need to be exactly true for the Poisson distribution to produce a good approximation to the observed data. However, if only a small set of people hit the Web site every day, and no one else accesses it, the observed data may not follow the Poisson distribution well. Other practical examples where the Poisson distribution is likely to apply would be the number of visual flaws in the finish of an automobile, the number of injuries in a factory each month, or the number of customer complaints received in a given time period.

A key distinction between the binomial and the Poisson is that the binomial counts both the "successes" and "failures," or "occurrences" and "nonoccurrences." If we sell 100 appliances, we can count both the number of times a service contract was sold and also the number of times that no contract was sold. On the other hand, it is impossible, or at least impractical, to count the number of nonoccurrences with the Poisson. The question "How many "non-hits" did we have on our Web site today?" doesn't make any sense. Similarly, it would be impractical to count the number of injuries that did *not* occur, the number of times we *didn't* get a complaint, or the number of locations on an automobile finish which do *not* have a visual flaw.

This distinction explains why we ask the question "Can we count nonoccurrences?" when determining the appropriate control chart in Chapter 5. The answer to this question helps differentiate between binomial ("Is a unit defective or not?") data and Poisson ("How many defects are present?") data. We then ask the question: "Are occurrences rare?" to determine how appropriate it might be to use the Poisson distribution to set the limits on the control chart. As noted above, a theoretical assumption is that the actual number of occurrences is quite small relative to the number of opportunities, which is typically assumed to be infinite. If occurrences are not rare relative to the opportunities, such as in the case of measuring production volume per day (in cases, tons, invoices, and so on), then the observed data will probably not follow the Poisson distribution closely. For example, if we are counting boxes of breakfast cereal produced in a factory, occurrence of a box being produced is not rare—but occurring continuously. We would therefore use a control chart for continuous data, perhaps individuals and moving range charts, even though the data are technically discrete. We do this because the Poisson distribution is not likely to work well in this case.

In Figure 9.4 we show several different Poisson distributions for different averages. Because the standard deviation of the distribution must equal the square root of the average, we do not need to specify the standard deviation. Because there is no absolute maximum value, the Poisson distribution theoretically goes from 0 to infinity. For example, there is theoretically no limit to the number of hits to our Web site, other than physical limits of the computer system. Note also that as the average increases, the shape of the distribution gets closer and closer to a bell-shaped, or normal, distribution. Just as in a binomial distribution, the central limit theorem applies. For averages greater than about 5, we can approximate the Poisson fairly accurately with the normal distribution, using an assumed standard deviation of $\sqrt{\mu}$. As with the binomial, one reason to do this is to quickly estimate the probability of getting a particular range of values, such as between 5,000 and 10,000 hits to our Web site, without having to calculate each individual probability and add them up.

The mathematical formula for the Poisson distribution is:

$$P(x) = \frac{e^{-\mu}\mu^x}{x!}$$

FIGURE 9.4
Sample Poisson Distributions

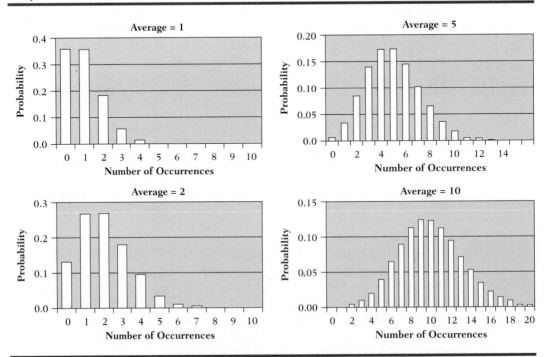

where

$P(x)$ means "the probability of observing exactly x occurrences"

e is the natural number e, which is approximately 2.72

μ is the average number of "defects," or occurrences

$x!$ means "x factorial," which is $x \times (x - 1) \times (x - 2) \times \cdots \times 2 \times 1$

The average of the Poisson distribution is μ, which appears in the formula above.

The standard deviation of the Poisson distribution is $\sqrt{\mu}$—that is, the square root of the average.

Note that there is no n in the Poisson formula: There is no limit on the number of possible occurrences.

Note that by definition $0! = 1$, so if we are calculating the probability of exactly 0 occurrences, the denominator of the formula above would be 1.

As with the binomial distribution, the variation is a function of the average. Therefore, models that use a Poisson count as the response will tend to have problems with nonconstant variation of the residuals. Again, transformations—typically a square root transformation—will often alleviate this problem. (See the discussion on transformations in Section 9.8.)

Continuous Distributions

Normal Distribution   The normal distribution is perhaps the most commonly applied probability distribution. Many statistical procedures, such as the *t* test, ANOVA, and regression analysis, are based on the normal distribution. We have also seen that other distributions start to look like the normal distribution as the sample size or (in a Poisson distribution) average increases. In practice, many continuous variables display distributions that can be well approximated by the normal distribution, although for reasons we will discuss below, no real variable has an exact normal distribution (again, "all models are wrong").

We first need to consider continuous distributions in general, and why they are noticeably different from discrete distributions. Continuous distributions, on an interval or ratio scale, can take on any value within some range. In other words, it is possible for cycle time to be 2.65472 hours, assuming we are capable of measuring time that precisely. Because an infinite number of outcomes are possible with a continuous variable, we do not represent their probabilities with bars, but will use a continuous curve instead. This changes our interpretation of what the height of the curve means.

Take another look at Figure 9.3. The height of each bar represents the probability of observing that particular number of contract sales. Note that, as the number of appliances sold increases, the number of bars increases, because there are now more possible outcomes. For 100 appliance sales, the distribution looks like a histogram of continuous data. Imagine what this distribution will look like as the number of appliances sold becomes larger. Eventually, the distribution would be composed of so many bars so close together that it would look like a continuous curve, instead of a sequence of discrete bars. Figure 9.5 is an

FIGURE 9.5   Normal Distribution With Average of 50, SD of 5

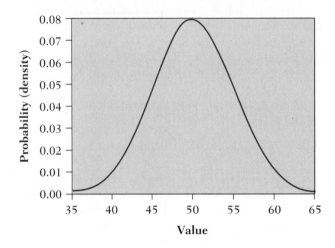

example of this: a normal probability distribution, for an average of 50 and standard deviation of 5.

The fact that the distribution is a continuous curve implies that there are an infinite number of possible outcomes. The higher the curve, or "density," at any point, the more likely it is to observe an outcome in that area. For example, the average of the distribution in Figure 9.5 is 50. Note that the distribution is highest at 50, indicating that values in this region are most likely. On the other hand, the height of the distribution is very low beyond about 60, indicating that it is rare to observe a value above 60. How would we use the normal curve to calculate specific probabilities?

Because there are an infinite number of possible outcomes on a continuum, the height of the distribution does not represent the probability of a specific observation, as it did with discrete distributions. Rather, we use *area under the curve* to represent probability. For example, to calculate the probability of observing a value between 50 and 60, we would calculate the area under the normal curve between 50 and 60, as shown in Figure 9.6. We use calculus to calculate this area. This methodology is beyond the scope of this text, but is easily handled by standard statistical software. In this case, it turns out to be about .48, meaning we could expect to observe a value between 50 and 60 almost 50% of the time.

A normal distribution is symmetric—that is, the left side looks just like the right side. The average is the most probable value (the "mode"), and divides the data in half: Half of the data fall above the average, and half below. The point at which this occurs is also called the "median," or 50th percentile. In a normal distribution, the mean, median, and mode are all the same. However, this is not the case with all continuous distributions. Theoretically, the normal distribution goes from minus infinity to plus infinity—that is, it never ends. This is another reason why no real variable is exactly normally distributed. However, many real data sets approximate the normal curve reasonably well, such as daily changes to the stock market, and this is why the normal curve is used so often in statistics.

FIGURE 9.6   Probability of Observing a Value Between 50 and 60 in a Normal Distribution With Average of 50, SD of 5

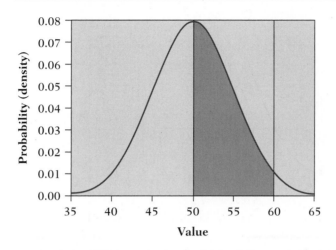

What Is the Probability of Getting Exactly 50 in a Continuous Distribution?   Theoretically, this probability is zero, because a single value has no area. Given an infinite number of decimal points, it is essentially impossible to get exactly 50. In practice, of course, when we round to a certain number of decimal points, it is possible to get exactly 50. This is why we say no real variable has exactly a normal distribution—we would have to measure to an infinite number of decimal points to accomplish this.

The specific equation for the normal distribution is:

$$f(x) = \frac{1}{\sqrt{2\pi\sigma^2}} \exp\left(\frac{-(x-\mu)^2}{2\sigma^2}\right)$$

where

  μ is the average
  σ is the standard deviation
  "exp" stands for exponent and means that the entire following expression is an exponent of the natural number e (approximately 2.72). Here, the expression is e to the power of

$$\frac{-(x-\mu)^2}{2\sigma^2}.$$

  π is the natural number pi (approximately 3.14)
  $f(x)$ is called the "probability density function" and represents the height of the normal curve, or the relative likelihood for any value of x.

FIGURE 9.7   Comparison of Normal Curve With Average of 50 and SD of 5
to Standard Normal (z) Distribution

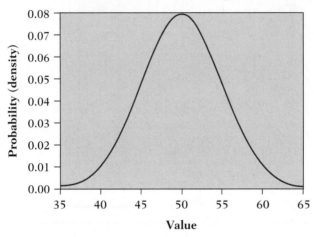

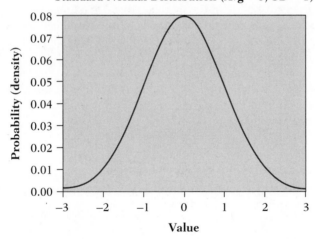

Standard Normal Distribution   The standard normal distribution is just a normal distribution with an average of zero and a standard deviation of 1. Figure 9.7 compares the normal curve seen in Figures 9.5 and 9.6 (which had an average of 50 and standard deviation of 5) to the standard normal curve. Note that the curves look exactly the same; only the scales are different. In other words, there is only *one* normal distribution; it just has many different scales. The standard normal distribution is also known as the z distribution.

FIGURE 9.8   Standard Normal Distribution (Avg = 0, SD = 1) With Confidence Levels Noted

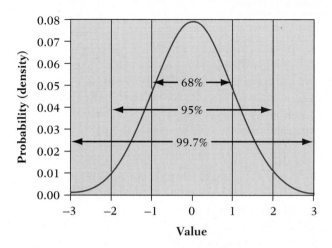

Why is the standard normal distribution important? Recall the discussion in Section 9.3 about creating a generic variable that allows us to reach one solution for a variety of similar problems, rather than having to solve each problem from scratch. The z distribution is another example of a generic variable. Virtually no real problems produce a z distribution, but it is very important in statistical applications.

One example of a z distribution is the confidence interval for an average. Figure 9.8 shows a standard normal distribution with the areas between −1 and +1, −2 and +2, and −3 and +3 marked with arrows. These areas correspond to the probability of an observation between these values with a standard normal, or z, distribution. This is called the empirical rule and will be discussed in greater detail below. Note that the probability of observing a value between −2 and +2 is about .95, or 95%. If we use −1.96 and +1.96, we get almost exactly 95%. For this reason, we use 1.96 in the formula for a 95% confidence interval when we have a large sample size. Recall that if we knew σ and did not have to estimate it from s, then we would not use t in the formula even for small sample sizes, but would replace it with 1.96, the z value.

For this confidence interval, we have calculated a generic variable, which has a z distribution if we know σ. The generic variable in this case is

$$z = \frac{\bar{x} - \mu}{\sigma / \sqrt{n}}$$

Note that by using standard algebra, the statement that we are 95% sure that

$$\bar{x} - 1.96 \frac{\sigma}{\sqrt{n}} < \mu < \bar{x} + 1.96 \frac{\sigma}{\sqrt{n}}$$

can be rewritten (by subtracting $\bar{x}$, then dividing by $\sigma/\sqrt{n}$, then multiplying by -1) as

$$-1.96 < z < 1.96$$

Mathematically, these two equations are identical. By converting any normal distribution to the generic z distribution, we can come up with one solution to developing confidence intervals for averages. Without using the z distribution, we would have to come up with a unique solution for every problem involving a confidence interval for an average. In the typical situation where we do not know the population standard deviation σ but have to estimate it from the sample standard deviation s, this introduces another source of uncertainty, and we use the t distribution instead of the z.

How do we convert a normally distributed variable to the z distribution? The formula is straightforward: If the variable x has a normal distribution with average μ and standard deviation σ, then the variable $z = (x - \mu)/\sigma$ has a z distribution. In other words, if you subtract the average and divide by the standard deviation, you convert any normal distribution to the standard normal, or z, distribution. This implies that the average becomes 0 in the z distribution, and the standard deviation becomes 1.0. A value greater than the mean takes on a positive z value, and a value less than the mean takes on a negative value. For example, a z value of -2 corresponds to an original value that is 2 standard deviations below the average. A z value of 1.5 corresponds to an original value that is 1.5 standard deviations above the average.

The Empirical Rule   The empirical rule (Wheeler & Chambers, 1992) refers to how much of a distribution will be within 1 or more standard deviations from the population average. It is referred to as the empirical rule because it has been observed empirically, that is, through data, with many distributions. It only holds precisely for a normal distribution, but it is approximately true for most continuous distributions. The empirical rule states the following:

- Approximately 68% of the distribution will fall within 1 standard deviation of the average. For distributions typically seen in practice, the exact percentage may be anywhere from about 60% to 75%.
- Approximately 95% of the distribution will fall within 2 standard deviations of the average. For typical distributions, the exact percentage may be anywhere from about 90% to 98%.
- Approximately 99% (99.7% for a normal distribution) of the distribution will fall within 3 standard deviations of the average. This rule tends to hold very well for virtually any distribution seen in practice.

This rule, illustrated for a normal distribution in Figure 9.8, is useful in helping interpret the meaning of a standard deviation in practical terms. For example, if students' test scores on a final exam averaged 80 with a standard deviation of 5, how good is a score of 90? Because 90 is 2 standard deviations above the average, we would expect very few students to score better than this. We expect about 95% of the students to have scored between 70 and 90 (plus or minus 2 standard deviations). Of the 5% we would expect to be outside this range, we would expect half (2.5%) to be below 70, and only 2.5% to be above 90. We would therefore expect a score of 90 to be at the 97.5th percentile of the exam scores. Of course, these calculations will not be exact if the test score distribution is not exactly normal or is not static over time. Even if the population distribution is normal, some deviation from an exact normal distribution will occur in the sample data, because of sampling variation—that is, variation in this particular sample from the population.

The Exponential Distribution   The exponential distribution is often used to approximate cycle time distributions—for example, the time before a piece of mechanical or electrical equipment fails. Suppose our business provides long-term service contracts on industrial motors and drives. In order to properly price the contract, we must estimate the frequency of repair. We have discovered from previously collected data that motor bearing failures are a key contributor to our costs to fulfill the service contracts. We have further determined that the time to failure of the bearings is stable and approximately follows an exponential distribution with an average of one year. Figure 9.9 shows the exponential distribution with an average of 1.0, which is the simplest case, because it simplifies the exponential probability distribution formula (explained below; with $\mu = 1$, the exponential formula becomes simply $f(x) = e^{-x}$). Note that the exponential distribution begins at 0 and is very skewed (asymmetric) to the right.

The exponential distribution has several unique properties. Because of its extreme skewness, the exponential deviates from the empirical rule more than most distributions. For example, theoretically about 86% of an exponential distribution will fall within 1 standard deviation of the mean. For the exponential distributions, we cannot observe negative values, and the mean and standard deviation are the

FIGURE 9.9   Exponential Distribution (Average = 1.0)

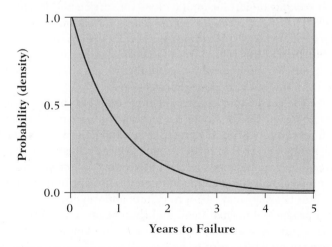

same. Therefore, "within 1 standard deviation of the average" implies "between 0 and 2 times the average." The theoretical value of 86% is considerably larger than the 68% given by the empirical rule and even well outside the typical range of 60% to 75%. This is because all of the distribution less than the average will be "within 1 standard deviation of the average," because we cannot observe negative values. However, 95% will theoretically be within 2 standard deviations of the mean, which does match the empirical rule. About 98% will fall within 3 standard deviations of the average, which is slightly less than the 99% value given by the empirical rule.

Over 60% of the exponential distribution will be less than the average. In other words, the average does not equal the median. For the industrial motor bearings discussed above, 60% will fail prior to the average of one year of service. This fact is due to the extreme skewness (lack of symmetry) of the exponential. Another unique feature of the exponential distribution is the "no memory" property, which states that for an exponential variable (motor bearing life in this case), the probability of failure in the next few hours is the same, regardless of the current age of the bearing. In other words, any equipment that follows an exponential distribution does not "wear out" at predictable times, but instead experiences "random failures," which occur at unpredictable times. This property can be seen graphically in Figure 9.9. If the portion of the distribution between 0 and 1 were discarded, the shape of the remaining distribution would essentially be the same as the original distribution—just a little smaller—and begin at 1 instead of 0. Therefore, if a specific bearing has lasted one year, the probability of failure in the second year is just 60%, the same as the original probability of failure in the first year.

The term "no-memory" is used to suggest that the bearing does not "remember" its age when "deciding" when to fail. This phenomenon may seem counterintuitive—certainly most equipment life does not follow an exponential distribution closely, but typically has some type of time-related wear out. However, some patterns follow the exponential distribution quite well: Certain types of bearing systems' failure tend to follow an exponential distribution, for example. The no-memory property has important implications for service and leasing, because we cannot accurately predict when the equipment will wear out. This is one example where understanding the distribution type provides insight into improving the system. If the bearing failures follow an exponential distribution, replacement of old equipment at periodic intervals is not cost effective, because it will not reduce the probability of failure, but will incur significant cost.

To reduce the servicing costs we must either obtain better bearings (with a longer average life) or find a way to detect that the bearing is beginning to fail. Even though we cannot accurately predict when the failure will occur, failure may be a gradual enough process to be detectable prior to catastrophic failure. For example, automobile components such as water pumps, alternators, or fuel pumps often make unusual noises indicating imminent failure prior to completely failing. Industrial motor bearings may begin to vibrate abnormally, then produce audible sounds, then produce abnormal heat, and then begin to smoke prior to actually "seizing." With sufficient advance warning, we can plan and implement preventive/predictive maintenance prior to failure.

The formula for the exponential distribution is

$$f(x) = \frac{1}{\mu}\, e^{-x/\mu}, \text{ for } x > 0$$

where

> x is some value of interest greater than zero
> $f(x)$ is the relative likelihood of observing x, called the "density"
> μ is the average of the distribution, as well as the standard deviation
> > (i.e., $\mu = \sigma$)
> e is the natural number e, which is approximately 2.72

9.6 Sampling Distributions

Sampling distributions are extremely important in business applications. By sampling distributions we mean probability distributions that result from taking a sample of data and calculating some statistic, such as the average. A normal distribution with an average of 50 and a standard deviation of 5 was illustrated in Figure 9.5. If we were to sample from this distribution for a long time and create a histogram of the raw data, it would look very similar to Figure 9.5, especially as the sample size increased. Suppose, however, that instead of sampling individual observations, we select samples of size five and calculate an average of these five observations. If we took several samples of five, calculated an average for each, and then plotted a histogram of the sample averages, what would it look like? The histogram of sample averages would be very different from the distribution in Figure 9.5. This is what we mean by a sampling distribution.

Why are sampling distributions important in statistical inference? For most real problems, the generic variable (discussed in Section 9.3) involves statistics calculated from samples, hence deriving the probability distribution of the generic variable requires an understanding of sampling distributions. For example, when performing a t test to compare two population averages, we calculate a t statistic that involves the two sample averages and standard deviations. To derive the t distribution, we need to understand the distributions of the two sample averages and standard deviations, which are sampling distributions. Similarly, the F test used in ANOVA is also based on sample averages and standard deviations. Virtually all of the confidence intervals and hypothesis tests discussed in Chapter 8 are based on some type of sampling distribution.

Following are some of the more common sampling distributions used in statistical inference applications.

The Sample Average

Perhaps the most commonly applied sampling distribution is the distribution of the sample average. This is used in the t test, ANOVA, and confidence intervals involving averages. For any distribution encountered in practice, the central limit

FIGURE 9.10 Sampling Distribution of Averages ($n = 4$) from a Normal Distribution With Average 50, SD 5

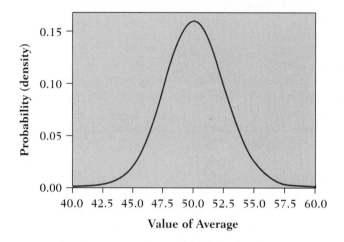

theorem tells us that this distribution will be approximately normal. The larger the sample size, the closer the distribution comes to exact normality. In addition, we can derive the exact average and standard deviation of the sampling distribution of the average for any sample size. For example, if a variable, say the cycle time for transferring funds in a bank, has a distribution with average μ and standard deviation σ, then sample averages of size n from that distribution will have a sampling distribution with average μ and standard deviation $\sigma/\sqrt{n}$. In other words, the average of the sample averages will be the same as for the raw data, but the standard deviation will be different. Specifically, the standard deviation of the sample averages decreases with the square root of the sample size. Therefore, if we take averages of samples of four transfers, these averages will have a standard deviation that is half of the standard deviation of the individual transfers. If we were to obtain a large number of these averages of four, their distribution would look like the sampling distribution shown in Figure 9.10. Note that the average is the same as for the individual data whose distribution is shown in Figure 9.5 (page 392), but the variation is much less. In fact, the standard deviation of the sample averages is 2.5—that is, $5/\sqrt{4}$. When we take averages, we do not change the average of the distribution, but we do reduce the variation.

The Central Limit Theorem

The central limit theorem says that if we sample from any distribution seen in practice and calculate sample averages, these sample averages will tend to have a normal distribution. The larger the sample size, the closer the actual distribution will be to an exact normal distribution. In addition, the long-term average of the sample averages ($\mu_{\bar{x}}$) will be the same as the average of the distribution from which we are sampling, and the long-term standard deviation of the sample aver-

What Does the "Average of the Averages" Mean? In the discussion of the sampling distribution of the sample average, we have noted what the average of this sampling distribution is. Because we are talking about the sampling distribution of the average, what does the average of the averages mean? It is the long-term average we would get if we defined a variable $\bar{x}$ that is the sample average of n observations. We could continue taking samples of size n for a long time period and calculate $\bar{x}$ each time. We would eventually create a probability distribution for the variable $\bar{x}$. This distribution would have an overall average, which is the "average of the averages."

ages ($\sigma_{\bar{x}}$) will be the long-term standard deviation of the distribution we are sampling from, divided by the square root of the sample size. In other words,

$$\mu_{\bar{x}} = \mu$$
$$\sigma_{\bar{x}} = \sigma/\sqrt{n}$$

where:

   μ is the average of the original distribution
   σ is the standard deviation of the original distribution
   n is the sample size

The central limit theorem is best appreciated graphically. Suppose we take samples of size 2 from the exponential distribution in Figure 9.9 and calculate the sample averages. What distribution would these sample averages have? How about averages of samples of size 5, or 10? Figure 9.11 shows histograms of averages of sample sizes 2, 5, and 10, as well as individual data from the original exponential distribution (i.e., random samples of size 1). In each case we have taken 1000 samples, the averages of which are plotted in the histograms. Of course, for $n = 1$, the "average" is just an individual observation sampled from this distribution. Note also that the scale has been kept constant for each graph, so that the degree to which the variation has been reduced by averaging is visible in the histograms.

Note that, even for averages of 2, the distribution is much more "normal" than for the individual data. For $n = 5$, the distribution of the averages is close to a normal distribution. For $n = 10$, the distribution is, for all practical purposes, normal. This is why Chapter 8 stated that some tests and confidence intervals (such as the t test) that theoretically assume a normal distribution are not sensitive to this assumption. Because we are actually performing the test on sample averages, we can be confident, based on the central limit theorem, that these averages will have a distribution very close to the normal, regardless of the original distribution. Therefore, we are not concerned about this assumption in practice. The central limit theorem also lets us use standard techniques—such as regression—when we are dealing with discrete data—such as percent defective. Even though such data may follow a binomial or Poisson distribution, the central limit theorem applies.

FIGURE 9.11

The Central Limit Theorem: Histograms of Sample Averages from an Exponential ($\mu = 1$) Distribution

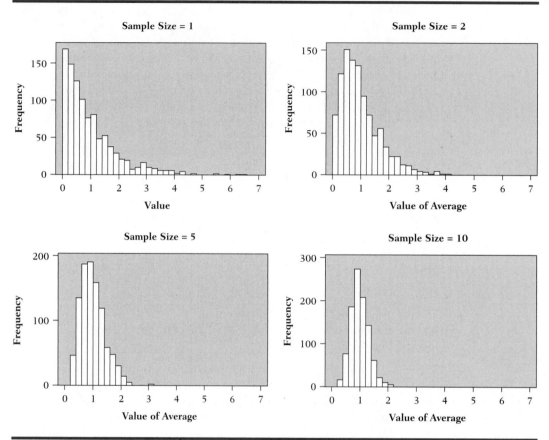

The exponential is perhaps the most nonnormal distribution seen in practice. Other common distributions will "become normal" faster as we take averages. For this reason, once we start taking averages of even size four or five, we tend not to worry about normality. Keep in mind, however, that we are not always primarily interested in the average. With prediction intervals, for example, we may be predicting individual observations, in which case the distribution is important from a practical point of view.

Although the central limit theorem has been theoretically known since the 19th century, the practical implications were not well understood until the middle of the 20th century. The main reason for this delay was the lack of computer technology, which eventually facilitated simulation studies like that shown in Figure 9.11. Prior to computers, the only way to determine how quickly the distribution of averages "became normal" was by manual calculation.

The Sample Variance (Standard Deviation)

Just as we can calculate the averages of samples, we can also calculate the sample variance or standard deviation. If we take several samples and calculate the standard deviation of each one, they will not be identical because of variation in the samples. There will be variation from one sample standard deviation to another, resulting in a sampling distribution for the standard deviation. It turns out that the sampling distribution of the variance (which is the standard deviation squared) is more easily quantified mathematically, hence we typically focus on the sampling distribution of the sample variance. Because the standard deviation is just the square root of the variance, we can use the sampling distribution of the variance to make any desired inferences concerning the standard deviation.

Quantifying the standard deviation, or variance, is useful in many statistical thinking applications. Chapters 1–4 discussed using it as an overall quantification of process variation. Chapter 6 discussed the standard deviation of model predictions as a measure of prediction accuracy. Chapter 8 explained how the standard deviation is required to perform many hypothesis tests, such as the *t* test. However, the ultimate intent of statistical thinking is to eliminate—or at least reduce—undesirable variation, not just to estimate it.

Figure 9.11 showed histograms of sample averages for samples of size 2, 5, and 10. Figure 9.12 shows histograms of sample variances from these same samples. Recall that the formula for the sample variance is:

$$s^2 = \frac{\sum (x - \bar{x})^2}{n - 1}$$

where

   $\bar{x}$ is the sample average
   Σ means "the sum of"

so we are summing the squared deviations of each data point from the average.

Recall also that the sample standard deviation, *s*, is just the square root of the variance—that is, $s = \sqrt{s^2}$. Note that the distributions of the sample variances in Figure 9.12 do not approach a normal distribution as the sample size increases. The central limit theorem does not apply to sample variances, only to sample averages and sums. These histograms reflect an approximate underlying χ^2 (chi-squared) distribution. Theoretically, the sample variance (times a constant based on the sample size) from a normal distribution has a χ^2 distribution. Because the distribution sampled here was the exponential, the sample variances will not have exactly a χ^2 distribution.

It is possible to derive the distribution of the standard deviation, the square root of the variance, from the χ^2 distribution. Because it comes from the square root of the χ^2 distribution, the distribution of the standard deviation is sometimes referred to as the χ (chi) distribution. Both of these distributions are dependent on the sample size, as is the distribution of averages.

Recall from Chapter 8 that when we compare two measures of variation, such as in ANOVA, or the *F* test, we utilize the *F* distribution. The *F* statistic is

FIGURE 9.12

Histograms of Sample Variances from an Exponential ($\mu = 1$) Distribution

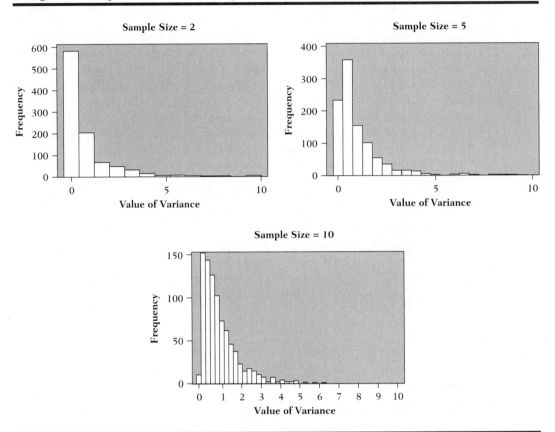

calculated by taking the ratio of two χ^2 distributions (times a constant based on the sample sizes), or alternatively, the ratios of two variances. Using the theory of statistical inference, one can derive the probability distribution of this ratio, which was named the F distribution by Snedecor (who discovered it) in honor of Sir Ronald Fisher, a pioneer in design of experiments and ANOVA.

The t Distribution

In Chapter 8 we discussed the t test for analyzing the difference between two averages. The distribution used in this test is a generic sampling distribution based on the sample average and standard deviation. It is generic in that we convert the actual distribution to one that does not depend on the specific average and standard deviation in the sample.

Recall that if we have a normal distribution, subtract the average, and divide by the standard deviation, the resulting variable has a standard normal, or z, distribution. In other words, $z = (x - \mu)/\sigma$. The standard normal is a generic distribution, in that it has an average of zero and standard deviation of 1, regardless of the value of μ and σ (that is, regardless of the original average and standard deviation of the variable x). Therefore, if x is the cycle time of approving a loan application and has a population average of 4 hours with a population standard deviation of .5 hour, then the variable $z = (x - 4)/.5$ has a standard normal distribution. Of course, with real applications we never know the true population average or standard deviation, μ or σ. The population average tends not to be a problem, because the null hypothesis often assumes that this is a specific value, such as zero. When comparing two averages, for example, we assume that the *average difference* between the two variables is zero.

The standard deviation is often an issue, however, particularly when we are testing averages and don't make assumptions about the value of the standard deviation. In practice, we often must estimate the population standard deviation using the sample standard deviation. In other words, instead of calculating $(x - \mu)/\sigma$, we actually calculate $(x - \mu)/s$. (This was done in our paper towel absorbency example in Chapter 8.) However, because there will be variation in the sample standard deviation s, the distribution we get for $(x - \mu)/s$ tends to deviate from the theoretical z distribution. For example, in the loan cycle time example, suppose that we don't know the true population standard deviation, but are testing the null hypothesis that the population average equals 4. We can no longer calculate the standard normal variable z, because we don't know the population standard deviation σ. Instead, we calculate the variable $t = (x - 4)/s$, where 4 is the null hypothesis value for the population average. The exact distribution we obtain is referred to as a t distribution:

$$t = (x - \mu)/s$$

Note that in many cases, the x variable of interest is a sample average, $\bar{x}$, which has an average of μ and a standard deviation of $\sigma/\sqrt{n}$ (estimated by $s/\sqrt{n}$). This results in the formula:

$$t = \frac{\bar{x} - \mu}{s/\sqrt{n}}$$

This distribution was developed by Gossett (see Box, Hunter, & Hunter, 1978). Gossett worked in a brewery in the United Kingdom and applied statistical analysis to improve the brewing process. He was frequently frustrated in trying to apply the standard normal distribution to small sample sizes, because his data generally had more variation than the theory predicted. He realized that this was due to an additional source of variation or uncertainty—the variation in *estimating* the standard deviation, versus *knowing* the true population value, σ. Using the theory of statistical inference, he finally succeeded in deriving the exact sampling distribution of the variable $(x - \mu)/s$, but his business did not allow employees to publish. He therefore published the results under the name "Student," and to this day the t distribution is often referred to as "Student's t."

This t distribution is not dependent on μ or σ—that is, it is "generic"—but it does depend on the sample size. (Recall the discussion of "degrees of freedom" from Chapter 8.) This is because the larger the sample size, the better we can estimate σ from s, or the closer the t distribution will be to the z distribution. In fact, once we get to a sample size of about 30, the t and z distributions are almost identical. Therefore, for 95% confidence intervals for the average, with large sample sizes we use the z distribution value of 1.96, rather than the exact t value. When the sample size is less than about 30, the t and z distributions are different enough to warrant use of the exact t value. (See Section 8.4 to review confidence intervals.)

9.7   Linear Combinations

When comparing two averages in hypothesis testing, we typically assume that the average difference between the two variables is zero. We do this by creating a third variable, say D, that is the difference between the two original variables of interest. In other words, we specify $D = x_1 - x_2$. The null hypothesis that the two population averages are the same can alternatively be stated as the population average of D is zero. For example, suppose we are comparing two express delivery services. We wish to test the hypothesis that the population average delivery times are the same for the two services. The null hypothesis could then be stated as $H_0: \mu_1 = \mu_2$.

Alternatively, we can define the variable D to be the difference between the delivery times of the two delivery services. In other words, we can define $\Delta = \mu_1 - \mu_2$, where Δ ("delta") is the population (long-term) average of D. The null hypothesis can then be stated as $H_0: \Delta = 0$. Why would we want to state the hypothesis this way? Quite simply, because this is an easier problem to solve. In the original statement of the problem, we have to worry about two random variables (each business's average delivery time), each of which is subject to variation and uncertainty. By creating a third variable that is the difference between the two original variables, we have simplified the problem so that it only involves one random variable. Testing that two average delivery times are different is now as simple as testing that one average is equal to zero. The only complication is that we have to derive the probability distribution of this generic variable, which is the difference between the two delivery times. This requires a special type of sampling distribution using linear combinations.

A linear combination is any variable that is formed by some linear equation of other variables. By linear equation, we mean that the linear combination can be written as follows:

$$y = c_0 + c_1 x_1 + c_2 x_2 + \cdots + c_n x_n$$

where

   y is the new variable, the "linear combination"
   x_1 is the first variable in the linear equation, x_2 is the second, and so on
   c_0 is an arbitrary additive constant
   c_1, c_2, and so on are arbitrary multiplicative constants

For example, in the delivery service case, $D = x_1 - x_2$, so c_0 is zero, c_1 is +1, and c_2 is −1. This is the same form as a linear regression equation. There are many other cases where creation of a linear combination simplifies statistical inference problems. For example, profit can be written as income minus expenses, and the cycle time of a sequential process can be written as the sum of the cycle times of each step in the process. The Consumer Price Index (CPI) is a weighted sum of prices for goods and services. The CPI is therefore a linear combination of individual prices, where the weights are the multiplicative constants (c_1, c_2, and so on).

How does the creation of a linear combination simplify the statistical inference problem? Fortunately, the average and standard deviation of the linear combination can be directly calculated from the averages and standard deviations of the variables in the linear combination. To be precise, it can be shown that

$$\mu_y = c_0 + c_1\mu_1 + c_2\mu_2 + \cdots + c_n\mu_n$$

and

$$\sigma_y = \sqrt{c_1^2\sigma_1^2 + c_2^2\sigma_2^2 + \cdots + c_n^2\sigma_n^2}$$

In other words, the average of the linear combination will just be the same linear combination of the individual variable averages. Note, however, that the standard deviation of y is *not* just the linear combination of the individual variable standard deviations. As this equation shows, the variances add, but the standard deviations do not. We add the variances (times the squared constants) and then take the square root of the final sum. This is sometimes called the "root sum of squares" formula, because we are taking the square root of the sum of the squared standard deviations. Note also that the additive constant c_0 has no impact on the standard deviation of the linear combination. We cannot determine the exact probability distribution of the linear combination from this formula, but according to the central limit theorem it will tend to be normal as the number of items in the sum increases.

The only theoretical assumption required for these formulas is that the variables x_1, x_2, x_3, and so on are approximately independent—that is, they are not correlated with one another. For example, we assume that if x_1 happens to be particularly large, this will not imply that other x's will be particularly large or small. (We can still directly calculate the average and standard deviation of the linear combination even if this assumption is not true by utilizing the correlation coefficients among the x's, but it is more complex than the equations noted above and beyond the scope of this text.) This will generally be a safe assumption, but will not always be valid. For example, if each of the x's is the price of a product or service in the U.S. economy (as is the case in the CPI), we would expect these prices to be highly correlated. If we know that the price of crude oil has increased sharply, it is likely that other prices will also be increasing. General economic dynamics, such as inflation or deflation, would likely cause these variables to move up and down together.

The formulas for the average and standard deviation of a linear combination greatly simplify the task of deriving the distributions of generic variables. For example, consider the situation where we wish to predict the average and variation of an interference fit in manufacturing bottles for medication. An interference fit means that the lid is actually smaller than the bottle opening it covers. The lid stretches and is held firmly in place by the friction between it and the bottle. The interference itself is very difficult to measure directly, but we can certainly measure the dimensions of the lid and the bottle. The interference can then be calculated as the difference between the bottle diameter and the lid diameter. In other words, $I = LD - BD$, where I is the interference, LD is the lid diameter, and BD is the bottle diameter. In this case, we want I to be negative—that is, we want the lid diameter (LD) to be less than the bottle diameter (BD) to ensure a snug fit.

By taking measurements of the bottles and lids that have been manufactured, we may determine that these processes are stable, with averages and standard deviations as follows:

	Average	Standard Deviation
Bottles (BD)	4.2 cm	.07 cm
Lids (LD)	3.9 cm	.12 cm

Given this information, what could we conclude about the average and variation of the interference fit? Given that $I = LD - BD$, I is a linear combination of the bottle and lid dimensions, with constants $+1$ and -1. It also appears reasonable to assume that bottle and lid dimensions are independent. If a particular bottle is large, this does not imply that the lid it will be combined with will be particularly large or small. We can therefore draw the following conclusions:

Average of I will be $3.9 - 4.2 = -0.3$

Standard deviation of I will be $\sqrt{(.07^2 + .12^2)} = 0.139$

We can now use this information to draw inferences about the strength of the interference fit, such as the probability that it is positive—that is, that there is no interference and the lids will fall off. This information could also enable us to set initial control limits on control charts of the interference.

Recall from the discussion of probability distributions in Section 9.5 that when we convert to a standard normal (z) distribution through the formula $z = (x - \mu)/\sigma$, z will have a zero mean and standard deviation of 1.0. We can derive this solution by applying linear combinations theory. Another way of expressing $z = (x - \mu)/\sigma$ is $z = -\mu/\sigma + x/\sigma$. These two expressions for z are mathematically identical, but the latter shows z to be a linear combination of the one variable x, where $c_0 = -\mu/\sigma$, and $c_1 = 1/\sigma$. This is because x is the random variable, and μ and σ are just constants. Therefore, the average of the variable z will be:

$c_0 + c_1\mu = -\mu/\sigma + \mu/\sigma = 0$

The standard deviation of z will be

$$\sqrt{([1/\sigma^2]\sigma^2)} = \sqrt{(\sigma^2/\sigma^2)} = 1.0$$

In this case, the z value for I is $[I - (-.3)]/.139$, or $(I + .3)/.139$.

9.8 Transformations

Why Do We Use Transformations?

The analysis of residuals from the model is an important step in the modeling process (see Chapter 6). Among other things, we look for serious departure from normality in the residuals. One reason to do this is that some of the analyses performed in conjunction with the model building, such as calculation of prediction intervals for a future observation, are dependent on the assumption of normality. In addition, conspicuous departure from normality is often a symptom of more serious issues, such as the existence of significant curvature that has not been accounted for in the model. One way to address nonnormality in the residuals is to measure the response (y) variable in another metric. This approach is called a "transformation," and the new variable is called the "transformed variable," or "transformed response."

Regression analysis is used as one example here, but the use of proper transformations is extremely useful in a variety of statistical applications. Transformations often enable us to develop a model that is closer to the physical relationships of the variables of interest and therefore provides more insight into the process. Some of the objectives of performing a transformation include the following:

- To achieve more linear relationships between variables
- To stabilize the variation, so it is more constant over the range of data
- To achieve a better approximation of the normal distribution

All of these objectives make the application of statistical inference more straight-forward.

Consider the common example of monitoring the level of the stock market, or the price of one particular stock. It is well known that the stock market does not increase linearly over time. Like most financial investments, even bank accounts, we expect our stock investments to increase *proportionally* to the stock's current value. In other words, although it may be acceptable for a $10 stock to increase $1 in a year, we would not be happy if it only increased $1 year after year. After 10 years, the $1 increase from $20 to $21 would only be a 5% return. After 90 years, the $1 increase from $100 to $101 would only be a 1% return. Obviously, we would expect the increase in the stock to be some proportion of the current price, such as 10% or 15% under normal conditions.

This is known as an exponential increase, and the equation for the value of an investment earning a fixed rate of return is an exponential equation:

$$V = P(1 + IR)^Y$$

where

V is the current value of the investment
P is the original principle
IR is the interest rate (expressed as a proportion, i.e., .1 for 10%)
Y is the number of years we have earned interest

For example, if we invest $100 at 10% interest for 5 years, the value of the investment after 5 years would be

$$\$100(1.1)^5 = \$161$$

Each year, the value increases by more dollars than it did the previous year because of the compounding of interest—that is, because we earn interest on all of the previous years' interest. This relationship is shown in graph A in Figure 9.13. Note that

FIGURE 9.13   Exponential Investment Growth on Original and Log Scales

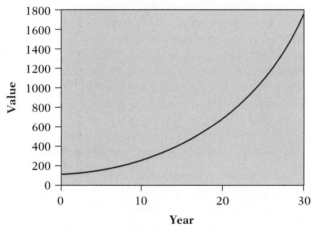

A. Value of a $100 Investment Growing 10% per Year (Original Scale)

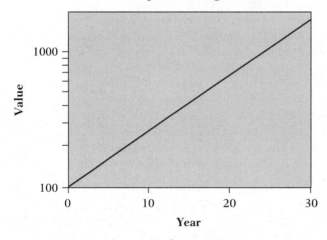

B. Value of a $100 Investment Growing 10% per Year (Log Scale)

it is clearly exponential, not linear. In many such cases, the variation in return that we observe about the regression line through the data is not constant, but proportional to the value of the response. In other words, the variation in return that we might observe at $100 is likely to be much more than at $1. This complicates our statistical inference because we must account for the unequal variation, not to mention nonlinearity and potential nonnormality of the residuals. (Note that "exponential" relationship does not imply an "exponential" probability distribution for either variable. These are separate concepts that use the same word.)

What Is a Logarithm? A logarithm, or log, is a transformation commonly used in business and engineering. It is frequently used in practice because, as we saw in the investment example, it makes an exponential relationship linear. The definition of the $\log(x)$ is: "that value, which when made an exponent of 10, produces the value x." For example, the log of 10 is 1, because $10^1 = 10$, $\log(100) = 2$ because $10^2 = 100$, $\log(1000) = 3$, and so on. This is for logarithms based on the number 10. It is also possible to base the logarithms on other numbers. For example, engineers often base logs on the natural number e (approximately 2.72), because this number is vital to the study and application of calculus, which is very important in engineering. Logarithms based on the number e are called "natural logs" and sometimes designated as "ln." Fortunately, logarithms based on any number all work the same way and differ only by a multiplicative constant. From a practical point of view, it does not matter which base we select. Note that logs are not defined for negative numbers and cannot be calculated for zero because, theoretically, the log of zero is minus infinity.

Mathematically, however, if we perform a logarithmic ("log") transformation on V in the above equation, our transformed equation becomes

$$\log(V) = \log(P) + \log(1 + IR)Y$$

This is because it can be shown mathematically that $\log(a^b) = b \log(a)$. This is a fundamental property of logarithms. Because P and the quantity $(1 + IR)$ are just constants and do not depend on the variable Y (years), similarly $\log(P)$ and $\log(1 + IR)$ will also be constants. For a given investment at a fixed interest rate, Y is the only part of the equation that changes—it is the only variable. This equation now looks like a standard linear regression equation (discussed in Chapter 6), where $\log(P)$ is the constant term, and $\log(1 + IR)$ is the slope. This is now a linear equation, and the transformed variable $\log(V)$ will increase by a constant amount each year, specifically, by $\log(1 + IR)$. If there is variation in Y, it is likely that the variation in $\log(V)$ will also be much more constant, and the distribution will probably be closer to a normal distribution than for V. Graph B in Figure 9.13 shows the same relationship, but with a transformed vertical (V) axis (log scale), resulting in a linear relationship.

Other Examples of Transformations

This phenomenon of investment growth is well known in the business community. Figure 9.14, for example, is a diagram from the *International Herald Tribune* (March 20/21, 1999) that shows the growth of the Dow Jones Industrial average from 1972 to 1999. Graph A shows growth in actual dollars plotted on linear scales. As noted above, however, we would expect the Dow to increase at an exponential, rather than linear, rate. Therefore, Graph B shows the same data, but on

FIGURE 9.14 Three Faces of the Dow

A. Dow Jones Industrial Average, Price Index

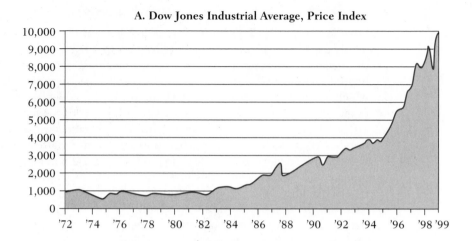

B. Log Scale

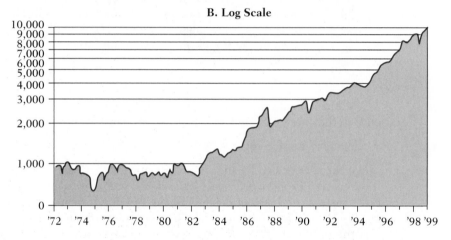

C. Inflation-Adjusted Performance: Prices in 1972 Dollars

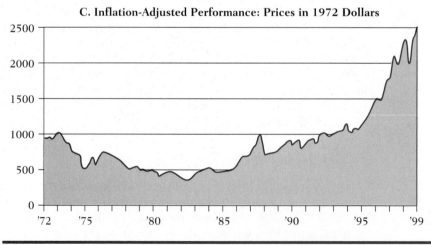

Source: "In the Age of Technology Stocks, Does the Dow Matter?" *International Herald Tribune,* March 20/21, 1999, pp. 9–10.

a transformed vertical scale that is not linear, but logarithmic. In other words, 1000, 2000, 3000, and so on are not equally spaced. Plotting the original variable on a transformed scale is mathematically equivalent to transforming the variable and then plotting on a standard linear scale.

When viewed on a log scale, the increase in the Dow appears to be approximately a linear increase, which is what one would expect of any investment when plotted on a log scale. Clearly, the increase in the Dow over this time period is not as impressive as one might think when viewing a plot of the raw data. Similarly, at the time of this writing, the Internet site CNBC.com and many others provide stock graphs on a logarithmic scale, to allow a more critical evaluation of stock performance.

Another use of transformations occurs with financial data, wherein the value of a unit of currency, such as the dollar or deutsche mark, is not constant, because of inflation. For example, a dollar in 1930 had much greater buying power than does a dollar in 2000. In order to eliminate the effect of inflation, financial figures are often reported in "constant dollars" based on a particular year. This transformed metric is obtained by taking the current year original figure and dividing by some measure of inflation between the base year and the current year, such as the Consumer Price Index. For example, in looking at financial data between 1970 and 2000, we might want to transform each year's dollar figure to constant 1970 dollars. This would be accomplished by calculating

constant dollars = original dollars/inflation

where "inflation" is a measure of the degree of inflation between the base and current year.

For example, if the CPI suggests that prices roughly tripled between 1970 and 2000, "inflation" for 2000 would be 3.0. Graph C in Figure 9.14 shows the same Dow data transformed to constant dollars. Again, we see that by eliminating the effect of inflation, the increase in the Dow is not as dramatic as one might have originally thought, at least prior to 1995.

A transformation is any change of metric for a variable of interest. Therefore, we could consider converting degrees Fahrenheit to degrees Celsius a transformation. This would be a linear transformation, because $C = (F - 32)/1.8$. Linear transformations do not change the shape of relationships, affect distribution, or stabilize variance, hence they are generally used for convenience, rather than for the three primary reasons noted on page 409 for using transformations in statistical inference. Other common transformations include converting money from one currency to another, using decibels to quantify sound volume (decibels involve a log transformation of sound energy), and using pH to quantify acidity (pH involves a log transformation of chemical particles involving hydrogen).

Common Transformations   Some of the common transformations used in practice are the log transformation, the square root, and the inverse. The log has already been discussed. The square root is just the same as the common square root button on a handheld calculator. The inverse transformation is 1 divided by the original variable, or $1/y$. For example, if y is the time to failure for a computer system, then $1/y$ is the rate of failure (failures per time period). Note that when taking the inverse, the scale is reversed—that is, a large value of y produces a small value of $1/y$. For this reason, a slight modification of the inverse transformation is often used in practice, such as $(1 - 1/y)$, or 1 minus the inverse. In this modification, a large value of y produces a large value of the transformed variable, which is easier to understand. Subtraction from 1 does not change the degree of skewness or normality, because it only involves use of a constant.

Reverse Transformations   The reverse transformation converts the transformed metric back to raw units—that is, it reverses the transformation (a useful practice when communicating with colleagues not trained in statistics). For example, if the transformation is a square root transformation, then the square is the reverse transformation: $y = \left(\sqrt{y}\right)^2$. Taking 10 to the power of the $\log(y)$ returns to the original metric for the log transformation: $y = 10^{\log(y)}$. Similarly, the inverse is the reverse of the inverse transformation: $y = 1/(1/y)$. If we have used the $1 - 1/y$ form of the inverse transformation, then the reverse transformation becomes $1/(1 - y)$, since $1/(1 - [1 - 1/y]) = 1/(1/y) = y$.

Transformations are fundamental to statistical inference. For example, the step of converting to generic variables, discussed in Section 9.3, always involves some type of transformation. A z (standard normal) distribution is just a normal distribution transformed to have an average of zero and standard deviation of 1.0. A t distribution is a transformation typically involving an average and a sample estimate of the population standard deviation. The F distribution is a transformation of two variances. In each case, the generic variable to which we have transformed allows us to develop a standard solution to the problem that is not dependent on the specific average and standard deviation of this particular data set. Note that the distribution of the transformed variable is typically different than the original variable, unless a linear transformation has been used.

The Goodwill Case

Recall Exercise 5.12 involving the amount of "goodwill" associated with acquisitions: Goodwill is the amount over and above the fair market value that we must pay to acquire a company. Some amount of goodwill is almost always necessary to provide incentive to the current owners to sell. Table 5.3, which included the amount of goodwill, the total assets of the company being acquired, and the term of the amortization of the goodwill, is repeated here as Table 9.3.

In Exercise 6.15 these data were to be analyzed in a regression model, and it was noted that a log transformation should be used for total assets and goodwill. Why was a transformation called for in the first place, and how do we know that the log transformation is appropriate? We will now reevaluate these data and walk through the process of detecting the need for transformations, determining the appropriate transformations, and presenting the conclusions appropriately. We will then formally present this overall process for applying transformations.

The model from Exercise 6.15 considered goodwill the response (y) and total assets and term of amortization the explanatory (x) variables. As explained in

TABLE 9.3 Goodwill Data (in millions, except Amortization, which is in years) from Table 5.3

Goodwill	Total Assets	Amortization	Goodwill	Total Assets	Amortization
9.5	3.8	10	32.1	18.8	15
5.2	4.5	5	36.3	7.2	30
0.8	1	10	1200	800	40
3.4	1.9	15	16.2	9.2	10
2.9	0	10	0.6	0.2	5
18	8	12	41	57	15
0.35	0.115	5	334	105	25
1	1.4	10	13	0.7	5
24.1	14.2	10	5.2	2.9	10
9	7.2	10	7.3	10.4	15
1.7	1.3	10	2.5	1.6	20
31	36.6	20	3.9	2.6	20
13.7	11.4	10	0.8	0.6	20
956	752	40	16.7	9.7	20
454	324	20	4.6	3.6	15
1266	601	40	29.3	6.4	10
2.2	1	10	8.1	2	10
2.1	0.3	10	5	14.8	7
1.9	0.8	15	43	13.8	10
1.2	1	10	14.6	9.7	10
46.4	28.8	15	6.1	5.1	7
422.4	281.9	30	27	18.4	10
181	83	30	11.9	10.2	10
1.4	1.2	5	186.5	336.6	30
11.3	8.5	15	20.3	18.1	10
13.5	10.7	10	2.3	1.8	10

Chapter 6, a logical first step to performing regression analysis is to examine diagnostic plots of the variables to help identify an appropriate model. Figures 9.15, 9.16, and 9.17 show a histogram of goodwill and scatter plots of total assets and amortization versus goodwill, respectively. Clearly, goodwill has a very skewed distribution, and the relationship between goodwill and amortization has significant curvature. In addition, although the relationship between goodwill and total assets appears relatively linear, most of the data are clustered in the lower left, and the remaining data appear to have more variation as we move toward the upper right of Figure 9.16. All of these issues point toward the need for some type of transformation. As a general rule of thumb, we anticipate having a problem with unequal variation whenever the range of a variable covers more than a factor of 10—that is, max/min > 10. For goodwill we have max/min = 1266/.35 = 3617, which is much greater than 10.

We will ignore these warning signs and proceed with a standard regression analysis, however, to illustrate how one might also use diagnostic plots from a regression analysis to identify the need for a transformation. The output of the

FIGURE 9.15   Histogram of Goodwill

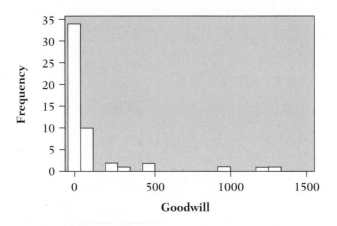

FIGURE 9.16   Scatterplot of Goodwill Versus Total Assets

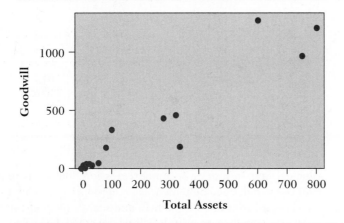

regression analysis utilizing these variables with goodwill as the response (y), and total assets and amortization as the explanatory (x) variables is given in Table 9.4, and the residual plots from that analysis are given in Figure 9.18. In the plots in Figure 9.18, we can clearly see outliers, nonnormality, and unequal variation. Close examination of Graph D, the plot of residuals versus predicted values (fits), indicates unexplained curvature as well. A transformation is clearly called for, but which one? A log transformation is a logical choice when the residuals versus predicted values plot produces a funnel or "megaphone" shape, indicating that the

FIGURE 9.17   Scatter Plot of Goodwill Versus Amortization

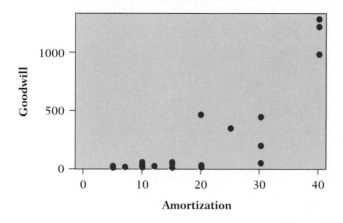

TABLE 9.4   Regression Output of Goodwill Using Total Assets and Amortization

The regression equation is
Goodwill = −23.2 + 1.41 Total Assets + 2.08 Amortization

Predictor	Coef	SD	T	P
Constant	−23.21	26.96	−0.86	0.393
Total Assets	1.4072	0.1024	13.74	0.000
Amortization	2.083	2.018	1.03	0.307

S = 77.90 R^2 = 92.5% R^2(adj) = 92.1%

variation is proportional to the average. In other words, as the predicted value increases, so does the variation in the residuals. If we take the logarithm of goodwill and repeat the histogram and scatter plots (shown in Figures 9.15, 9.16, and 9.17) with total assets and amortization period in Figures 9.19, 9.20, and 9.21, we see that log(goodwill) looks more normally distributed, and the relationship between amortization period and log(goodwill) now appears linear.

However, the relationship between log(goodwill) and total assets now has a great deal of curvature. Because this had been an approximately linear relationship prior to transforming goodwill (see Figure 9.16), and given that total assets is also highly skewed, performing the same log transformation on total assets seems to make sense. The plot of log(goodwill) versus log(total assets) in Figure 9.22 now looks not only linear, but the variation now appears to be constant over the entire range of the data. This is called a "log-log" plot. (Note that we eliminated the data point shown in Table 9.3 that had total assets of 0.0, because we cannot take a log

FIGURE 9.18

Residual Model Diagnostics: Plots for Goodwill Using Total Assets and Amortization as *x* Variables*

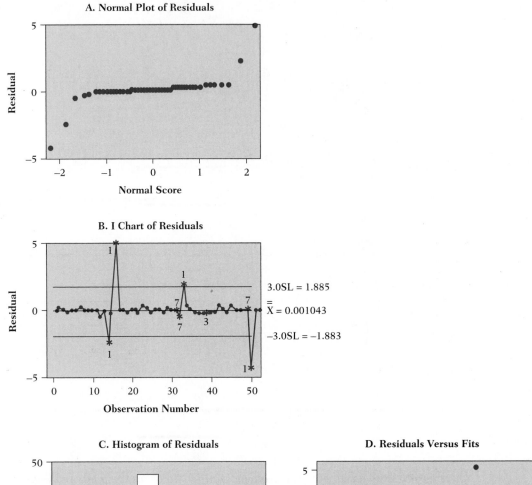

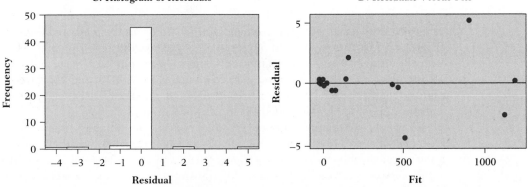

*Residuals plotted for this example are "standardized" by dividing the raw residual by its standard deviation. They should therefore have approximately a *z* distribution since the raw residuals' average is zero and we have a large sample size (52).

FIGURE 9.19   Histogram of log(Goodwill)

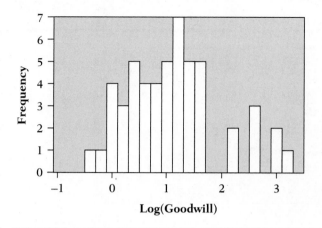

FIGURE 9.20   Scatter Plot of log(Goodwill) Versus Total Assets

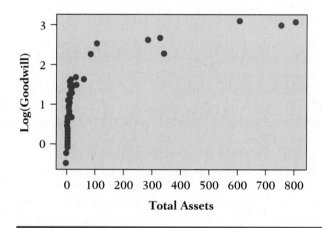

FIGURE 9.21   Scatter Plot of log(Goodwill) Versus Amortization

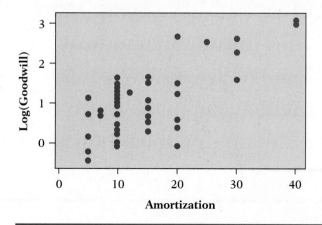

FIGURE 9.22   Scatter Plot of log(Goodwill) Versus log(Total Assets)

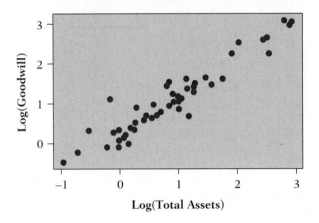

of zero [theoretically, it is minus infinity].) The regression analysis is now repeated using log(goodwill) as the *y* variable, and amortization period and log(total assets) as the *x*'s. The output is given in Table 9.5, and the residual plots in Figure 9.23. Although abnormal behavior remains in some of the residual plots, particularly an apparent outlier, the overall behavior of the residuals is much better. Because amortization period is not quite statistically significant (its *p*-value is slightly greater than .05), we might choose to drop this variable from the analysis. This would suggest that, if we know the total assets of the acquisition, we don't really need the amortization period to predict goodwill. The new model produces statistical output and residuals given in Table 9.6 and Figure 9.24, respectively.

Note that we are not concerned that the adjusted R^2 has slightly decreased in the transformed metric. The current model is much more appropriate because it

TABLE 9.5   Regression Output of log(Goodwill) Using log(Total Assets) and Amortization

The regression equation is
log(Goodwill) = 0.210 + 0.833 log(Total Assets) + 0.0104 Amortization

51 cases used, 1 case contains missing values

Predictor	Coef	SD	T	P
Constant	0.21011	0.07418	2.83	0.007
log(Total Assets)	0.83290	0.06141	13.56	0.000
Amortization	0.010391	0.006120	1.70	0.096

S = 0.2697 R^2 = 90.8% R^2(adj) = 90.4%

FIGURE 9.23
Residual Model Diagnostics: Plots for log(Goodwill) Using log(Total Assets)
and Amortization as *x* Variables

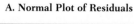

A. Normal Plot of Residuals

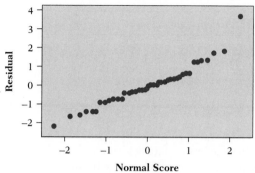

B. I Chart of Residuals

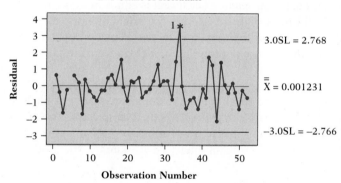

C. Histogram of Residuals

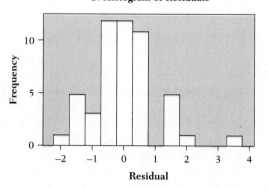

D. Residuals Versus Fits

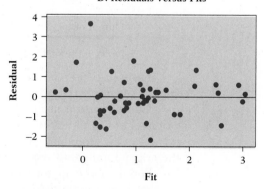

TABLE 9.6   Regression Output of log(Goodwill) Using log(Total Assets)

The regression equation is
log(Goodwill) = 0.298 + 0.909 log(Total Assets)

51 cases used, 1 case contains missing values

Predictor	Coef	SD	T	P
Constant	0.29800	0.05414	5.50	0.000
log(Total Assets)	0.90925	0.04263	21.33	0.000

$S = 0.2749$ $R^2 = 90.3\%$ $R^2(\text{adj}) = 90.1\%$

better satisfies our assumptions and fits the entire data set much better. The normal probability plot (A) and histogram (C) of residuals in particular look more normal in Figure 9.24 than in Figure 9.23. All four charts are much better in Figure 9.24 than in Figure 9.18, which was done for goodwill in original units. The larger R^2 for the original model is misleading because it is almost totally caused by fitting the extreme goodwill values slightly better. In the original metric these values dominate the analysis because of their magnitude, to the near exclusion of the other goodwill values. Finding a more appropriate model via transformation will usually result in improved summary statistics, but this is not always the case.

Now that we have found a more appropriate formulation of the model, the next question is how to present the results. Financial managers are not likely to understand what log(goodwill) means. To present the results in terms that make practical sense, we can calculate the "reverse transformation"—that is, the transformation that brings us back to the original metric. For example, the model we currently have is:

log(Goodwill) = 0.298 + 0.909 log(Total Assets)

Using the laws of logarithms, this is mathematically equivalent to:

$$\text{Goodwill} = 10^{0.298 + 0.909 \log(\text{Total Assets})}$$
$$= 10^{0.298} \times 10^{0.909 \log(\text{Total Assets})}$$
$$= 1.99 \times 10^{0.909 \log(\text{Total Assets})}$$

However, graphs of the variables in original units are likely to be the best method to convey the relationship. Using modern statistical software we can fit the model in transformed units, but present the results in the original metric. For example, in Figure 9.25 we show a plot of total assets versus goodwill, with the regression equation and confidence/prediction intervals included. The plot is in the original units, but the regression line and confidence/prediction bands are based on the more appropriate model in log units. Financial managers can now see the analysis in units they understand, but with statistically accurate bands of uncertainty. Because the variation is not constant, the confidence/prediction bands are increasing for higher values of goodwill or total assets.

FIGURE 9.24
Residual Model Diagnostics: Plots for log(Goodwill) Using log(Total Assets) as x Variable

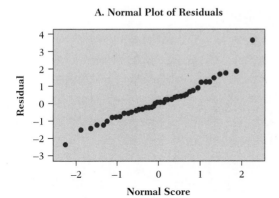

A. Normal Plot of Residuals

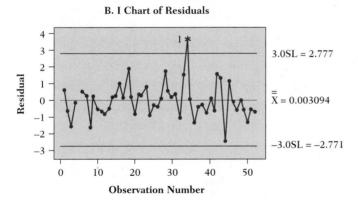

B. I Chart of Residuals

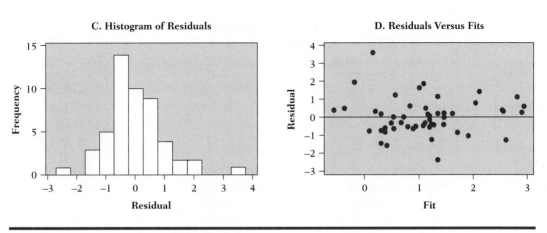

C. Histogram of Residuals

D. Residuals Versus Fits

FIGURE 9.25   Plot of Goodwill Versus Total Assets With Regression Line and Confidence/Prediction Intervals from Log-Log Regression

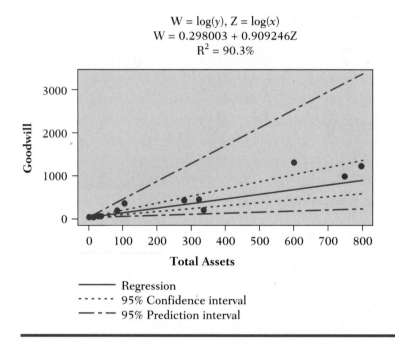

$$W = \log(y), Z = \log(x)$$
$$W = 0.298003 + 0.909246Z$$
$$R^2 = 90.3\%$$

———— Regression
- - - - - 95% Confidence interval
— - — 95% Prediction interval

The Process of Applying Transformations

Transformations add complexity to our analysis, hence we generally do not apply them unless there is a particular reason to do so. Like other activities, applying transformations involves a process. The overall process for applying transformations is shown in Figure 9.26. The first step is to identify an issue, with analysis of the original data, that a transformation may help address. One example would be the identification of unexplained curvature in residual plots from a regression analysis. Attempting to model a response that is known to be highly skewed, such as cycle time, would be another example. Both of these were issues in the goodwill case.

The next step is to determine what transformation—log, inverse, or square root—is likely to resolve the issue. The type of issue we are facing may give insight into the transformation that is most likely to be successful. For example, it is well known that if we have a Poisson random variable, such as the number of visual defects in the finish of a new car, the variation of number of defects will not be constant, but will vary with the average (recall that for the Poisson, the standard deviation equals the square root of the average). Therefore, if the number of visual defects is a response variable in a regression analysis, the assumption that the variation in the response is constant will be violated. However, if we analyze the square root of a Poisson-distributed variable (the number of visual defects in this

FIGURE 9.26   The Process of Applying Transformations

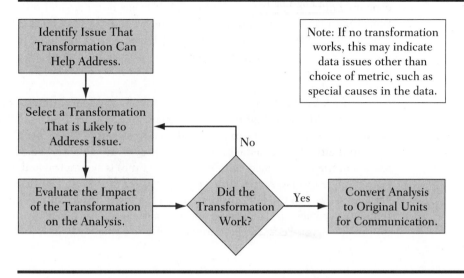

case), the variation of this transformed variable will be approximately constant, regardless of the level of the average. This result also holds, although to a somewhat lesser degree, for binomial random variables, such as percent defective.

Each of the three common transformations (described on page 413) reduces skewness in a right-skewed variable (i.e., a distribution where the right "tail" is longer than the left "tail"). The square root is least dramatic, followed by the log, and then the inverse, which is most severe. Therefore, for extreme skewness, the inverse transformation would be a likely first choice. This is often used when analyzing cycle time if we believe that the time to occurrence is highly skewed or nonlinear, but the rate of occurrence (1/time) is likely to be fairly linear. For discrete counts, which may have a Poisson distribution, or even a binomial, the square root is a logical first choice. Where exponential relationships are expected, such as investment performance over time (as in Figure 9.14), the log is a likely choice. When a plot of residuals versus predicted values shows levels of variation increasing proportional to the average (a "megaphone" effect), the log is a logical transformation. This was the situation in the goodwill case.

Of course, there is no limit to the number of possible transformations; these are just three of the more commonly applied transformations. A more rigorous approach to identifying appropriate transformations is to apply the Box-Cox family of power transformations, which makes calculations and produces graphs that indicate the most appropriate transformation. (For more information on this method, see Box, Hunter, and Hunter, 1978.)

The next step is to evaluate the impact of the selected transformation by repeating the analysis in the transformed metric and determining if the original issue has been successfully addressed. For example, if systematic curvature was detected in residual plots from a regression analysis, repeat the regression analysis using the transformed variable and look at the new residual plots to see if the un-explained curvature has been addressed. If it has not, consider other transformations that might work. If no transformation works, this may be an indication of a more serious problem than choice of metric, such as an unstable process or poor choice of a model. In the goodwill case, the extreme nonnormality, curvature, and unequal variation in the residuals were almost totally eliminated by taking logs of goodwill and total assets.

Figure 9.27 shows the skewed histogram of a random variable—cycle time in this case—and the histograms of the square root, log, and inverse of cycle time. Note that the square root only partially addresses the skewness: There is still a skew to the right. The histogram of the log seems to have worked, in that the histogram of the transformed variable log(cycle time) is symmetric and approximately normal. The inverse seems to have gone too far, however. The histogram is now actually skewed to the left. For these data, the log transformation appears best for reducing skewness and achieving approximate normality. It is likely this would also minimize curvature in the relationship of y to other variables, and stabilize the

FIGURE 9.27

Histograms of Cycle Time and Several Transformations

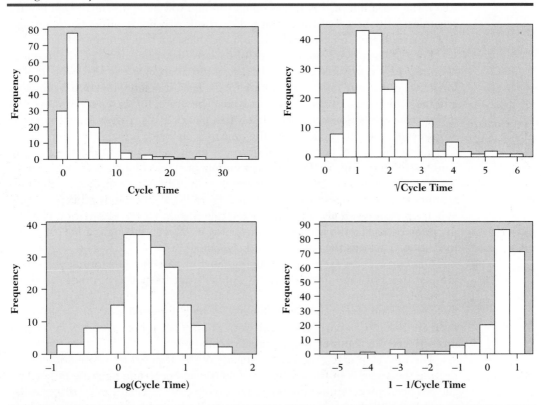

variation. Recall from Chapter 6 that in regression analysis the residuals are assumed to be normally distributed, not the original y data. However, it is the original data that we transform, typically beginning with the y variable.

Communicating the Results If the selected transformation has addressed the original issue, we will no doubt obtain a more meaningful and theoretically valid analysis. We do have a new problem, however, in communicating the results of the analysis to others. For example, if we are working on a cycle time reduction project, business leaders will not be interested in how much we have reduced log(cycle time); they will want to know how much we have reduced "real" cycle time. Similarly, it would be confusing to announce that we had increased square root of income; naturally people are more interested in income than the square root of income. Therefore, present graphs and summary results in the original metric: Perform all statistical analyses in the most appropriate transformed metric and then convert the final predicted responses or graphs into original units. (This was done for the goodwill analysis in Figure 9.25.)

Suppose we are comparing the cycle time to approve credit applications between three branches of a financial institution. In plots of the data we detect extreme skewness and find in residual plots from an analysis of variance (ANOVA) that the variation of cycle time increases with the average. Because this lack of constant variation violates one of the assumptions of ANOVA, we decide to transform cycle time. A log transformation is selected. Analysis of log of cycle time indicates minimal skewness, and the variation of the residuals for the transformed response appears constant and equal for all three branches. We are therefore convinced that the log is an appropriate transformation and perform our ANOVA on log(cycle time). This analysis demonstrates a difference between the three branches (p-value is less than .05), with Branch 1 having a particularly long cycle time, and Branch 2 having a particularly short cycle time. How should we present the results to the management of the three branches, who perhaps have not been trained in statistics?

It would be wise to show plots of the raw data, perhaps histograms, on the same scale. These plots will no doubt indicate a difference in average cycle time. We might then go on to note that formal analysis proves beyond a reasonable doubt that the long-term average cycle times of the three branches are in fact different. If asked how different the branches are, we would report their average cycle times in the original units. If we had done a regression analysis of log of cycle time, we would note which x variables had the most dramatic impact on log(cycle time) in the analysis. To convey these relationships, we might show scatter plots of specific x's versus cycle time (in the original metric—this was done in the goodwill case).

If we made any predictions of what cycle time would be for specific combinations of x's—that is, predictions and prediction intervals—we would use the regression model in log units to obtain the predictions and prediction intervals. Before communicating them to others, however, we would convert the predictions and prediction intervals back into raw units with the reverse transformation. Therefore, we would take 10 to the power of the predicted log response and 10 to the power of the endpoints of the prediction interval. Because we are 95% sure that log(y) will be within the prediction interval, we are also 95% sure that y will be within the interval formed by taking 10 to the power of the original endpoints.

For example, suppose we had predicted that for a specific combination of x values, $\log(y)$ would be 1.4, and the 95% prediction interval was 1.2 to 1.6. In this case, the prediction for the cycle time we would observe at this combination of x values would be $10^{1.4} = 25.12$, and the 95% prediction interval for cycle time would be from $10^{1.2} = 15.85$ to $10^{1.6} = 39.81$. Note that in the original metric of cycle time, this interval is not symmetric because of the extreme skewness of cycle time. In Figure 9.25 we plotted the regression line and confidence/prediction intervals from the analysis in log units on a scatter plot in the original units. This accomplishes the same objective in graphical form.

9.9   Summary

- The theory of statistical inference is the theoretical methodology used to derive the formulas for inference tools used in practice, such as confidence intervals and hypothesis tests.
- Although much of the theory is not directly important to practitioners, certain aspects have important practical implications, such as the central limit theorem.
- A probability distribution quantifies the probability of observing any particular value, or set of values, for the variable in question. Understanding probability distributions is the key to statistical inference.
- Different types of variables (discrete, continuous, etc.) have different types of probability distributions.
- Understanding the probability distribution can provide insight into the process and how to improve it.
- Statistics calculated from samples, such as averages and standard deviations, are themselves random variables that have probability distributions, referred to as sampling distributions.
- The central limit theorem states that, when we take averages, the averages will tend to have a normal distribution, regardless of the original distribution.
- Linear combinations involving random variables have well-defined averages and standard deviations.
- In many analyses, transforming the data to another metric better satisfies statistical assumptions and results in a more meaningful analysis. The log, square root, and inverse are common transformations applied in practice.

9.10   Project Update

You may not be required to apply the underlying theory of statistical inference to your project. However, several aspects of this theory may enhance the analyses you have performed. For example, it would be appropriate to go back to formal analyses previously performed to determine if some type of transformation would enhance them. If so, note any modifications to your previous conclusions. How will you present the results of analyses involving a transformed variable?

What probability distributions have you encountered in your project? Has there been a combination of discrete and continuous, or just one or the other? If continuous, have they been approximately normal, or perhaps skewed? What practical implications does an understanding of the relevant probability distributions have for your project?

Are there any linear combinations that are meaningful for your project, such as the sum of cycle times or an interference fit? If so, can you derive the average and standard deviation of the linear combination? Are there any sample statistics for which it would make sense to derive a sampling distribution?

In addition to applying the underlying theory of statistical inference, you should be wrapping up the project, ensuring that you have completed each step of the process improvement or problem-solving strategy. You should describe not only the final results of the project, but also the process you used to get there. What did you learn from this experience that you could apply to future applications of statistical thinking? (Refer to Appendix B on written presentation skills, which will provide advice on preparing a final, written report.)

EXERCISES

1. Explain why the underlying theory of statistical inference is important to statistical thinking applications.

2. The data below consist of cycle time for performing manual account reconciliations. Do the data appear to be normally distributed? Suppose we wish to perform some analysis that requires normality, such as calculating a prediction interval. Find an appropriate transformation that will accomplish this objective. Obtain a 95% prediction interval for the next observed cycle time in the transformed metric and convert it back to the original units. What is this interval in original units? How should it be interpreted?

0.99	0.83	0.98	1.96	0.81
1.58	9.38	0.52	3.23	1.13
0.47	1.03	1.97	1.58	4.26
2.59	1.52	0.58	1.15	3.07
1.17	2.79	0.67	0.27	5.71
1.54	0.50	3.84	0.28	3.16
0.59	2.65	4.54	3.27	0.92
0.70	1.88	0.83	0.14	0.39
5.50	2.16	3.21	1.68	1.52
0.75	1.65	2.20	1.18	13.21

3. On what scale would we expect to measure each of the following variables? Consider nominal, ordinal, integer, interval, and ratio scales.

   ■ Scores on SAT exams (*Hint:* Missing all questions does not produce a zero score.)

- A student's rank in a graduating class (1st, 10th, and so on)
- Earnings in dollars
- Types of accounts a customer has at a bank (savings, checking, IRA, and so on)
- The amount of time spent with a potential customer during a sales call
- A survey response on a 1–5 scale, with 1 = Strongly Disagree, 5 = Strongly Agree
- The number of sales transactions done over the Internet
- The percentage of customers who are classified as minorities (*Hint:* Assume a large number of customers and remember the central limit theorem.)

4. Suppose we are making sales calls on two customers today, and two more tomorrow. With each customer, we will either make a sale or not, and we believe that we have approximately a 50–50 chance of making the sale. Derive and draw the probability distribution for the number of sales we will make

- Today only
- Today and tomorrow total

How likely is it that we will make at least three sales?

5. Explain in your own words why the theoretical assumptions of normal distributions are not important in practice for the *t* test and ANOVA. Explain why the assumption of equal variance in ANOVA is important, even when the normality assumption is not.

6. Suppose we have estimates for the cycle time distributions of several steps in our insurance claim settlement process, as follows:

Step	Average	Standard Deviation
Receive and Input Claim	1.2	.4
Process Claim	2.4	.6
Investigate Claim	3.5	1.1
Make Decision	1.4	.2
Communicate Decision	2.1	.4
Make Payment	5.2	.7

If these steps make up the overall process, what estimates would you make for the average and standard deviation of the overall process, assuming a payment is made? Without additional information, what can you say about the shape of the distribution for the overall process? What assumptions would you have to make?

7. During the evaluation of income from individual fast-food franchises, a sales manager has determined that income is reasonably stable, averages about $74,000 per week with a standard deviation of $30,000, and has a skewed distribution. If the sales manager is compensated on the basis of average income for the 14 franchises in her territory, what can you conclude about the distribution, average, and standard deviation of average weekly sales for the territory? If each of the 14 franchises had a different average and standard deviation, how would this change your answer?

8. A colleague is performing a regression analysis on customer complaints. Initial analysis indicates that the number of complaints received approximately follows a Poisson distribution. What issues might this analysis present, based on the typical

assumptions made in regression analysis? What methods from this chapter might be employed to help address this situation?

9. List three practical situations *not* mentioned in this text where you might expect the binomial distribution to apply. Explain why the binomial should apply.

10. Repeat Exercise 9 for the Poisson distribution.

11. Repeat Exercise 9 for the normal distribution.

12. List one practical situation where the data produced might be reasonably well approximated by the binomial, Poisson, and also normal. (*Hint:* Consider when the binomial would be similar to the Poisson, and when these would be similar to the normal.)

13. Check on the Internet or local library to find the specific equation that is used to calculate the Consumer Price Index (CPI). Is this a linear combination? Is it reasonable to assume that the individual items behave independently? Why or why not?

14. Repeat Exercise 13 for the Dow Jones Industrial Average.

15. Randomly generate 100 values for each of four variables—x_1, x_2, x_3, and x_4—such that each value has an average of 10 and a standard deviation of 1. Use a normal distribution or any continuous distribution that can have an average of 10 and standard deviation of 1 (the exponential is not possible in this case). Using linear combinations, predict the standard deviation of the sum of these four variables. Now sum the simulated data for the four x variables and calculate the standard deviation of the sum (i.e., calculate 100 sums, one for each set of four x values, and calculate the standard deviation of these 100 sums). How close is the actual standard deviation to the predicted value?

16. Randomly generate 100 values from three Poisson distributions, with averages of .5, 1, and 1.5. Plot the histograms for each variable and note the degree of skewness of the histograms. Now create a new variable that is the average of the three Poisson variables. That is, let the first average be the average of the first value of the three Poisson variables, let the second average be the average of the second value of the Poisson variables, and so on. Based on a histogram of the averages, compare the skewness of the averages to the skewness of the original Poisson variables. What do you see? Why did this occur?

17. Check the price of a stock of your choice on any Web site. Obtain a graph of the history of the stock price. What type of scale does this graph use, linear or logarithmic? See if you can obtain a graph on a linear scale if the original was logarithmic, and vice versa. Does your impression of the stock's performance change in any way? Why did this occur? Which graph do you feel is more informative to an investor?

18. Table 7.21, which presented the results of an experiment on popcorn, is reproduced here as Table 9.7. One of the key responses a food products business might be interested in is the percentage of unpopped kernels, called "bullets." If we develop a model to predict bullets based on the levels of the five explanatory variables listed here, would we expect the variation to be constant over the full range of bullets? Why or why not? If not, what type of transformation might be appropriate to deal with this problem? Develop models using original units (percent unpopped kernels) and also in the transformed metric selected. Which model appears more appropriate? Why? Is there another transformation that works better?

TABLE 9.7   Microwave Popcorn Experiment from Chapter 7

Test	Run Order	A	B	C	D	E	Bullets	Taste
1	12	−1	−1	−1	−1	1	1.5	7.5
2	9	1	−1	−1	−1	−1	1.4	8.0
3	6	−1	1	−1	−1	−1	1.9	9.0
4	18	1	1	−1	−1	1	0.6	6.5
5	1	−1	−1	1	−1	−1	1.8	7.0
6	14	1	−1	1	−1	1	0.3	7.5
7	7	−1	1	1	−1	1	0.2	2.5
8	5	1	1	1	−1	−1	0.9	1.0
9	17	−1	−1	−1	1	−1	1.7	7.0
10	15	1	−1	−1	1	1	0.8	6.0
11	3	−1	1	−1	1	1	0.6	4.5
12	16	1	1	−1	1	−1	0.9	4.0
13	4	−1	−1	1	1	1	0.6	9.0
14	13	1	−1	1	1	−1	1.3	7.5
15	NA	−1	1	1	1	−1	Missing	—
16	NA	1	1	1	1	1	Missing	—
17	2	−1	−1	−1	1	−1	3.2	8.5
18	8	1	0	1	1	1	0.1	4.0
19	11	1	0	1	−1	−1	0.8	5.0
20	10	−1	0	1	1	−1	1.6	5.5

A = Price (generic, brand), B = Time (4 minutes, 6 minutes), C = Power (medium, high), D = Preheat (no, yes),
E = Elevate (no, yes).
Source: Anderson & Anderson, 1993.

REFERENCES

Agresti, A. (1990). *Categorical data analysis.* New York: John Wiley.

Anderson, M. J., & Anderson, H. P. (1993, July–August). Applying DOE to microwave popcorn. *PI Quality,* 1–3.

Box, G. E. P., Hunter, W. G., & Hunter, J. S. (1978). *Statistics for experimenters.* New York:
Wiley/Wiley Interscience.

Grant, E. L. (1952). *Statistical quality control* (2nd ed.). New York: McGraw-Hill.

Moubray, John. (1992). *Reliability-centered maintenance.* New York: Industrial Press.

Walpole, R. E., & Myers, R. H. (1993). *Probability and statistics for engineers and scientists*
(5th ed.). Englewood Cliffs, NJ: Prentice Hall.

Wheeler, D., & Chambers, D. (1992). *Understanding statistical process control* (2nd ed.).
Knoxville, TN: SPC Press.

10

Summary and Path Forward

I don't know what the data say about whether or not my diabetes improved, but I can tell you for sure I know it worked because I feel a lot better now than I did before we started!

Carolyn Pohlen

10.1 Overview

Chapters 1–9 discussed using statistical thinking and methods in the improvement of business processes. The statistical thinking approach to business improvement has three key elements: process, variation, and data. All work is a process, and business processes provide the context and focus for business improvement. Roadmaps (models) to follow and tools to use when solving problems and improving processes were presented. Process data and designed experiments can be used to develop process models and in turn improve processes. The tools of statistical inference and probability that are useful in process improvement and problem solving were also presented and discussed. Along the way you used these methods on your own project. Hopefully, this activity deepened your understanding of statistical thinking and methods and their use in process improvement.

To determine where we go from here, consider: What other aspects of statistical thinking and methods should we explore and utilize? Following are two case studies that may suggest answers to this question. The first is a personal health care experience that illustrates the use of statistical thinking and methods in everyday life, with potential application to the health care field. The second is a study of business improvement tactics. These will be followed by a text summary and discussion of potential next steps to delve deeper into the field of statistical thinking and methods and, finally, guidance on how to debrief your project.

10.2   A Personal Case Study by Tom Pohlen*

I was introduced to statistical thinking in October, 1988 when I attended Heero Hacquebord's course "Statistical Thinking for Leaders." I went into the course thinking that I already knew everything I needed to know about SPC. I came out of the course with a whole new perspective on statistics, looking upon SPC and other statistical applications more as a way of thinking about processes so we can learn how to improve them. I also found that I could never again be satisfied with looking at numbers without graphical analysis.

I immediately began to think of all sorts of ways to apply this new knowledge at work and at home. One of my first applications was to control chart my weight, and I have been doing so ever since. My enthusiasm, however, wasn't shared by my wife, Carolyn. She had been an insulin-dependent diabetic since November 1976. I could see an obvious application of control charts to her blood glucose level, which she was testing daily. No matter what my argument was, she wasn't interested.

Then in June 1994, something changed. Somewhere between being tired of pricking her finger up to four times per day to get blood, having laser surgery in November 1993 to repair eye damage caused by diabetes, and having been sick so frequently during the winter of 1993/1994, she finally said she was ready to be my guinea pig for applying statistical thinking.

The Objectives

We started out with a few simple objectives. For Carolyn it was to reduce the pain and inconvenience of diabetes, and in particular, to regain her health and reduce the blood testing to no more than once per day. My objective was to get her what she wanted. To do so I knew that we had to understand her "process" variation, gain control of it, and reduce that variation. It seemed like a simple problem; after all it should be just like a production process! It wasn't!

The "Process"

Carolyn is a "brittle" diabetic, due to the fact that she has little or no insulin production. In June 1994, her swings in blood glucose levels were high. The blood glucose range considered normal for non-diabetics is 70–120 milligrams/deciliter. It was not uncommon for Carolyn to vary from over 300 mg/dl to under 70 mg/dl (usually accompanied by an insulin reaction) in a very short time, sometimes within 24 hours. I ran a regression analysis to try to determine the effect of insulin on her blood glucose and re-learned the futility of analyzing "production" records for correlations. The analysis indicated that "increasing insulin increased blood glucose," an obvious error! What was actually happening was that we had to increase insulin because blood glucose was up.

---

*Tom Pohlen is a Senior Quality Engineering Specialist with 3M in Hutchinson, MN. Originally published in Britz, Emerling, Hare, Hoerl, & Shade, 1996.

The Goal

Our main goal early in the project was to reduce glucose variation with an emphasis on reducing the over-control that was occurring. This was not a simple task because with diabetes there is no choice but to eat (usually sugar) when blood glucose drops (toward 70 mg/dl in Carolyn's case), and there is very little choice but to take extra insulin when blood glucose gets high (> 250 mg/dl). There was no doubt that we were dealing with a truly "chaotic" process. This indicated the need to find a control condition for Carolyn's diabetes that would make her "process" more robust (insensitive to sources of variation).

Understanding Variation

In June 1994 before we could really get going with our efforts to control her diabetes, Carolyn had a bout with high blood glucose that we could not contain and she ended up in intensive care. She came out of the hospital June 26 still not very healthy. She continued to have high blood glucose problems and by August was back in intensive care.

This second failure was a big letdown, but we recommitted ourselves and decided it had to be done and it could be done. Carolyn long ago had learned that she had to take control of her own sickness and learn as much as she could about it. We dug into books and magazines and learned much about diabetes, including some of the mechanisms and key causes of blood glucose variation (food types, exercise, illness, infections, emotional stress, etc.). A key book (Bernstein, 1984) identified a very important fact: The effect of any insulin on BG (blood glucose) usually diminishes as BG rises. This indicates blood glucose has a nonlinear response to insulin, a characteristic of a chaotic process.

After Carolyn came out of the hospital August 24, 1994, we began to make progress. Around this time we shared our data with Carolyn's physician, Dr. John Zenk, doctor of internal medicine. He was pleased to see our active interest in controlling Carolyn's blood glucose and was particularly satisfied with our charting of her numbers. As we gained better control, Carolyn had more low blood glucose reactions, a very undesirable complication. Low glucose tends to cause severe headaches and creates the potential for unconsciousness. The latter never happened, but the possible brain cell damage is very undesirable. Her reactions tended to occur when sleeping in the early morning.

To deal with this we developed a theoretical quadratic model of how her insulin levels would build up during a normal day. She was taking multiple shots per day of two kinds of insulin: Ultralente (long acting, typically over 24 hours) and Regular (short acting, typically over 3–6 hours, taken to handle the immediate effect of meals). Of her total insulin, about 65–75% was Ultralente (initially 36 total units of insulin). Our graphical model indicated we could improve the overall uniformity of insulin if she delayed her last shot of Ultralente until about 8 P.M. rather than taking it at 6 P.M. with dinner. To coincide with the change we also delayed our dinnertime. The result was a reduction in problems with early morning insulin reactions.

Finding Solutions

Knowing that we could not eliminate the causes of high blood glucose, we realized we needed to develop a control strategy to bring high blood glucose down when it occurred. In the spirit of statistical thinking our first reaction to out-of-control signals was to ask, "Why?" and investigate. Then based on that investigation, if an increase in insulin was required, we needed an adjustment plan.

The theory that seemed to fit best was that the blood glucose level is really trying to be at some average level based on all the competing factors (insulin level, stress, illness/health, food intake, and so on). Due to the dynamics of the situation, all these factors either were rising or falling simultaneously in amount. At any moment in time no single value of either the insulin dose or blood glucose level really has meaning. Thus, it was decided to use 24–48 hour averages of each of these two key parameters for determining the amount of change to make in the insulin doses to bring down high blood glucose. An initial crude formula was developed that worked well in dealing with high blood glucose levels in November 1994 and February 1995.

In January 1995 a more fundamental theory was developed incorporating the nonlinear effect of insulin on blood glucose. This was combined with an approach to apply differential equations, further theorizing that the problem was similar to a population growth model. The final theory adopted the idea that the derivative of blood glucose with respect to insulin dose was an exponential function of blood glucose.

Successful Results

The use of the equation was successful in dealing with high blood glucose levels that occurred in May, June, July, August, and December 1995. We learned that in applying the equation it was important to keep the relative doses of Ultralente and Regular at about the same ratio (about 65–75% Ultralente). Usually, more than one iteration of the formula was needed to counteract high blood glucose due to the dynamic nature of each situation. Using the equation in cases of low blood sugar tended not to be as effective since these instances often were accompanied by insulin reactions which generally required large doses of extra sugar to regain blood glucose levels.

The control charts (Figure 10.1 and Figure 10.2) show the results of Carolyn's blood glucose between the visits to the hospital June–August 1994 and after August 1994. The limits are based on one of the best periods of control: November 29, 1994 through January 3, 1995 ($\bar{x} = 107$ mg/dl; $s = 34.8$ mg/dl). The early data were taken usually as three to four readings a day. Beginning in January 1995 most data were taken once a day in the morning except during periods of high blood glucose, when more readings were needed. In general, analysis of the days with multiple readings failed to show any specific pattern in blood glucose levels throughout the day; thus, the data are graphed all together.

FIGURE 10.1   Blood Glucose, Summer 1994

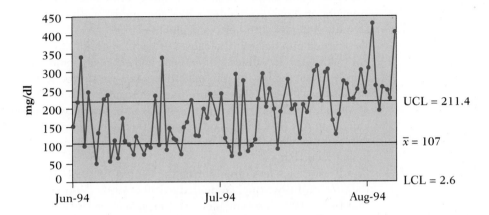

Limits based on 11/29/94–1/3/95—a future period of good control

FIGURE 10.2
Blood Glucose with New Control Plan

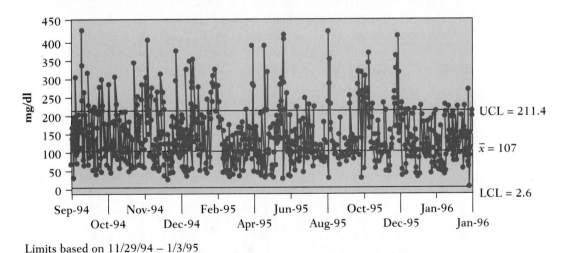

Limits based on 11/29/94 – 1/3/95

Benefits

There is clearly still plenty of room for improvement in controlling the complex process of Carolyn's blood glucose. However, we have already obtained many benefits, most of which we believe can be directly attributed to our use of statistical thinking with control charts:

1. We met our objective of safely getting down to one blood glucose test a day. This was accomplished because we successfully reduced the variation in her "process" making it possible to "see" and predict within limits what was actually going on with her blood glucose.

2. We learned much about the causes of Carolyn's high blood glucose (different foods, exercise, illness, infections, and especially emotional stress).

3. Carolyn is taking less insulin now than when we started in June 1994 and she also has lower blood glucose (current insulin = 26 units; June 1994 = 36 units). My theory is that her body has now gotten used to lower blood glucose and is now more robust against fluctuations. In fact, Carolyn used to feel jittery when she was at near-normal blood glucose levels; she no longer feels that way.

4. Carolyn was sick less often during the winters of 1994–1995 and 1995 to the time of this writing, compared to1993–1994. (Winter 1994–1995: sick four times. 1995–present: sick two times. Winter 1993–1994: No hard data, but memory recalls a frequency of an illness every one–three weeks.) In addition Carolyn has been much more energetic. High blood glucose tended to make her lethargic, whereas with lower blood glucose she clearly has had more energy.

5. A successful control strategy was developed for appropriately correcting high blood glucose with minimal over-control.

6. The time between insulin reactions has improved; that is, the time between these reactions increased when compared to where it was when we first started to control the blood glucose. (Figure 10.3. Note: log of time-between-reactions was used due to the highly skewed nature of the time.)

FIGURE 10.3 Time Between Insulin Reactions

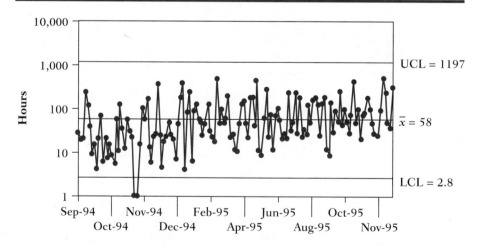

Limits based on log(Hours) 1/95–1/96

FIGURE 10.4 Glycosylated Hemoglobin Levels

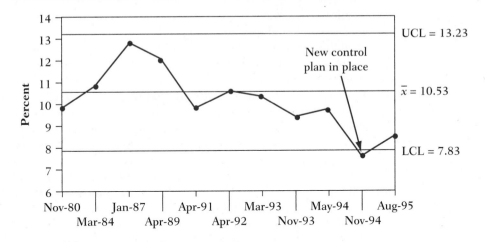

Limits based on November 1980–May 1994

7. Carolyn has had a significant drop in her Glycosylated Hemoglobin (GH) levels since we started this program (Figure 10.4). This test estimates the long-term average blood glucose over approximately the last two months prior to performing the test. It is an indicator of future complications due to diabetes. Recent studies have indicated that people with GH levels below 8.1% have significantly lower risk of kidney disease.
8. Carolyn was anticipating having a second laser surgery on her eye in November 1994; however, her eye exam in November indicated no need to have the surgery.
9. We have also developed some additional ideas for further control of Carolyn's diabetes. One thought to reduce the chance of insulin reactions in the night is to have Carolyn take a small amount of food (a sugar tablet) in the early morning (e.g., 2 or 3 A.M.). Of course, that idea is not very popular with the subject and has been rejected at this time.

Lessons Learned

Beyond the tangible benefits of the control plan for Carolyn's health, we have also learned or reaffirmed some additional valuable lessons regarding statistical thinking and science. These include:

1. A process has no regard for the "specifications"; it just does what it is capable of doing. There is no better evidence than this case of diabetes. The human internal insulin regulatory system controls blood glucose to within ±25 mg/dl. The best we've been able to get Carolyn's variation is ±104 mg/dl. Tampering only increased the variation.

2.   Statistical thinking should be used to help us develop theories as well as to test them. To do so we should remember to use all of our scientific/mathematical background to try to explain the patterns in data.
3.   Chaos is alive and well and living in diabetes. Recognizing the presence of conditions in systems that can result in chaos can help us become conscious of the importance of creating robust processes that are less sensitive to sources of variation.
4.   The real value of control charts and statistical thinking is to help us learn about our processes. It is a serious fallacy to avoid introducing control charts to a process due to lack of adequate knowledge of how to control it. Failure to introduce the charts essentially guarantees that one will continue to be ignorant of how to control the process.
5.   The human body is a marvelous creation that is extremely robust.
6.   While I was preparing this case study my wife pretty well summed up our results when she said, "I don't know what the data say about whether or not my diabetes improved, but I can tell you for sure I know it worked because I feel a lot better now than I did before we started!" As my friend, the late Ken Kotnour, once said in quoting Deming: "The customer doesn't always know what they need, but they will treasure it if you give it to them."

Once Tom and Carolyn had a better understanding of the variation in the process, they were able to apply techniques to reduce variation. The results were heartwarming—and, perhaps, life-saving.

10.3   CASE STUDY: Newspaper Accuracy

This case study focuses on reducing errors in newspaper publishing, but it illustrates many common elements of reducing errors in business processes. Defects—as measured by errors, accuracy, completeness, and so on—are a key measure of the performance for many business processes. This case study illustrates a common approach to reducing the errors (defects) of a process. This project followed the "Six Sigma" five-phase process improvement strategy of Define, Measure, Analyze, Improve, and Control (DMAIC) (discussed in Chapter 4 and also in Appendix D).

Introduction

A major newspaper reported the promotion of a new CEO to a major U.S. corporation and misspelled the name of the newly promoted executive. The newspaper received a call from a very unhappy reader—not the new CEO, or the public relations officer of the company, or the CEO's spouse, but rather the new executive's mother! This points out that the demand for accurate information comes from many different, and sometimes surprising, sources.

One of the key customer requirements of a newspaper is accurate reporting. Nothing is more important to a newspaper. It is essential that the names, facts, numbers, and other information reported in newspaper articles be accurate. Newspaper errors greatly affect the newspaper's reputation as well as increase costs incurred by fixing the mistakes that are caught before the paper hits the streets. In extreme cases, mistakes result in lawsuits. Although reducing costs is important, the paper's reputation is critical to its long-term success. If you get the facts wrong, spell a name incorrectly, or make an arithmetic error, readers notice it and you lose their trust.

Project Definition

The newspaper editor decided to form a project team to reduce the number of errors in the newspaper. The newsroom was very busy, and it was decided that the team would meet for 2 hours once per month. The individual team members would of course do project work on their own or in small groups between the monthly meetings. The team consisted of the editor, two copy editors, two graphics editors, one reporter, and four supervisors. A person skilled in process improvement was assigned to guide the team through the process. Some preliminary data suggested that as many as 30–40 errors could be caught at the copy desk on a given day. Errors not caught at the copy desk would likely make it into the newspaper unless detected in the production process, where errors were often caught because of the diligence of the printing staff and the small monetary reward they received for finding errors.

The team recognized that a goal should be set for the total number of errors at the completion of the project: The goal for errors found at the copy desk would be less than 10 per day, approximately a 50% reduction from the current levels. The long-term goal for the process was, of course, zero errors; hence additional studies would be needed to eliminate errors completely. Other objectives for the project included elimination of rework throughout the story creation process, the freeing up of copy editors to do more value-added work and less checking, and the creation of a culture that emphasized "doing it right the first time" and taking personal responsibility for accuracy.

An analysis estimated how much an error cost the organization. It was found that, at a minimum, an error caught at the copy desk cost $62, if detected in the composing room $88, if a page had to be redone $768, and $5000 if the presses had to be stopped and restarted. Of course, the cost of an error being published is "unknown and unknowable."

To measure errors and reduce them required an agreed-upon definition of what constitutes an error. The team decided that an error would be (1) any deviation from truth, accuracy, or widely accepted standards of English usage or (2) a departure from accepted procedures that causes delay or requires reworking a story or a graphic. An operational definition is used to define the characteristic being measured, so that different people will come to the same conclusion. Examples of errors included incorrect information; a hole in a story; mistakes in spelling, grammar, or punctuation that cause confusion; mistakes in grammar, spelling, or punctuation that put readers off; and departures from standard procedures.

Process Measurement

Once the project was well defined, the team moved on to "Measure," and began quantifying the extent of the problem. The first step was to measure each day's total errors detected at the copy desk and classify the errors by type. The following suspected types of errors were tracked:

Misspelled Word   Word Missing
Wrong Number   Duplicated Word
Wrong Name   Fact Wrong
Bad Grammar   Other
Libel

A high-level process map showing the workflow was also created (Figure 10.5).

During considerable discussion of the sources of errors, the following variables emerged as prime candidates (theories of causes):

- Size of the paper (number of pages)
- Number of employees absent each day
- Major change in front-page story (yes, no)

The first two variables are self-explanatory; the third, a major change in the front-page story, occurs when a late-breaking story takes place and the lead story has to be written from scratch under tight time constraints. Several cause-and-effect diagrams were constructed to display these three variables and the categories of errors noted above. The most useful diagram is shown in Figure 10.6.

FIGURE 10.5   Newspaper Writing and Editing Process

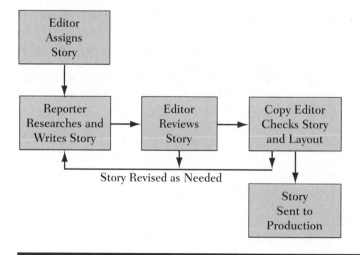

FIGURE 10.6   Causes of Newspaper Errors: Cause-and-Effect Diagram

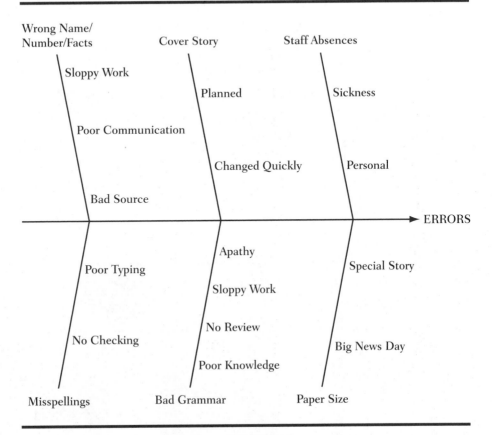

Process Analysis

First, the team analyzed the first month's data using a run chart and a Pareto chart. The run chart showed no trends or obvious special causes, and that the process was averaging around 20 errors per day. The Pareto chart showed that the main sources of errors were misspelled words; wrong names, facts, and numbers; and bad grammar. It was decided to collect another month of data to confirm these trends and to increase the sample size, which would enable the use of regression analysis to assess the effects of the size of the paper, absent employees, and front-page story changes on the total errors. The additional data confirmed the trends seen in the first month of data. Run and Pareto charts of the combined data set are given in Figures 10.7 and 10.8, respectively.

Investigation found that although the reporters writing the stories had spellcheckers, the software wasn't being used. It was conjectured that many didn't take the time to spellcheck their articles because they knew that the copy editors would catch any errors they made.

FIGURE 10.7   Newspaper Errors, March–April: Run Chart

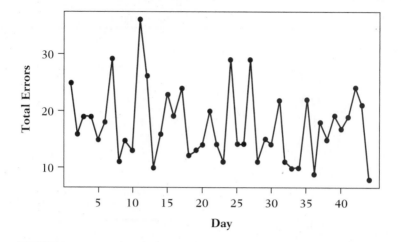

FIGURE 10.8   Newspaper Errors, March–April: Pareto Chart

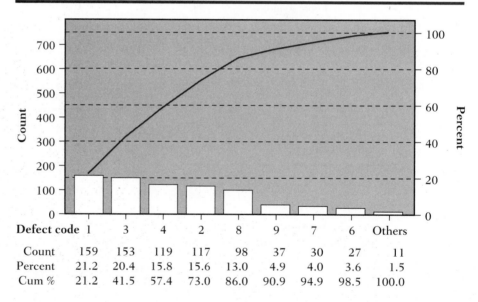

Defect code	1	3	4	2	8	9	7	6	Others
Count	159	153	119	117	98	37	30	27	11
Percent	21.2	20.4	15.8	15.6	13.0	4.9	4.0	3.6	1.5
Cum %	21.2	41.5	57.4	73.0	86.0	90.9	94.9	98.5	100.0

Defect code
1 = **Misspelling**
2 = **Wrong number**
3 = **Wrong name**
4 = **Poor grammar**
5 = **Libel**
6 = **Word missing**
7 = **Duplicate word**
8 = **Wrong fact**
9 = **Other**

TABLE 10.1   Newspaper Error Data: Weekly Averages

Month	Week 1	Week 2	Week 3	Week 4	Week 5*	Average
March	19.8	16.8	20.2	18.0	17.4	18.4
April	14.2	13.4	16.8	17.6	n/a	15.5
May	8.6	12.8	16.0	9.2	16.5	12.6
June	15.4	15.0	13.0	**	n/a	14.5
July	14.0	13.4	14.6	13.6	11.6	13.4
August	13.2	12.2	16.6	16.4	n/a	14.6
September	15.5	8.2	12.2	7.0	n/a	10.7
October	6.2	7.0	8.0	5.6	5.8	6.5
November	5.2	5.2	5.8	4.6	n/a	5.2
December	8.2	**	**	**	**	—

*Where applicable.
**Data not available for this week.

A regression analysis was conducted to evaluate the effects of size of paper, absent employees, and lead article changes, as well as

- Day of week (M, T W, Th, F)
- Month of year (March through December)

Inclusion of discrete variables into a regression analysis involves the creation of "dummy variables," which is beyond the scope of this text. This technique is explained in any of the regression references in Chapter 6. A regression analysis of the data indicated that absent employees and day of the week had no effect on total errors. The number of pages (size of paper) significantly increased total errors, as did changes to the front cover. Unexpectedly, there was a reduction in total errors between April and May. In Table 10.1 we see that May and June total errors appear lower than those in March and April. This effect was attributed to the Hawthorne effect, in which things often get better when people know that management is working on improving the process.

The regression analysis also showed that the effect of changing the front-page cover story at the last minute was smaller in June than in March, April, or May.

Process Improvement

The improvement strategy adopted in the "Improve" phase was for all staff to be particularly alert when changes were made to the front-page lead story, or when the number of pages in the paper increased. It was found to be likely that more errors would occur on these days, and the staff prepared for both possibilities.

Attention was focused on the largest categories of errors: spelling errors; wrong facts, names, and numbers; and bad grammar. In addition to not using the spellchecker, the reporters were not routinely checking the accuracy of their articles. It was made mandatory that the reporters verified that they had checked the accuracy of their articles. Three job aids were created: a "Spellchecker How-To,"

FIGURE 10.9 Newspaper Errors, March–December: Weekly Averages

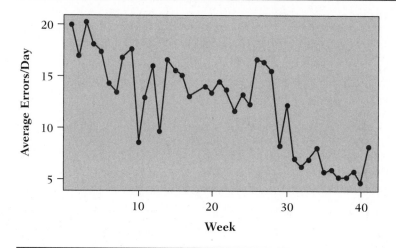

a list of "Ten Rules of Grammar," and the "Pyramid of Trust," which detailed what sources were to be trusted to produce accurate names, facts, and numbers.

An all-hands meeting was held at the end of July, and everyone was shown the results of the study, the new job aids, and the requirement that all reporters check their articles for accuracy to reduce errors. The interim goal of reaching fewer than 10 errors per day was reviewed.

Mandating Versus Implementing Change One month went by, and the data for the month of August were analyzed: The total errors had not improved. The management team was assembled and the situation reviewed. Why were the errors still high? It was learned that the new procedures were not being used. Many employees did not feel that management was serious about the changes and therefore did not take the changes seriously. The editor reiterated that the new procedures were mandatory and that the management team would lead this new way of working. Another all-hands meeting was held to address the issue.

One month later when the latest data (September) were analyzed, the total errors had dropped significantly. By year end, the total errors had dropped by approximately 65%—as compared with the goal of 50% reduction (see Table 10.1 and Figure 10.9). The new procedures were clearly working.

Process Control

The errors were significantly reduced, but were not yet at zero, and more work was needed to achieve further improvement. Statistical thinking is an iterative process, and more than one pass through the improvement cycle is typically needed. In the meantime, a control plan was instituted to maintain the gains of the work done to date and to keep the errors at the reduced level. This is the purpose of the "Control" phase in Six Sigma and is similar to the "Standardize" phase in the problem-solving strategy. As part of the control plan, the team decided to monitor the following measures using control charts:

- Total errors
- Errors by category
- Percentage of articles checked by the author
- Percentage of articles spellchecked

The latter two measurements were found to be particularly useful in detecting when the reporters were not following procedures and thus producing errors.

Checklists were also created for reporters and others; roles and responsibilities, including backups, were defined to reduce handoff problems between departments. This enabled people to view their own work processes as part of an overall system. Work was also initiated to find the sources of errors in the graphics.

Results

In addition to reducing the errors by 65%, the process produced other benefits, including:

- Fewer missed deadlines, including increased ability to deal effectively with extremely tight deadlines
- Improved morale at the copy desk: Copy editors were freed up to make better use of their talents and training.
- Re-keying of names (rework) reduced
- More efficient and less costly sources of information were found, resulting in reduced errors and less number input time (less keying of data). News assistants were freed up to do more valuable work.

Fewer errors resulted in less non-value-added work and a more streamlined and effective process. Such effects are typical when the errors and defect levels of business processes are reduced. The processes work more effectively and efficiently, costs are reduced, employee morale improves, and customer satisfaction increases.

10.4   Review of the Statistical Thinking Approach

The thought process behind the statistical thinking approach used in both case studies is summarized in Table 10.2. Any organization, profit or nonprofit, serves customers with the goal of delighting, or at least satisfying, those customers. In a business environment, such as the newspaper's, a key "customer" of the overall organization is the stockholder, who expects improvement in dividends (profits) and stock price. In the personal case study, Carolyn Pohlen was the customer of Tom's statistical thinking application. Improvement is more than an option in most cases—it is a requirement for growth, and occasionally survival. (Recall that General Electric is the only surviving member of the companies included in the original Dow Jones Industrial Average.)

TABLE 10.2   Using Statistical Thinking to Improve Processes

- Our goal is to improve results.
- In order to improve results, we must satisfy, if not delight, our customers and do so efficiently.
- Processes generate the results; we must improve the process to improve the results.
- Process performance varies, which makes our customers unhappy and produces waste and rework.
- Using a principle of good business and science, we measure process performance and generate data in the form of process measurements.
- We use statistical tools and methods to analyze process data in the context of our existing hypotheses.
- This analysis leads to better understanding of process variation—that is, improved hypotheses—which allows us to take informed action to improve the process.
- Often several rounds of data collection, hypothesis revision, and improvement are needed to achieve the desired results.

To improve results, we must improve the processes that produce the results. In the personal case study, these processes included Carolyn's eating habits and the processes of taking medication. The newspaper case focused on processes for writing and editing. All work occurs in a system of processes, and an organization touches its customers through the processes it operates. Therefore, to improve results, such as profitability or customer satisfaction, analysis should focus on the processes that generate value for customers. Improving the effectiveness of these processes to better satisfy customers typically produces more revenue (top-line growth), and improving the efficiency of these processes produces cost savings (bottom-line improvements).

The overall statistical thinking approach requires merging data-based methods with subject matter knowledge in an iterative fashion. We gather data to verify or refute existing hypotheses and, based on the data analysis, we typically modify those hypotheses. This leads to further questions that need to be verified or refuted, continuing the iterative cycle of hypothesis suggesting data, data modifying hypothesis. Recall that data have no meaning in themselves; they take on meaning only in the context of some conceptual model of the process being studied. This cycle is clearly illustrated in the personal case study, where Tom and Carolyn began with some knowledge of diabetes and the chemistry of insulin affecting blood glucose, then gradually refined and improved these hypotheses using data from Carolyn's "process." Several rounds of data gathering and analysis were required to obtain the ultimate improvements. Similarly, people working for the newspaper had some hypotheses as to the potential root causes of errors. Some of these turned out to be true, while others were refuted by the data, resulting in refined hypotheses.

A key cause of poor customer satisfaction and process inefficiency is excessive process variation. We recall that variation is a fact of life: all around us and present in everything we do. Understanding and reducing variation are key to process improvement. Excessive variation in Carolyn's blood glucose level was life-threatening. As is often the case in business, the average level was not the issue. Variation was the real culprit in her diabetes. In the newspaper case, although the main emphasis was on reducing the average level of errors, process variation had to be reduced through standardized procedures to accomplish this objective. To aid process improvement, adopt a principle of good science (and business) and measure the process performance. This permits quantifying, understanding, and hopefully reducing the process variation. Careful measurement and analysis of Carolyn's data were key to her breakthrough. Output and process measurements were also critical to the improvements at the newspaper.

This is where statistical thinking is particularly valuable, perhaps invaluable. The concepts and tools of statistical thinking, the science of statistics, were developed to deal with variation. Statistical thinking and methods help characterize, quantify, analyze, control, and reduce variation. In the final steps of the thought process, statistical tools and methods are used to analyze the data so we can understand the variation and take action to improve the process—which leads to improved results. Control charts were used to detect and diagnose changes in Carolyn's blood glucose levels and to monitor and diagnose several key output and process measures at the newspaper.

Statistical thinking can be applied to improve processes in many ways. The process improvement strategy improves processes subject to common-cause variation, and the problem solving strategy is effective at dealing with special-cause variation. The best approach depends on the type of variation we are dealing with. Carolyn's diabetes and the total newspaper errors were generally stable, but unacceptable, performance. In both case studies, the approach to improvement tended to follow the process improvement strategy more closely, although not exactly. There is no single "correct" approach to applying statistical thinking.

The key themes of the personal case study are summarized in Table 10.3. The themes consist of the objectives and goal, the process studied, the magnitude and effects of variation, a key process measurement (blood glucose), statistical analysis and modeling done, actions taken, and statement of customer satisfaction—all the aspects of statistical thinking discussed above and summarized in Table 10.2.

TABLE 10.3 Personal Case Study Themes

- Carolyn's goals were to reduce the pain and inconvenience of diabetes and reduce blood testing to no more than once a day.
- The processes studied included the mechanism by which the body produces insulin and how it is affected by insulin injection and other factors.
- Variation was the key problem, as stated in the goal to "reduce glucose variation and the overcontrol that was occurring."
- Process performance was measured through blood glucose levels.
- Various statistical tools and analyses were used in an iterative fashion to understand the process variation and refine predictive models.
- A successful control strategy (action) resulted in less sickness, lower blood glucose levels, and no need for planned eye surgery.
- A happy customer resulted.

10.5   Text Summary

The following paragraphs summarize the key points of each of the chapters of the book. These provide a useful overview of the book, and refresher of the main points discussed.

Chapter 1: The Need for Business Improvement

Changes in the business environment over the last few decades have required organizations to be able to make significant changes to survive and prosper. One key change is that continuously improving how we work must now be a routine part of everyone's job. Numerous methods of improvement have been implemented and produced significant results, such as benchmarking, total quality management, reengineering, and Six Sigma. The key concepts of statistical thinking—process, variation, and data—are integral to most of these improvement initiatives.

Chapter 2: The Overall Statistical Thinking Approach

The objective of statistical thinking applications is to improve results. To do so, we must improve the business processes that produce the results. Improving processes typically requires us to improve the average and also reduce variation. Improving dynamic business processes is not easy! Success often requires a sequential approach combining existing subject matter knowledge with new information gained through collecting and analyzing process data. Our knowledge guides our data collection, and the subsequent data analysis improves our understanding of the process. Several iterations of this process are typically required to achieve improvement objectives.

Chapter 3: Understanding Business Processes

All activities in a business are performed through a system of processes, although these processes are not always obvious to the casual observer. Therefore, a key preliminary step toward improvement is to map the process in question via the SIPOC model. This analysis will often identify non-value-added activities, such as the "hidden plant," which could be eliminated through appropriate process improvements. Part of analysis should view the process from the larger context of the system in which it operates. This ensures that efforts will improve the overall business system, rather than push the problem from one area to another. The process of creating data—the measurement process—is critically important and should not be overlooked.

Chapter 4: Process Improvement and Problem-Solving Strategies

In some cases, our process is "broken," and some aspect of the process needs to be fixed to return it to its typical performance level. In such cases, the problem-solving strategy provides one overall approach that integrates various tools in a logical sequence to solve the problem. In other cases, however, the process is performing consistently, but at an unsatisfactory level. There is nothing broken to fix; we need instead to fundamentally change the process to improve its performance. For these cases, the process improvement strategy provides a different sequence of steps and tools. Run and control charts are key tools to diagnose whether the process is stable. This information helps us determine the most appropriate improvement strategy.

Chapter 5: Process Improvement and Problem-Solving Tools
Various types of tools are utilized in the improvement strategies. Data collection tools help obtain the data needed for improvement. Sampling tools ensure that we have the right quality and quantity of data. Data analysis tools extract information from numerical data. Knowledge-based tools analyze qualitative data—that is, ideas. These tools are even more dependent on participation from the people with the most appropriate knowledge of the process. Best results are obtained when these various types of tools are used in conjunction with each other in a logical sequence.

Chapter 6: Building and Using Models
Developing models (equations) to quantify the impact of key input or process variables on key output variables is another important statistical thinking tool. Models help us understand, predict, and modify the key output variables, thereby improving the process. Developing a good model is an iterative process involving subject matter knowledge, data, and critical examination of the preliminary model. When using models, we are reminded that "all models are wrong, but some are useful." In other words, even good models are just approximations of the true relationships between the variables. We should therefore be cautious with our use of the model, particularly when predicting outside the range of our data.

Chapter 7: Using Process Experimentation to Build Models
Model building is typically limited by the quality and quantity of the data available. One strategy for obtaining the right amount of the right data is to proactively determine the data we need and then collect them. When we use statistical principles to do this, the strategy is called statistical design of experiments. Statistically designed experiments are both effective and efficient and thereby minimize the cost of obtaining the needed information. They are particularly powerful in experiments with several process variables or in studies of interactions. Often, experimentation and model building are used together in a sequential approach, with each new design resolving questions from the previous model.

Chapter 8: Applications of Statistical Inference Tools
Statistical inference tools allow us to draw conclusions about an entire population or process based on only a sample of data. Predicting election results based on sample polls is a common application. Confidence intervals quantify the uncertainty in estimates of the population average, standard deviation, proportion, and so on. Hypothesis tests determine whether we have sufficient evidence to draw definitive conclusions about the process, such as concluding that the average cycle time for processing checks is longer at one bank than another. Sample size formulas determine the amount the data required to achieve a desired level of uncertainty in our estimates.

Chapter 9: The Underlying Theory of Statistical Inference
Understanding the theory underlying inference tools provides additional insight into how and why the tools work. Certain aspects of this theory, such as the central limit theorem and linear combinations, are directly applicable. Probability distributions are the foundation of the theory of inference. All variables—continuous, discrete, and even statistics calculated from samples, such as averages and standard deviations—have probability distributions that quantify their behavior. For some distributions, we obtain a more meaningful analysis by transforming to a different metric.

10.6 Potential Next Steps to Deeper Understanding of Statistical Thinking

So where do we go from here? Figure 10.10 summarizes some of the potential next steps we may consider. It shows that statistical thinking, together with basic mathematics, economics, and quantitative and analytical thinking and computing skills, provide the foundation for studying more statistically advanced subjects. This text presented basic skills in statistical thinking. However, one text cannot possibly develop detailed knowledge of all the tools used in a field as broad as sta-

FIGURE 10.10   *Statistical Thinking: Improving Business Performance*—The Foundation for Progress

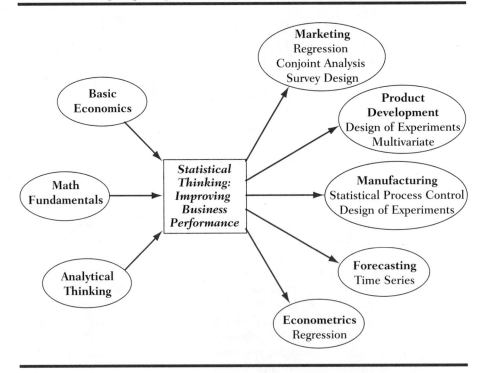

tistical thinking and methods. You may therefore choose to develop more in-depth knowledge of particular statistical methods relevant to your career. This is the beginning of your education in statistical thinking, not the end.

The subjects you pursue for deeper knowledge depend on your career direction.

- Marketing and market research require skills in regression modeling, conjoint analysis (a type of designed experiment and subsequent analysis), and survey design. Marketing professionals can also benefit from knowledge of multivariate methods.
- Research and product development workers will need to know more about design of experiments (DOE) and multivariate statistics, such as principal component analysis, factor analysis, and cluster analysis.
- Those who pursue operations management or other manufacturing fields will need to know statistical process control (SPC) as well as design of experiments. SPC makes heavy use of control charts.
- Business strategists, planners, and econometricians will likely focus on learning more about regression modeling and forecasting using time series models.

In all of these fields, statistical thinking and its elements—process, variation, and data—are fundamental to deeper understanding and use of the associated statistical methods. Further discussion of the advanced statistical techniques mentioned above can be found in the references at the end of this chapter.

10.7 Project Summary and Debriefing

The purpose of the project debrief is to review and summarize the project work and document what was accomplished and learned. This is done in the form of a project report or presentation. The key items to be covered in the report or presentation include:

- Project background, purpose, objectives, goals, and sources of information and assistance utilized, including any team members
- Executive summary listing key conclusions
- Description of work done, including project planning, data collection, data analysis, and actions taken during each phase of the statistical thinking process
- What was learned about the process being studied
- Barriers encountered
- Recommended actions
- Recommended future studies
- Key learnings regarding the use of the statistical thinking approach to a real problem

The project summary and debriefing not only helps communicate the project, but also helps develop an integrated view of statistical thinking and what you have learned regarding its use in the improvement of business processes.

EXERCISES

Relate your answers in the following exercises to real situations such as those discussed in this book (e.g., Sections 2.3, 4.5, 7.6, and 10.2), your project, or some other real situation with which you are familiar.

1. **The Two-Minute Summary.** Define and discuss the value, benefits, strengths, and limitations of the statistical thinking approach to business process improvement. Condense your discussion to a two-minute speech and be prepared to give the speech to anyone who asks.
2. **Newspaper Accuracy.** Summarize the key themes of the Newspaper Accuracy Project discussed in Section 10.3. Compare these themes to the key aspects of process improvement summarized in Table 10.2 and the key themes of the personal case study listed in Table 10.3.
3. **Uses of Statistical Thinking.** Why and how is statistical thinking useful in solving problems and improving processes?
4. **Relationship Between Statistical Thinking and Statistical Methods.** Discuss the roles of and relationships among statistical thinking, statistical methods, process, variation, and data.
5. **What If I Don't Have Data?** Can statistical thinking be used if you don't have data? If so, how? If not, why not?
6. **Effect of Variation on Me Personally.** Write an essay on how variation affects your life, short term and long term. Refer to the questions at the end of Chapter 2.
7. **Learning More.** What steps will you take to broaden and deepen your knowledge, skills, and use of statistical thinking?
8. **Using Statistical Thinking.** How do you see yourself using statistical thinking to solve problems and improve processes in the future?
9. **Projects to Reduce Errors.** Using the newspaper accuracy project as a frame of reference, what processes in your organization or field of interest do you believe should be studied to reduce the error/defect rates? Be specific, giving the name of the process(es), what errors will be measured, how the errors will be measured, the study objectives, and how the study should be designed and conducted.

REFERENCES

Anderson, T. W. (1984). *Introduction to multivariate statistical analysis* (2nd ed.). New York: John Wiley.

Bernstein, R. K. (1984). *Diabetes: The glucograf method for normalizing blood sugar.* Los Angeles: Jeremy P. Tarcher.

Box, G. E. P., Hunter, W. G., & Hunter, J. S. (1978). *Statistics for experimenters.* New York: Wiley-Interscience, John Wiley and Sons.

Box, G. E. P., & Jenkins, G. M. (1970). *Time series analysis, forecasting and control.* San Francisco: Holden-Day.

Britz, G., Emerling, D., Hare, L., Hoerl, R., & Shade, J. (1996). *Statistical thinking.* Milwaukee, WI: American Society for Quality Statistics Division. Available from the Quality Information Center, American Society for Quality, P.O. Box 3005, Milwaukee, WI 53201-3005, 800-248-1946. Publication number S07-07.

Draper, N. R., & Smith, H. (1981). *Applied regression analysis* (2nd ed.). New York: Wiley-Interscience, John Wiley and Sons.

Jackson, J. E. (1991). *A user's guide to principal components.* New York: John Wiley.

Johnson R., & Wichern, D. (1992). *Applied multivariate statistical methods.* Englewood Cliffs, NJ: Prentice Hall.

Montgomery, D. C. (1991). *Design and analysis of experiments* (3rd ed.). New York: John Wiley.

Montgomery, D. C. (1991). *Introduction to statistical quality control* (2nd ed.). New York: John Wiley.

Morrison, D. (1967). *Multivariate statistical methods.* New York: McGraw-Hill.

APPENDIX A

Effective Teamwork

It's easy to get the players. Getting 'em to play together is the hard part.

Casey Stengel

A.1 Introduction

As soon as you begin to think about doing an improvement project, you need to think about who will be involved and how the work will get done. Working in teams is the most effective way to do such work, both in the classroom and in the business world. In class projects, teams bring different interests and skills to the problem, as well as reduce the amount of work to be done by students and instructors alike. Simply put, better projects are produced by teams. But as Casey Stengel (Hall of Fame manager of the New York Yankees) points out, it's not easy to get teams to work together.

People in any organization will be a part of many teams, both formal and informal, throughout their careers. It is rare that a piece of work is successfully completed by a single individual working alone and without interacting with anyone. When teams are appropriate, the work gets done more quickly with higher quality. This appendix provides general suggestions to increase the probability that teams will be successful. For more detailed information on effective teams, see the references on page 464.

A.2 Benefits of Using Teams

Effective teamwork provides significant benefits to both the organization and the individual team members. The organization gets a complete entity that collectively has all the knowledge and skills needed to improve the process. Thus an effective team can often make fundamental, lasting process improvement, rather than just pushing the problem from one department to another—as often happens without effective teams. The individuals also benefit by learning from fellow team members about other aspects of the process or better approaches to problem solving and process improvement. The pooling of information may eliminate long-standing headaches in organizational processes that individuals have been unable to solve on their own.

When forming a team, consider what each team member will get out of the project and how the organization will benefit. Teams members may want to know "What's in it for me?" In a healthy organization, however, problem solving and process improvement will be considered part of the job and not "extra work" (see Chapter 1). Teams are most productive when all team members get something significant out of the effort. New experiences, new skills, new acquaintances/friends, broader responsibility, and leadership experience—as well as financial rewards—are benefits that often accrue from working on teams.

A.3   When to Use a Team

Of course, forming a team is not always the best approach. Some activities, such as writing poetry or a novel, are best done by individuals. Rarely has a great work of art been produced by a team. A team should be used in any situation in which no single person has the collective knowledge, skills, and experience needed to get the job done effectively in the desired time frame. Following are key issues that identify when a team is needed.

When to Use a Team

- The task is complex.
- Creative ideas are needed.
- The problem crosses functions.
- Broad buy-in is needed.
- Implementation involves many people.
- The issue is controversial.
- The path forward is unclear.

A team should not be used when one person can do the job, unless the objective is to give employees new experiences. People may ask to be on a team if they feel that their views will not be adequately represented by the current members of the team. In other cases, everyone may wish to join the team working on a "hot" project. One company paid employees a bonus depending on how many teams they joined. Predictably, the result was a proliferation of large teams—which typically do not work well. Success results when the team is small and includes a mechanism to see that all relevant views are heard. In major projects, such as building a new facility, a "core team" might oversee the entire project, while numerous other teams work on individual tasks associated with the project. The core team's job is to ensure that each individual team is working toward the same common objective—that is, to ensure overall "system optimization" versus functional suboptimization within each team (see Chapter 3). This tiered structure of teams allows major projects to be accomplished with fairly small teams.

A.4 Forming a Team

Business process improvement is typically a team activity because different groups are involved and are affected, requiring different backgrounds and skills to make the needed improvements. Rarely is one person the expert on the entire business process. Student projects work best with two to four members; three is an ideal team size. Student teams that form naturally, such as among existing friends or those in the same department or housing, work well. When this is not possible, the instructor provides guidance on team formation as well as the topic on which the team should work. Topic selection is discussed in the next section.

In the business world, teams of four to six people work best. Teams that include more than six people often slow the improvement process: It becomes difficult to get the entire team together for meetings, and consensus is typically harder to achieve. Sometimes purely volunteer teams are used. In general, this is not a good practice because management support is required to get the needed personnel, time, and funds allocated. The makeup of the team should be based on the problem to be addressed and on the skills and knowledge that will be required to get the job done.

A.5 Selecting Team Projects

A team can be doomed to failure by selecting an inappropriate project. For example, often teams select a problem scope or objective that is much too large for them to actually accomplish in a reasonable amount of time. Such projects are often referred to sarcastically as trying to "solve world hunger" or "boil the ocean." Careful thought needs to go into selection of the improvement project itself, to give the team a chance to succeed.

For student projects, where we may select the team prior to selecting the project, some key criteria for the project are:

- Topic is of interest to all team members.
- Situation can actually be affected by the team.
- Scope can be completed in allotted time.
- Data are available in allotted time.
- Issue is important—people who are *not* team members care about it.

A written proposal for the project should be submitted to the instructor for approval prior to beginning the project.

In the business world, projects are generally selected first, and then team members identified. It is hard to pick the right team if we don't know what problem we are addressing. Projects that can be completed in 4–6 months work best.

Longer projects often drag on too long, with little tangible benefit to show management given the staff hours used. It is important to attach a financial value to the project at an early stage so that management can determine whether the predicted benefits of the project are worth the cost of doing the project.

A.6   Ingredients for a Successful Team

Scholtes et al. (1996) discuss the ingredients of a successful team. An adaptation of their model is shown below.

Ingredients of a Successful Team

- Clear goals
- Clear roles
- Project plan
- Use of scientific methods including data
- Well-defined decision procedures
- Knowledge of the process being improved
- Problem-solving skills
- Productive team dynamics
  - Clear communication
  - Balanced participation
  - Follow ground rules
  - Group process awareness

Paying attention to these ingredients increases the probability that the team will be successful. All of these elements need to be present for success. Ignoring even one or two can greatly affect the team's success. For example, if everyone does not have the same goals, then hidden agendas, politics, and conflict are likely to result.

A.7   Stages of Team Growth

Teams typically go through four generic stages of growth: forming, storming, norming, and performing. The team must first form and begin to function as a team. It may take a while before team members feel comfortable working together. As the team begins to operate together, it will likely have numerous issues to resolve (or "storm"): dominant personalities, lack of understanding of team processes, disagreement over the best approaches to use for the project, and so on. In the norming phase, the team works through these issues and starts to function effectively. By the time it has reached the performing phase, the team is functioning like a well-oiled machine. Much of the frequent resentment over working on committees and participating in team projects is due to teams that never get out of the storming phase. This is more likely if the team does not consciously focus on each of the elements of a successful team listed above. Scholtes et al. (1996) used a swimming metaphor to summarize the characteristics of each stage of team growth.

Stages of Team Growth

Forming

- Hesitant swimmers
- Standing beside the pool
- Sticking toes in the water
- Thinking about getting involved

Storming

- In the water thrashing about
- Overwhelmed by the amount of work
- Little time to do the work
- Doing project work in addition to regular duties
- Panic can set in

Norming

- Team members get used to working together.
- Team members start helping each other.
- Common spirit and goals develop.
- Team believes that they can be successful.

Performing

- Team members become comfortable with each other.
- Everyone works in concert.
- The team becomes an effective unit.
- They understand each other's strengths and weaknesses.
- There is a close attachment to the team.
- They're satisfied with their accomplishments.

Assessing the stage of growth that a team is in can greatly help one understand the behavior the team is exhibiting and identify ways to get the team back on the road to success. Teamwork is a process that can be studied and improved. Don't expect your process to start out functioning perfectly.

A.8   Running Effective Meetings

People generally dislike meetings because they take too much time and accomplish too little. These problems can be avoided, however, with appropriate meeting management techniques. Effective meetings are well designed with a purpose, set of desired outcomes, agenda, and well-defined roles, listed below:

Roles for Running Effective Meetings

Meeting Leader

- Is a fully participating team member
- Sets the agenda, with input
- Leads team through meeting agenda
- Opens meeting
- Opens discussion of each agenda item
- Closes meeting
- Is accountable for the meeting "results"

Facilitator

- Facilitates (means "to make easier")
- Is neutral—does not participate in meeting "content"
- Helps team come to closure on each agenda item
- Helps team choose appropriate decision methods
- Keeps discussion focused
- Keeps team on the agenda
- Owns the meeting "process"

Scribe (may also be the Note Taker)

- Records key points and ideas on flipchart to keep discussion focused
- Is a fully participating team member
- Writes large and legibly
- Is brief, using fewest words possible
- Uses words of the speaker whenever possible (no editing)
- Documents agreements
- Updates action list and "parking lot" as needed

Note Taker/Recorder

- Is a fully participating team member
- Keeps records of key occurrences and commitments
- Works with team leader to develop and distribute minutes

Timekeeper

- Is a fully participating team member
- Makes sure team keeps to its agenda
- Has a clock or watch available
- Notifies meeting leader 5–10 minutes before a specific agenda item is due to end
- Notifies meeting leader when time has run out so team can reach consensus to adjust agenda if necessary

In small meetings these roles are sometimes combined, with the meeting leader acting as the scribe, note taker, and/or timekeeper (although taking on too many roles can result in this person dominating the meeting). The scribe sometimes also serves as the note taker/reporter. When an organization becomes more familiar with using these meeting roles, the meeting participants will often volunteer for the needed roles. A neutral facilitator is often needed to help the team get through the storming stage. Once it reaches the performing stage, this role is often dropped.

During meetings, it is helpful to have an "action list" posted to capture assignments, including action items, responsible person, and time for completion or next report. This ensures that something tangible comes out of the meeting. The next meeting typically begins with a review of the status of the action items from the last meeting. A "parking lot" is a way to capture important ideas that are off the agenda, but are nonetheless worthy of the team's attention. The parking lot items are usually addressed at the end of the meeting. Paying attention to these roles usually results in very productive meetings and in willing participation by team members, because so much useful work is accomplished. A more detailed discussion of how to run effective meetings can be found in *The Team Memory Jogger* (GOAL/QPC, 1996) or Doyle and Straus (1982).

A.9   Dealing With Conflict

Conflict often occurs among team members, especially in the storming stage. Some ways for dealing with conflict—from *The Team Memory Jogger* (GOAL/QPC, 1996)—are listed below.

Dealing With Conflict

Avoid Conflict by Being Objective
- Stay focused on the subject, not the people involved.
- Try to understand the other person's point of view.
- Avoid judgmental and inflammatory language.

Handling Disagreements
- Decisions are built on a series of small agreements.
- Work to find areas of agreement.
- Build on areas of agreement.
- Identify areas of agreement and disagreement.
- Listen carefully and check for understanding.

Dealing with Feuds
- Keep the meeting focused on the work of the team, not the feud.
- Recognize that a feud may have started long ago.
- Help the team move forward in spite of the feud.
- Work to keep the feuding parties from dominating the meeting.
- Ensure that the feud is handled outside the meeting.

The key to avoiding conflict is to stay focused on the work of the team, listen and identify areas of agreement, and avoid personal issues—stay objective. Build on the areas of agreement, however small, that help the team be successful. Decisions are usually built on such small agreements.

Of course, people will often have differences of opinion, even if the team follows all this good advice. After all, diversity of opinion and knowledge was one of the reasons we formed a team in the first place. However, one can disagree without being disagreeable. This avoids feuds between members that become distractions to the team. Often if two knowledgeable, intelligent people disagree, different assumptions or beliefs are at the root of their disagreement. Probing for these fundamental differences may help focus the disagreement and allow the team to either make an objective decision or decide to obtain more data to resolve the assumptions.

A.10 Why Project Teams Fail

The reasons for project team failure (identified by Snee et al. [1998] from a study of more than 70 projects) are summarized below. Note that the majority of the reasons identified are management related.

Reasons for Team Failure

- Team is not supported by management.
    - No champion identified.
    - Champion does not meet with team.
- Project scope is too large.
- Project objectives are not significant.
- No clear measure of success is identified.
- Team is not given enough time to work on project.
- Team is too large.
- Team members have multiple agendas.
- Team is not trained.
- Data are not readily available.

Attention to these items increases the probability that your project will be successful. The ability to use teams to manage and improve an organization is a skill that must be developed by both managers and employees. Teams require guidance, coaching, counseling, training, recognition, and reinforcement. However, successful teamwork can radically improve organizational performance and employee growth, so the effort required is well worth making.

REFERENCES

Doyle, M., & Straus, D. (1982). *How to make meetings work.* New York: Jove Books.

GOAL/QPC & Joiner Associates. (1996). *The team memory jogger.* Madison, WI: Author.

Scholtes, P. R., Joiner, B. L., & Streibel, B. J. (1996). *The team handbook* (2nd ed.). Madison, WI: Joiner Associates.

Snee, R. D., Kelleher, K. H., & Reynard, S. (1998, May). Improving team effectiveness. *Quality Progress,* 43–48.

APPENDIX B

Presentations and Report Writing

The job isn't complete until the paperwork is done

B.1 Reporting Is Important and Necessary

Process improvement studies are of no value unless the process changes identified by the studies are implemented. This typically requires that the study results be presented to management and others to build support for the proposed changes and obtain the needed resources. These presentations can take many forms, such as informal one-on-one discussions, formal presentations to various groups, and written reports. Key requirements for any presentation or report are that it be clear, concise, and accurate. The following information will help you prepare for such interactions.

B.2 Presentations to Individuals or Small Groups

The simplest and most frequent communication is a presentation to a single person or to a small group. One should not take such interactions lightly. As John Wooden, renowned UCLA basketball coach, pointed out, "Failing to plan is planning to fail." Careful preparation can make the difference between getting and losing the support you need. First, identify the purpose for the meeting and your expected outcomes (i.e., what you would like to happen as a result of the meeting). Next, construct an agenda for your meeting that will produce your desired outcomes. A typical agenda might include:

- Introductions
- Meeting purpose and desired outcomes

- Project description
- Study design and data collection and analysis
- Results, interpretation, and conclusions
- Accomplishments to date, or since last review
- Progress toward goals as reflected in the key project metrics
- Recommendations, needed resources, and help
- Key learnings and issues
- Next steps and meeting conclusion

An agenda for a shorter presentation to management might include:

- Project description
- Key results
- Accomplishments to date, or since the last review
- Progress toward goals as reflected by the key project metrics
- Issues, needs, and next steps

The agenda tells your audience what problem you worked on, the work you did, and your recommendations based on your work. You will find that the positive outcome of the meeting will make the thorough preparation well worthwhile. Such a presentation may use an LCD projector or an overhead projector for computer-generated visuals or slides and may involve a handout and use of a flipchart to make your points. Handouts and flipcharts often work well for small groups.

B.3   Presentations or Project Reviews for Large Groups

The preparation for presentations and project reviews for large groups is similar to that for small groups. The key difference is that you will be more dependent on projected visual aids to make your points. The discussion following your presentation may also be more formal because of the large audience size. The content of the presentation can be similar to that of the small group.

It is particularly important that the slides be easy to read and understand when the audience is large. Some guidelines for creation of visuals (overhead or LCD projector) are

- Use one slide for each 2 minutes of presentation.
- Use no more than 30 words per slide.
- Use no more than 8 lines per slide.
- Do not use acronyms or abbreviations.
- Use 20-point or larger font size.
- Use high contrast between lettering and background—dark lettering on a light background or vice versa.

These guidelines will help keep the information content per slide reasonable. However, in some instances you may choose to violate these guidelines. You can get away with more information per slide in smaller groups from a readability standpoint but understandability may still be an issue. Remember that, when these guidelines are violated, you may waste valuable meeting time explaining your slides instead of your project.

Pitfalls

Some common presentation mistakes to avoid include:

- Talking to the projection screen (always face the audience when speaking).
- Saying "uh" between sentences (a very common nervous habit).
- Speaking too fast or using slang when addressing an international audience.
- Reading a speech (boring and insulting to the audience—speak in your own words, even if you must memorize what you want to say).
- Speaking in a monotone (try to vary the inflection of your voice).
- Unreadable visuals (discussed above).

One-Paragraph Summary or Abstract

You will often need to create a one-paragraph summary or abstract of your project. Such a paragraph typically contains three parts: problem/issue description, work done, and results/impact/implications. Depending on the required length, it is usually appropriate to include a two- or three-sentence description for each of the three areas—that is, a total of six to nine sentences. It is particularly important in the business world to include the results, impact, and implications. A mere description of the work done is not sufficient and may irritate some of your audience.

B.4 Written Reports

You may often need to follow up the presentation with a written report. Reports take time to prepare. As a compromise in many instances, a copy of the presentation slides if they are complete will be sufficient for this purpose. When a written report is needed, its contents should include the following items (in order):

- Cover letter if appropriate
- Key conclusions in an executive summary
- Project background
- Study design
- Data collection and analysis
- Results and interpretation
- Discussion, conclusions, recommendations

The contents and style of the report should always match the needs and culture of the intended audience. The executive summary is a concise statement of the key conclusions, recommendations, and take-aways from your project. Keep it general and short—don't present details in this section.

Use of Graphics

A graphic should be included to illustrate each of the key points of your report. Graphics should be clear and understandable. Graphics can contain too much information (i.e., a "busy" graph) in the same way that a slide can contain too many words. See Tufte (1983) for general advice on the use of graphics.

Graphics may have been used during your research for data exploration, analysis, and communication. In presentations and reports, graphics will help communicate your results. However, the graphics used in the exploration and analysis phases of the project are not always appropriate for communication of results. Plan to revise or replace charts that are too obscure or complicated. In addition to the display of summary results, graphics can be used to display process flows, graphs of models (e.g., *x* versus *y*), and procedures.

Presenting Statistical Results

A graph is the best way to present statistical results. Include statistical results in the text of the report or in tables when you need to support decisions, conclusions, and recommendations. Readers of your report will want to know what data you used as a basis for your conclusions. When possible, supporting statistics should be accompanied by some measure of uncertainty such as confidence limits.

Tables are another effective way of reporting data and statistical results. Tables should be clear, concise, and as simple as possible. Keep in mind that the objective is to help the reader understand what analyses you did and how you reached your conclusions. Clearly label the table title and the names of the rows and columns. Use table footnotes where necessary to help the reader understand and use the tables. As much as possible, each table should stand on its own and not require reference to the text to understand the table contents. The table should be constructed so that it is easy to make comparisons of interest. For example, it is easier to see trends in columns of numbers than in rows of numbers. Further discussion of the effective construction of data tables is contained in Ehrenberg (1975, Chapter 1).

Pitfalls

Some common mistakes to avoid in written reports include:

- Burying the key conclusions at the end (see advice above concerning the executive summary).
- Explaining how you did what you did before explaining why you did it and what you actually did (discussed above).
- Using technical language beyond the understanding of the intended audience. (KISS: keep it simple, stupid! The object is to communicate, not to impress.)
- Getting bogged down in details, such as a complex financial analysis (state the conclusions in the body and include the details as an appendix).
- Making the report a one-sided "position paper" (objectively state the results and provide data to back up key recommendations).

REFERENCES

Ehrenberg, A. S. C. (1975). *Data reduction.* New York: Wiley Interscience, John Wiley.

Tufte, E. R. (1983). *The visual display of quantitative information.* Cheshire, CT: Graphics Press.

APPENDIX C

More on Surveys

Effective management always means asking the right questions.

Robert Heller

This information was prepared by a group of professionals in survey methodology and is a good introduction to the topic. Another useful reference for using surveys to better understand the "voice of the employee" is Snee (1995), which discusses how employee surveys can be used both as a communication tool and an improvement tool. It includes a case history, as well as general guidelines for designing and implementing sound surveys.

What Is a Survey?*

It has been said the United States is no longer an "industrial society" but an "information society." That is, our major problems and tasks no longer mainly center on the production of the goods and services necessary for survival and comfort.

Our "society," thus, requires a prompt and accurate flow of information on preferences, needs, and behavior. It is in response to this critical need for information on the part of the government, business, and social institutions that so much reliance is placed on surveys.

Then, What Is a Survey?

Today the word *survey* is used most often to describe a method of gathering information from a sample of individuals. This "sample" is usually just a fraction of the population being studied.

---

*Reprinted with permission from the American Statistical Association Section on Survey Research Methods, Fritz Scheuren editor.

469

For example, a sample of voters is questioned in advance of an election to determine how the public perceives the candidates and the issues . . . a manufacturer does a survey of the potential market before introducing a new product . . . a government entity commissions a survey to gather the factual information it needs to evaluate existing legislation or to draft proposed new legislation.

Not only do surveys have a wide variety of purposes, they also can be conducted in many ways—including over the telephone, by mail, or in person. Nonetheless, all surveys do have certain characteristics in common.

Unlike a census, where all members of the population are studied, surveys gather information from only a portion of a population of interest—the size of the sample depending on the purpose of the study.

In a bona fide survey, the sample is not selected haphazardly or only from persons who volunteer to participate. It is scientifically chosen so that each person in the population will have a measurable chance of selection. This way, the results can be reliably projected from the sample to the larger population.

Information is collected by means of standardized procedures so that every individual is asked the same questions in more or less the same way. The survey's intent is not to describe the particular individuals who, by chance, are part of the sample but to obtain a composite profile of the population.

The industry standard for all reputable survey organizations is that individual respondents should never be identified in reporting survey findings. All of the survey's results should be presented in completely anonymous summaries, such as statistical tables and charts.

How Large Must the Sample Size Be?

The sample size required for a survey partly depends on the statistical quality needed for survey findings; this, in turn, relates to how the results will be used.

Even so, there is no simple rule for sample size that can be used for all surveys. Much depends on the professional and financial resources available. Analysts, though, often find that a moderate sample size is sufficient statistically and operationally. For example, the well-known national polls frequently use samples of about 1,000 persons to get reasonable information about national attitudes and opinions.

When it is realized that a properly selected sample of only 1,000 individuals can reflect various characteristics of the total population, it is easy to appreciate the value of using surveys to make informed decisions in a complex society such as ours. Surveys provide a speedy and economical means of determining facts about our economy and about people's knowledges, attitudes, beliefs, expectations, and behaviors.

Who Conducts Surveys?

We all know about the public opinion surveys or "polls" that are reported by the press and broadcast media. For example, the Gallup Poll and the Harris Survey issue reports periodically describing national public opinion on a wide range of current issues. State polls and metropolitan area polls, often supported by a local newspaper or TV station, are reported regularly in many localities. The major broadcasting networks and national news magazines also conduct polls and report their findings.

The great majority of surveys, though, are not public opinion polls. Most are directed to a specific administrative, commercial, or scientific purpose. The wide variety of issues with which surveys deal is illustrated by the following listing of actual uses:

Major TV networks rely on surveys to tell them how many and what types of people are watching their programs.

Statistics Canada conducts continuing panel surveys of children (and their families) to study educational and other needs.

Auto manufacturers use surveys to find out how satisfied people are with their cars.

The U.S. Bureau of the Census conducts a survey each month to obtain information on employment and unemployment in the nation.

The U.S. Agency for Health Care Policy and Research sponsors a periodic survey to determine how much money people are spending for different types of medical care.

Local transportation authorities conduct surveys to acquire information on commuting and travel habits.

Magazines and trade journals use surveys to find out what their subscribers are reading.

Surveys are conducted to ascertain who uses our national parks and other recreation facilities.

Surveys provide an important source of basic scientific knowledge. Economists, psychologists, health professionals, political scientists, and sociologists conduct surveys to study such matters as income and expenditure patterns among households, the roots of ethnic or racial prejudice, the implications of health problems on people's lives, comparative voting behavior, and the effects on family life of women working outside the home.

What Are Some Common Survey Methods?

Surveys can be classified in many ways. Two dimensions are size and type of sample. Surveys also can be used to study either human or nonhuman populations (e.g., animate or inanimate objects—animals, soils, housing, etc.). While many of the principles are the same for all surveys, the focus here will be on methods for surveying individuals.

Many surveys study all persons living in a defined area, but others might focus on special population groups—children, physicians, community leaders, the unemployed, or users of a particular product or service. Surveys may also be conducted with national, state, or local samples.

Surveys can be classified by their method of data collection. Mail, telephone interview, and in-person interview surveys are the most common. Extracting data from samples of medical and other records is also frequently done. In newer methods of data collection, information is entered directly into computers either by a trained interviewer or, increasingly, by the respondent. One well-known example is the measurement of TV audiences carried out by devices attached to a sample of TV sets that automatically record the channels being watched.

Mail surveys can be relatively low in cost. As with any other survey, problems exist in their use when insufficient attention is given to getting high levels of cooperation. Mail surveys can be most effective when directed at particular groups, such as subscribers to a specialized magazine or members of a professional association.

Telephone interviews are an efficient method of collecting some types of data and are being increasingly used. They lend themselves particularly well to situations where timeliness is a factor and the length of the survey is limited.

In-person interviews in a respondent's home or office are much more expensive than mail or telephone surveys. They may be necessary, however, especially when complex information is to be collected.

Some surveys combine various methods. For instance, a survey worker may use the telephone to "screen" or locate eligible respondents (e.g., to locate older individuals eligible for Medicare) and then make appointments for an in-person interview.

What Survey Questions Do You Ask?

You can further classify surveys by their content. Some surveys focus on opinions and attitudes (such as a pre-election survey of voters), while others are concerned with factual characteristics or behaviors (such as people's health, housing, consumer spending, or transportation habits).

Many surveys combine questions of both types. Respondents may be asked if they have heard or read about an issue . . . what they know about it . . . their opinion . . . how strongly they feel and why . . . their interest in the issue . . . past experience with it . . . and certain factual information that will help the survey analyst classify their responses (such as age, gender, marital status, occupation, and place of residence).

Questions may be open-ended ("Why do you feel that way?") or closed ("Do you approve or disapprove?"). Survey takers may ask respondents to rate a political candidate or a product on some type of scale, or they may ask for a ranking of various alternatives.

The manner in which a question is asked can greatly affect the results of a survey. For example, a recent NBC/*Wall Street Journal* poll asked two very similar questions with very different results: (1) Do you favor cutting programs such as social security, Medicare, Medicaid, and farm subsidies to reduce the budget deficit? The results: 23% favor; 66% oppose; 11% no opinion. (2) Do you favor cutting government entitlements to reduce the budget deficit? The results: 61% favor; 25% oppose; 14% no opinion.

The questionnaire may be very brief—a few questions, taking five minutes or less—or it can be quite long—requiring an hour or more of the respondent's time. Since it is inefficient to identify and approach a large national sample for only a few items of information, there are "omnibus" surveys that combine the interests of several clients into a single interview. In these surveys, respondents will be asked a dozen questions on one subject, a half dozen more on another subject, and so on.

Because changes in attitudes or behavior cannot be reliably ascertained from a single interview, some surveys employ a "panel design," in which the same respondents are interviewed on two or more occasions. Such surveys are often used during an election campaign or to chart a family's health or purchasing pattern over a period of time.

Who Works on Surveys?

The survey worker best known to the public is the interviewer who calls on the telephone, appears at the door, or stops people at a shopping mall.

Traditionally, survey interviewing, although occasionally requiring long days in the field, was mainly part-time work and, thus, well suited for individuals not wanting full-time employment or just wishing to supplement their regular income.

Changes in the labor market and in the level of survey automation have begun to alter this pattern—with more and more survey takers seeking to work full time. Experience is not usually required for an interviewing job, although basic computer skills have become increasingly important for applicants.

Most research organizations provide their own training for the interview task. The main requirements for interviewing are an ability to approach strangers (in person or on the phone), to persuade them to participate in the survey, and to collect the data needed in exact accordance with instructions.

Less visible but equally important are the in-house research staffs who, among other things, plan the survey, choose the sample, develop the questionnaire, supervise the interviews, process the data collected, analyze the data, and report the survey's findings.

In most survey research organizations, the senior staff will have taken courses in survey methods at the graduate level and will hold advanced degrees in sociology, statistics, marketing, or psychology, or they will have the equivalent in experience.

Middle-level supervisors and research associates frequently have similar academic backgrounds to the senior staff or they have advanced out of the ranks of clerks, interviewers, or coders on the basis of their competence and experience.

What About Confidentiality and Integrity?

The confidentiality of the data supplied by respondents is of prime concern to all reputable survey organizations. At the U.S. Bureau of the Census, for example, the data collected are protected by law (Title 13 of the U.S. Code). In Canada, the Statistics Act guarantees the confidentiality of data collected by Statistics Canada, and other countries have similar safeguards.

Several professional organizations dealing with survey methods have codes of ethics (including the American Statistical Association) that prescribe rules for keeping survey responses confidential. The recommended policy for survey organizations to safeguard such confidentiality includes:

> Using only number codes to link the respondent to a questionnaire and storing the name-to-code linkage information separately from the questionnaires.
>
> Refusing to give the names and addresses of survey respondents to anyone outside the survey organization, including clients.
>
> Destroying questionnaires and identifying information about respondents after the responses have been entered into the computer.
>
> Omitting the names and addresses of survey respondents from computer files used for analysis.
>
> Presenting statistical tabulations by broad enough categories so that individual respondents cannot be singled out.

What Are Other Potential Concerns?

The quality of a survey is largely determined by its purpose and the way it is conducted.

Most call-in TV inquiries (e.g., 900 "polls") or magazine write-in "polls," for example, are highly suspect. These and other "self-selected opinion polls (SLOPS)" may be misleading since participants have not been scientifically selected. Typically, in SLOPS, persons with strong opinions (often negative) are more likely to respond.

Surveys should be carried out solely to develop statistical information about a subject. They should not be designed to produce predetermined results or as a ruse for marketing and similar activities. Anyone asked to respond to a public opinion poll or concerned about the results should first decide whether the questions are fair.

Another important violation of integrity occurs when what appears to be a survey is actually a vehicle for stimulating donations to a cause or for creating a mailing list to do direct marketing.

Where Can I Get More Information?

In addition to the pamphlets in this series, the American Statistical Association (ASA) also makes other brochures available upon request:

> *Ethical Guidelines for Statistical Practice*
> *Surveys and Privacy,* produced by the ASA Committee on Privacy and Confidentiality
> *What Is a Survey?: How to Plan a Survey*
> *What Is a Survey?: How to Collect Survey Data*

For the above brochures or other pamphlets, contact:

Section on Survey Research Methods
American Statistical Association
1429 Duke Street
Alexandria, VA 22314-3402 USA
(703) 684-1221/fax: (703) 684-2037
E-mail: asainfo@amstat.org

REFERENCE

Snee, R. D. (1995, January). Listening to the voice of the employee. *Quality Progress*, 91–95.

APPENDIX D

More on the Six Sigma Improvement Approach[*]

With Six Sigma permeating much of what we do, it will be unthinkable to hire, promote or tolerate those who cannot, or will not, commit to this way of work.

Jack Welch, CEO, General Electric

D.1 The Basic Concept of Six Sigma

The *Financial Times* (Tomkins, 1997) defines the Six Sigma initiative as "a programme aimed at the near-elimination of defects from every product, process and transaction." This concept was introduced at, and popularized by, Motorola in their quest to reduce defects of manufactured electronics products. When used as a metric, Six Sigma technically means having no more than 3.4 defects per million opportunities in any process, product, or service. Statisticians may notice that having specification limits 6 standard deviations away from the average of an assumed normal distribution will not produce 3.4 defects per million, and they are correct. The 3.4 figure assumes that the specification limits are 6 standard deviations away from the target, but that the process average may drift over the long term by as much as 1.5 standard deviations, despite our best efforts to control it. This results in a probability of being outside the specifications of about 3.4/1,000,000, assuming that the specifications are 4.5 standard deviations away from the average.

More important than the technical definition is the concept of Six Sigma as a disciplined, quantitative approach for improvement of defined metrics in manu-

---

[*]Adapted from R. W. Hoerl, "Six Sigma and the Future of the Quality Profession," *Quality Progress*, June 1998, 35–42. Further discussion of Six Sigma as it relates to the process improvement and problem-solving strategies and the statistical thinking model is contained in Chapter 4.

facturing, service, or financial processes. This approach drives the overall processes of (1) selecting the right projects based on their potential to improve performance metrics, and (2) hiring and training the right people to get the desired results. Improvement projects follow a disciplined process of four macro phases: Measure, Analyze, Improve, and Control (MAIC). Sometimes a preliminary Define step is added, which relates to appropriate selection of projects and problem definition. The purpose of each step is as follows:

- **Measure:** Select the appropriate responses (the "*y*'s") to be improved based on customer input, ensure that they are quantifiable and that we can accurately measure them. Determine what is unacceptable performance (i.e., a "defect"). Gather preliminary data to evaluate current performance.
- **Analyze:** Analyze the preliminary data to document current performance (baseline process capability) and also to begin identifying root causes of defects (i.e., the "*x*'s," or independent variables) and their impact.
- **Improve:** Determine how to intervene in the process to significantly reduce the defect levels. Several rounds of improvements may be required.
- **Control:** Once the desired improvements have been made, put some type of system into place to ensure the improvements are sustained, even though additional Six Sigma resources may no longer be focused on the problem.

Following are the major elements of Six Sigma implementation:

- The initiative is driven by leaders at the highest levels of the organization—often, the CEO—and permeates through *all* levels of management and operations. CEO-led initiatives were led by Jack Welch of General Electric, Bob Galvin at Motorola, and Larry Bossidy at AlliedSignal. This aspect is not just corporate PR and is perhaps the most important reason for success.
- The main focus was initially on manufacturing, and specifically on cost and waste reduction, yield improvement, and operations where opportunity exists to improve capacity without major capital expenditure. Currently, strong emphasis is placed on understanding and satisfying customer needs. In addition, as organizations realize how large the financial impact could be, nonmanufacturing processes have also begun to receive similar attention.
- Performance metrics are established that directly measure the improvement in cost, quality, yield, and capacity. Contrary to some TQM initiatives, financial figures are required both to select projects and evaluate success, and performance metrics are tracked rigorously.
- Typical projects are targeted for at least $50,000 per year contribution to the bottom line. At AlliedSignal, the initial projects typically exceeded $1 million in benefits, and many new projects are still of this magnitude.
- Practitioners (engineers, accountants, computer scientists, and so on) are identified to contribute 50–100% of their time to these projects, with help from other team members. These practictiners are called various names in different companies, such as "Black Belts" (at GE and Motorola), "Process Improvement Masters" (at AlliedSignal), or "Variability Reduction Leaders" (at Polaroid). For consistency, we will refer to them as Black Belts (BBs) here.

- The BBs take up to 4 to 5 weeks of intensive, highly quantitative training, roughly corresponding to the four macro steps of the Six Sigma methodology. For example, the first week covers measurement tools; emphasizes documenting customer needs, process mapping, measurement system evaluation; and begins process capability analysis. The Analyze step continues process capability; stresses basic tools to seek root causes and relationships, such as scatter plots, run charts, Pareto charts; and introduces more formal statistics, such as hypothesis testing and ANOVA. Improve typically stresses designed experimentation, and Control discusses control charting, mistake proofing, and standard operating procedures. See Table D.1 for a typical course outline. Soft tools, such as effective communication and team leadership skills, are typically part of the curriculum. The "price of admission" is a significant project affecting the business's bottom line.
- The initial training courses in the company are usually taught by an external expert (e.g., at GE by the Six Sigma Academy's Mikel Harry). The audience includes future BBs, managerial "champions," and a group of carefully selected Master Black Belts (MBBs), including some internal statisticians. MBBs are the "Six Sigma technical experts" and have Six Sigma training, experience, and skills beyond that of BBs. The MBBs have the responsibility for training the Black Belts and for providing overall leadership. The MBB role subsequently evolves into a mentoring, teaching, and reviewing role.

TABLE D.1
Typical Six Sigma BB Training Curriculum

Week 1	*Week 3*

- Six Sigma Overview and Process Improvement Planning; The MAIC Road Map
- Process Mapping
- QFD (Quality Function Deployment)
- FMEA (Failure Mode and Effects Analysis)
- Organizational Effectiveness Concepts
- Basic Statistics Using Minitab
- Process Capability
- Measurement Systems Analysis
	- DOE (Design of Experiments)
	  - Factorial Experiments
	  - Fractional Factorials
	  - Balanced Block Designs
	  - Evolutionary Operation (EVOP)
	  - Response Surface Designs
	- ANOVA
	- Regression (Multiple)
	- Facilitation Tools

Week 2	*Week 4*
- Review of Key Week 1 Topics - Statistical Thinking - Hypothesis Testing (*F, t,* and so on) - Correlation - Passive Multivariate Analysis and Regression (Simple) - Team Assessment	- Control Plans - SPC/Advanced Process Control - Mistake-Proofing - Team Development - Wrap-up of Tools

Notes:
1. Project reviews are done each day in weeks 2–4.
2. Hands-on exercises are done on most days.
3. Three weeks of applied time between sessions.

- Upon completion of the first project, preferably within 4 months, the BB moves on to a new project repeating the deployment of the tools in the MAIC sequence. Most BBs work on several projects simultaneously and have to formally report on these projects to management.
- Although the tools used are not new, the Six Sigma approach adds considerable value to them. Its advantages include:
  - Formalizing the use of statistical tools versus having isolated individuals use them in a disconnected way.
  - Providing an overall "road map," or a multistep approach (MAIC) to integrating the tools appropriately. Many people have commented that their college statistics courses left them confused about why they needed to learn various statistical tools and how they fit together. The Six Sigma road map has finally made this clear and enabled practitioners to actually apply the tools to real problems.
  - Stressing the need to understand and reduce variation versus only estimating it.
  - Emphasizing a data-based approach to management versus gut feel or intuition. Six Sigma requires that everything be quantified, even "intangibles" such as customer perceptions.
  - Developing standardized vocabulary, metrics, and tools throughout highly diverse companies.

In addition to BBs and MBBs, Green Belts (GBs) receive similar or, in some places, reduced training. Like the BBs, they enter the training with a chartered project important to their operation's success. However, unlike the Black Belts, the Green Belts are not expected to spend the preponderance of their time on Six Sigma projects. Many hourly workers in financial operations, factories, and the like have also been trained, using the titles of Yellow Belts (YBs). They may receive a total of four days of introductory training on the MAIC tools to assist on BB or GB teams. These additional training efforts are intended to get everyone involved in Six Sigma. The obvious long-term objective is for every employee of the company to make improvement of their work processes a normal, everyday part of the job. To illustrate the possible scope of this effort and one company's level of commitment, GE insisted that all of its professional workforce be Green Belt or Black Belt trained and do projects by the end of 1998, including the senior executives of the company. Obviously, filling in for these resources, particularly the full-time MBBs and BBs, required a significant investment by companies embarking on Six Sigma.

D.2 The Evolution of Six Sigma Quality Beyond Manufacturing and Traditional Quality

Although most of the initial emphasis of Six Sigma was on quality improvement in manufacturing, it is rapidly spreading to key areas beyond manufacturing and beyond what would traditionally be considered "quality." This includes transactional and administrative processes such as employee recruitment, billing, order entry, distribution, purchasing, and other parts of the supply chain. A special form of Six Sigma, known as Design for Six Sigma, is used in new product development and other research and development processes. AlliedSignal has developed its Technical Excellence thrust around Six Sigma concepts, value chain analysis, and customer satisfaction. The focus is on getting good data on customer requirements and on reducing failure modes and variation in product design, scale-up, and commercialization. Allied also has significant Six Sigma initiatives in financial and business services.

Similarly, General Electric has placed a major focus on Commercial Quality Six Sigma, which emphasizes both the needs of GE's service businesses (GE Information Services, NBC, and the massive GE Capital Services) and the non-manufacturing operations of GE's manufacturing businesses (software systems and development, billing, human resources, and so on). In addition, recognizing that long-term success is highly dependent on product and service design, GE has begun a significant drive for Design for Six Sigma (DFSS) and, most recently, for Design for Reliability (DFR) as a key element of the Six Sigma initiative.

Each of these extensions has built on the foundation provided by the basic Six Sigma concepts and training, including high emphasis on successful projects and quantification of results, thus expanding application of the Six Sigma concepts beyond the traditional manufacturing arenas to engineering, reliability, financial, and human resource processes, where there are many opportunities to make significant improvements. Moreover, opportunities for broader application abound in society: banking, health care, government, and teaching (including curriculum design) are just a few areas that come to mind.

D.3 What Have Been the Results?

Six Sigma has worked extremely well and is building further momentum. Some key reasons for this are as follows.

Big Financial Impact Between 1995 and the first quarter of 1997, AlliedSignal reported results exceeding $800 million in cost savings from its Six Sigma–related Operational Excellence initiative (per the company's first quarter 1997 report). Similarly, GE's third quarter 1997 rise in operating margin from 13.8% to 14.5%, worth about $600 million, "stemmed from the Six Sigma quality initiative" (Tomkins, 1997). Chairman Jack Welch has indicated that Six Sigma may add $8 to $12 billion in five years, if utilized successfully (Clifford, 1997). These results have been the accumulation of numerous individual projects. For example, one team of three BBs at an AlliedSignal site returned over $25 million in cost savings and capacity improvement on one project alone. This involved identification of critical process variables that affected capacity on a chlorofluoro-carbon-substitute process. In his address at the GE company 1997 annual meeting (Welch, 1997), CEO Jack Welch described in some detail typical Six Sigma projects at GE Lighting, GE Capital Mortgage Corporation, GE Plastics, and GE Power Systems. He added that "Six Sigma has gone in less than two years from being an alien concept, full of complex calculations and unfamiliar jargon, to a consuming passion sweeping across the company." He added, "In 1996 (the first full year of Six Sigma quality), we spent about $200 million on projects and training—and got somewhere in the neighborhood of $150 million in returns. In 1997 we'll invest over $400 million, and we'll get $550 million to $600 million back . . . and by 2000 . . . the cumulative bottom-line impact of Six Sigma quality will be measured . . . in billions."

Continued Top Management Support and Enthusiasm   The support by top company leaders for the Six Sigma quality initiative remains unabated. For example, GE unambiguously announced in 1997 that only those who have been Six Sigma trained and have actually done projects are eligible for promotion to executive levels. To make the point further, business leaders are being evaluated on their progress on Six Sigma: 40% of each GE executive's bonus in 1997 was based on Six Sigma implementation and results. In addition, all salaried employees must be trained and have project experience by the end of 1998 as a condition of employment. Jack Welch has made it clear, by word and deed, that if you are not enthusiastic about Six Sigma, GE is simply not the right company for you. Similar management enthusiasm was required at companies like AlliedSignal and Motorola in order to achieve significant financial impact.

The Quantitative and Disciplined Approach to Process Improvement   Using the MAIC approach consistently across diverse businesses and processes has allowed GE to achieve significant tangible business results. Six Sigma has become a common language for different business units to talk to one another, share successes and failures, and in general, learn from one another. Senior leaders recognize what may seem obvious to quality professionals: To improve results, improve the process that generates the results. In addition, awareness is growing that improving the process will require a disciplined, data-based approach, rather than the more traditional "ready, fire, aim" method. This leads directly to more effective use of statistical tools. Similarly, vague statements, such as "We think we have that under control" or "Recent performance seems to be improving" are no longer accepted at any level. The typical response to these kinds of statements is now: "Show me the data!"

The Value Placed on Understanding and Satisfying Customer Needs   Although most companies claim this as a value already, it is amazing how little most businesses really know about their customers prior to Six Sigma. With the emphasis on metrics, customer needs and current performance in meeting them must be quantified and documented. Formal customer interactions and evaluations have replaced anecdotal information from the sales force.

Combining the Right Projects, the Right People, and the Right Tools   In the past, statisticians may have focused too narrowly on the statistical tools. The tools are clearly powerful, but only if used by the right people on the right projects. When companies have stressed tools per se, the typical result has been a lot of unimportant, perhaps even trivial, applications. Tremendous synergy can be generated by carefully selecting important and appropriate projects and getting the most talented people properly trained to apply relevant statistical (and non-statistical) tools to these projects. Training is tremendously enhanced by directly and immediately applying the material learned to a real project. Project-based training has had such a dramatic effect that GE businesses generally require a project as a prerequisite for attending Six Sigma training. It is also important to note that GE has insisted that candidates for significant Six Sigma roles be the most highly regarded people in the business, not simply "warm bodies" who were available. Project and people selection have been just as important to achieving results as use of the proper tools.

REFERENCES

Clifford, H. (1997, November). Six Sigma. *Continental Airlines Magazine,* 64.

Hahn, G. J., Hill, W. J., Hoerl, R. W., & Zinkgraf, S. A. (1999). The impact of Six Sigma improvement—a glimpse into the future of statistics. *The American Statistician,* 53 (3), 208–215.

Tomkins, R. (1997, October 10). GE beats expected 13% rise. *Financial Times,* p. 22.

Welch, John F. (1997). A learning company and its quest for Six Sigma. Presented at the General Electric Company 1997 Annual Meeting, Charlotte, North Carolina, April 23. Reprints available from Executive Speech Reprints, Radius Group Inc., 705 Corporation Park, Scotia, NY 12302.

APPENDIX E

More on Design of Experiments

Experiment, and it will lead you to the light.

Cole Porter

E.1 Experiments Involving Four or More Variables

Chapter 7 discussed the use of statistically designed experiments to collect data for building models, focusing on experiments involving three variables. As you use this approach, you may encounter situations in which you have four or more variables to study. This appendix presents some designs that enable you to investigate the effects of four or more variables and provides guidance on how to analyze the results. The designs discussed are summarized in Table E.1 and shown in detail in Tables E.2–E.7. The references at the end of this appendix present further details on the use and analysis of these designs.

Table E.1 shows the number of variables that each design is typically used to study. The designs in Tables E.5, E.6, and E.7 are used to study multiple numbers of factors. These designs should be used to estimate the effects of "L" factors using the first L columns of the designs. For example, the first ten columns of Table E.7 should be used to estimate the linear (main) effects of ten factors.

E.2 Analysis of Two-Level Experiments

Two-level factorial designs are analyzed either by computing the factor effects as discussed in Chapter 7 or by using regression analysis to compute coefficients (i.e., effect = 2 × regression coefficient of the factor) as discussed in Chapter 6.

TABLE E.1
Some Useful Experimental Designs

Design Code	Table	Design Name	Number of Variables	Number of Runs	Permits Estimation of
A	2	Two-level factorial	4	16	All linear (main) effects and interactions
B	3	Two-level factorial	5	32	All linear (main) effects and interactions
C	4	Fractional-factorial	5	16	All linear (main) effects and two-factor interactions
D	5	Fractional-factorial	6, 7, or 8	16	All linear (main) effects and two-factor interaction "chains"
E	6	Plackett-Burman	4, 5, 6, or 7	12	Linear (main) effects only
F	7	Plackett-Burman	8 to 15	20	Linear (main) effects only

TABLE E.2   Two-Level Four-Factor Factorial Design

	Variable			
Run	x_1	x_2	x_3	x_4
1	−1	−1	−1	−1
2	1	−1	−1	−1
3	−1	1	−1	−1
4	1	1	−1	−1
5	−1	−1	1	−1
6	1	−1	1	−1
7	−1	1	1	−1
8	1	1	1	−1
9	−1	−1	−1	1
10	1	−1	−1	1
11	−1	1	−1	1
12	1	1	−1	1
13	−1	−1	1	1
14	1	−1	1	1
15	−1	1	1	1
16	1	1	1	1

TABLE E.3
Two-Level Five-Factor Factorial Design

Run	Variable					Run	Variable				
	x_1	x_2	x_3	x_4	x_5		x_1	x_2	x_3	x_4	x_5
1	−1	−1	−1	−1	−1	17	−1	−1	−1	−1	1
2	1	−1	−1	−1	−1	18	1	−1	−1	−1	1
3	−1	1	−1	−1	−1	19	−1	1	−1	−1	1
4	1	1	−1	−1	−1	20	1	1	−1	−1	1
5	−1	−1	1	−1	−1	21	−1	−1	1	−1	1
6	1	−1	1	−1	−1	22	1	−1	1	−1	1
7	−1	1	1	−1	−1	23	−1	1	1	−1	1
8	1	1	1	−1	−1	24	1	1	1	−1	1
9	−1	−1	−1	1	−1	25	−1	−1	−1	1	1
10	1	−1	−1	1	−1	26	1	−1	−1	1	1
11	−1	1	−1	1	−1	27	−1	1	−1	1	1
12	1	1	−1	1	−1	28	1	1	−1	1	1
13	−1	−1	1	1	−1	29	−1	−1	1	1	1
14	1	−1	1	1	−1	30	1	−1	1	1	1
15	−1	1	1	1	−1	31	−1	1	1	1	1
16	1	1	1	1	−1	32	1	1	1	1	1

TABLE E.4 Two-Level Five-Factor Fractional-Factorial Design

Run	Variable				
	x_1	x_2	x_3	x_4	x_5
1	−1	−1	−1	−1	1
2	1	−1	−1	−1	−1
3	−1	1	−1	−1	−1
4	1	1	−1	−1	1
5	−1	−1	1	−1	−1
6	1	−1	1	−1	1
7	−1	1	1	−1	1
8	1	1	1	−1	−1
9	−1	−1	−1	1	−1
10	1	−1	−1	1	1
11	−1	1	−1	1	1
12	1	1	−1	1	−1
13	−1	−1	1	1	1
14	1	−1	1	1	−1
15	−1	1	1	1	−1
16	1	1	1	1	1

TABLE E.5 Two-Level Fractional-Factorial Design for Six, Seven, or Eight Variables

				Variable				
Run	x_1	x_2	x_3	x_4	x_5	x_6	x_7	x_8
1	−1	−1	−1	1	1	1	−1	1
2	1	−1	−1	−1	−1	1	1	1
3	−1	1	−1	−1	1	−1	1	1
4	1	1	−1	1	−1	−1	−1	1
5	−1	−1	1	1	−1	−1	1	1
6	1	−1	1	−1	1	−1	−1	1
7	−1	1	1	−1	−1	1	−1	1
8	1	1	1	1	1	1	1	1
9	−1	−1	−1	−1	−1	−1	1	−1
10	1	−1	−1	1	1	−1	−1	−1
11	−1	1	−1	1	−1	1	−1	−1
12	1	1	−1	−1	1	1	1	−1
13	−1	−1	1	−1	1	1	−1	−1
14	1	−1	1	1	−1	1	1	−1
15	−1	1	1	1	1	−1	1	−1
16	1	1	1	−1	−1	−1	−1	−1

Chains of Two-Factor Interactions
12 + 37 + 48 + 56
13 + 27 + 46 + 58
14 + 28 + 36 + 57
15 + 26 + 38 + 47
16 + 25 + 34 + 78
17 + 23 + 68 + 45
18 + 24 + 35 + 67

Note: With this design, if any two-factor interactions do exist, they will not bias estimates of the linear (main) effects.

TABLE E.6
Twelve-Run Plackett-Burman Design

						Variables					
Run	x_1	x_2	x_3	x_4	x_5	x_6	x_7	x_8	x_9	x_{10}	x_{11}
1	1	−1	1	−1	−1	−1	1	1	1	−1	1
2	1	1	−1	1	−1	−1	−1	1	1	1	−1
3	−1	1	1	−1	1	−1	−1	−1	1	1	1
4	1	−1	1	1	−1	1	−1	−1	−1	1	1
5	1	1	−1	1	1	−1	1	−1	−1	−1	1
6	1	1	1	−1	1	1	−1	1	−1	−1	−1
7	−1	1	1	1	−1	1	1	−1	1	−1	−1
8	−1	−1	1	1	1	−1	1	1	−1	1	−1
9	−1	−1	−1	1	1	1	−1	1	1	−1	1
10	1	−1	−1	−1	1	1	1	−1	1	1	−1
11	−1	1	−1	−1	−1	1	1	1	−1	1	1
12	−1	−1	−1	−1	−1	−1	−1	−1	−1	−1	−1

TABLE E.7
20-Run Plackett-Burman Design

	Variables																		
Run	x_1	x_2	x_3	x_4	x_5	x_6	x_7	x_8	x_9	x_{10}	x_{11}	x_{12}	x_{13}	x_{14}	x_{15}	x_{16}	x_{17}	x_{18}	x_{19}
1	+	+	−	−	+	+	+	+	−	+	−	+	−	−	−	−	+	+	−
2	−	+	+	−	−	+	+	+	+	−	+	−	+	−	−	−	−	+	+
3	+	−	+	+	−	−	+	+	+	+	−	+	−	+	−	−	−	−	+
4	+	+	−	+	+	−	−	+	+	+	+	−	+	−	+	−	−	−	−
5	−	+	+	−	+	+	−	−	+	+	+	+	−	+	−	+	−	−	−
6	−	−	+	+	−	+	+	−	−	+	+	+	+	−	+	−	+	−	−
7	−	−	−	+	+	−	+	+	−	−	+	+	+	+	−	+	−	+	−
8	−	−	−	−	+	+	−	+	+	−	−	+	+	+	+	−	+	−	+
9	+	−	−	−	−	+	+	−	+	+	−	−	+	+	+	+	−	+	−
10	−	+	−	−	−	−	+	+	−	+	+	−	−	+	+	+	+	−	+
11	+	−	+	−	−	−	−	+	+	−	+	+	−	−	+	+	+	+	−
12	−	+	−	+	−	−	−	−	+	+	−	+	+	−	−	+	+	+	+
13	+	−	+	−	+	−	−	−	−	+	+	−	+	+	−	−	+	+	+
14	+	+	−	+	−	+	−	−	−	−	+	+	−	+	+	−	−	+	+
15	+	+	+	−	+	−	+	−	−	−	−	+	+	−	+	+	−	−	+
16	+	+	+	+	−	+	−	+	−	−	−	−	+	+	−	+	+	−	−
17	−	+	+	+	+	−	+	−	+	−	−	−	−	+	+	−	+	+	−
18	−	−	+	+	+	+	−	+	−	+	−	−	−	−	+	+	−	+	+
19	+	−	−	+	+	+	+	−	+	−	+	−	−	−	−	+	+	−	+
20	−	−	−	−	−	−	−	−	−	−	−	−	−	−	−	−	−	−	−

487

TABLE E.8   Process Control Study: Five-Factor Fractional-Factorial Design

			Variable			
Run	x_1 Solvent	x_2 Catalyst	x_3 Temp	x_4 Purity	x_5 pH	y Color
1	−1	−1	−1	−1	1	−0.63
2	1	−1	−1	−1	−1	2.51
3	−1	1	−1	−1	−1	−2.68
4	1	1	−1	−1	1	−1.66
5	−1	−1	1	−1	−1	2.06
6	1	−1	1	−1	1	1.22
7	−1	1	1	−1	1	−2.09
8	1	1	1	−1	−1	1.93
9	−1	−1	−1	1	−1	6.79
10	1	−1	−1	1	1	6.47
11	−1	1	−1	1	1	3.45
12	1	1	−1	1	−1	5.68
13	−1	−1	1	1	1	5.22
14	1	−1	1	1	−1	9.38
15	−1	1	1	1	−1	4.31
16	1	1	1	1	1	4.05

The effects or regression coefficients are subsequently analyzed by hypothesis tests or normal probability plots to identify important variables. Normal probability plots are not particularly useful when estimating fewer than ten or eleven effects, hence they have not yet been discussed. Dot plots can also be used to identify the factors that have large effects.

The data in Table E.8 are from a process control study whose purpose was to identify the significant factors affecting the color of the product produced by the process (Snee et al., 1985). The experiment utilized a half-fraction of a two-level five-factor factorial design (see Table E.4), which permits the estimation of the five linear (main) effects and all ten two-factor interactions. The factor effects are summarized in Table E.9 and shown graphically in Figure E.1, which shows that factors A, B, D, and E have large linear effects and need to be tightly controlled if the process is to produce a product with consistent color (i.e., small variation in color). As with normal probability plots of residuals, a straight line in the plot indicates random variation. An effect that stands well off the line indicates a real effect. There are no significant interactions. This simplifies the process control methods needed. Factor C, temperature, has a very small effect in the temperature range studied and needs little control when kept within the low and high levels studied in the experiment.

The 16-run design in Table E.5 does not allow the experimentor to estimate each interaction. It does permit the estimation of "chains" (linear combinations) of two-factor interactions as shown at the bottom of Table E.5. For example, if we used the first seven columns of Table E.5 to estimate the effects of seven factors,

TABLE E.9   Process Control Study: Estimated Effects and Coefficients for Color

Term	Effect	Coefficient
Constant	—	2.876
Solvent	1.644	0.822
Catalyst	−2.504	−1.252
Temperature	0.769	0.384
Purity	5.586	2.793
pH	−1.744	−0.872
Solvent × Catalyst	0.109	0.054
Solvent × Temperature	0.126	0.063
Solvent × Purity	−0.191	−0.096
Solvent × pH	−0.611	−0.306
Catalyst × Temperature	0.084	0.042
Catalyst × Purity	−0.089	−0.044
Catalyst × pH	0.371	0.186
Temperature × Purity	−0.626	−0.313
Temperature × pH	−0.576	−0.288
Purity × pH	0.001	0.001

FIGURE E.1   Process Control Study: Normal Probability Plot of the Effects
(response is color)

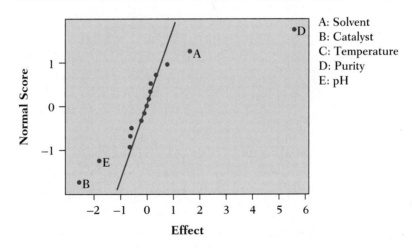

A: Solvent
B: Catalyst
C: Temperature
D: Purity
E: pH

the x_1x_2 effect would actually estimate the sum of the x_1x_2, x_3x_7, and x_5x_6 interactions (we ignore x_4x_8 because we have no x_8 in this case). If there is a large interaction, mathematically we can't be sure which of these three potential interactions is the source. We can get a clue to which variables are causing the interactions by studying the magnitudes of the linear effects. Experience has shown that factors that have large linear effects often have large interactions. One strategy is to run additional experiments to estimate the two-factor interactions associated with large linear effects, as discussed in Box, Hunter, and Hunter (1978)—an approach that should be familiar to serious students of DOE. Another approach is to delete the nonsignificant factors, if there are any, and reestimate the effects including the suspected two-factor interactions. It is also helpful to review Table 7.4 in Snee et al. (1985), which describes a 16-run, seven-factor experiment in which a significant interaction is identified by examining chains of two-factor interactions.

E.3   Analysis of Mixed-Level Factorial Experiments

The design, analysis, and interpretation of two-level experiments have been emphasized because of their wide applicability. Of course, many studies involve more than two levels for one or more factors (several examples of which were shown in Chapter 7). A $2 \times 3 \times 4$ factorial design is shown in Table E.10.

TABLE E.10   Days Sales Outstanding (DSO) Study

Run	SBU	Billing Organization	Customer Size	DSO
1	1	1	1	32
2	1	1	2	38
3	1	1	3	39
4	1	1	4	52
5	1	2	1	51
6	1	2	2	57
7	1	2	3	53
8	1	2	4	52
9	1	3	1	41
10	1	3	2	43
11	1	3	3	46
12	1	3	4	49
13	2	1	1	43
14	2	1	2	43
15	2	1	3	47
16	2	1	4	53
17	2	2	1	55
18	2	2	2	54
19	2	2	3	58
20	2	2	4	56
21	2	3	1	48
22	2	3	2	50
23	2	3	3	51
24	2	3	4	60

Variables studied were:

Strategic business unit: SBU1 and SBU2

Billing organization: Org 1, Org 2, and Org 3

Customer size (in \$1K annual sales); 1 250; 2 = 250 − 500; 3 = 500 − 750; 4 750.

FIGURE E.2   Linear (Main) Effects Plot for DSO Data

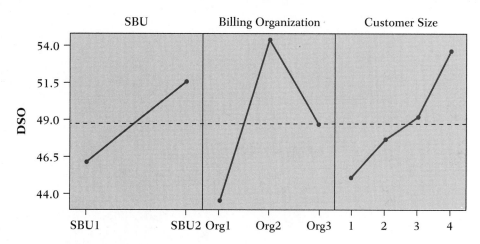

Variables plotted are
Strategic business unit: SBU1 and SBU2
Billing organization: Org 1, Org 2, and Org 3
Customer size (in \$1K annual sales): $1 \leq 250$; $2 = 250-500$; $3 = 500 - 750$; $4 \geq 750$

An effective initial analysis of a mixed-level factorial experiment is to construct plots of the means for the levels of each variable. Of equal importance are plots of all possible pairs of interactions between the variables in the design. The study summarized in Table E.10 investigated the effects of three variables on the Days Sales Outstanding (DSO) for the accounts receivable (i.e., the number of days it takes for the customer to pay the bill) of two strategic business units. The variables studied are shown at the bottom of the table. The data in Table E.10 are the average DSO of all invoices sent out in a 3-month period. The objective is for DSO to be around 35 days (to allow for mail delays) because customers are given 30 days to pay an invoice. The main-effects plots (Figure E.2) show that (1) the DSO for SBU1 is less than that of SBU2, (2) there are large differences among the three organizations issuing the bills, and (3) DSO increases with the sales volume of the customer.

The main effects are easily interpreted only if there are no interactions between the factors (see Figure E.3). In these plots, parallel lines indicate no interaction. Nonparallel lines indicate interaction. Note that in the plot of billing organization versus customer size at the lower right, the lines are not parallel. This suggests that the effect of billing organization depends on customer size. An analysis of variance (ANOVA) indicated that this interaction is statistically significant ($p < .05$) and that there was a significant difference between the DSO of the two business units. Further details on the use of analysis of variance in the analysis of mixed-level experiments can be found in the references.

FIGURE E.3   Interaction Plots for DSO Data

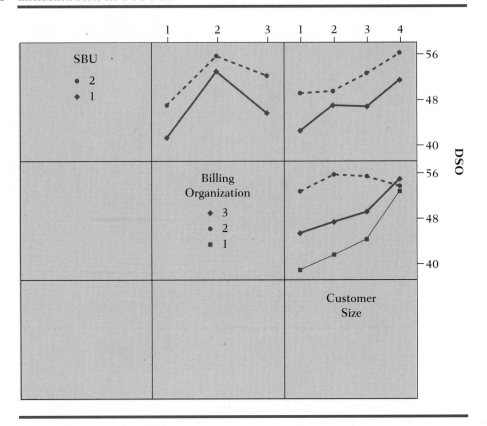

E.4   Response Surface Experiments

Response surface experiments are run to develop a prediction equation (model) that will predict the output (*y*) of the process anywhere within the design region of the process input (*x*'s) variables. Such models are used to control the process, develop optimum operation conditions, identify key operating variables, and better understand the process. An example of a chemical yield study that utilized the response surface approach is shown in Table E.11. The objective of the experiment was to identify the combination of reaction time and temperature that would give the maximum yield. Time and temperature were varied using a central composite design (Figure E.4) that consists of 9 points. Note that like all response surface designs, this has more than two levels of each *x*—five. The center point of the design was replicated 4 times, giving a total of 12 runs.

The following two-variable, six-coefficient quadratic regression model (see Chapter 6)

$$\text{Yield} = b_0 + b_1x_1 + b_2x_2 + b_{12}x_1x_2 + b_{11}x_1x_1 + b_{22}x_2x_2 + e$$

TABLE E.11 Chemical Yield Study

Run	Time	Temperature	Yield (%)
1	10	50	79
2	30	50	84
3	10	100	91
4	30	100	77
5	5	75	83
6	35	75	81
7	20	40	81
8	20	110	80
9	20	75	88
10	20	75	87
11	20	75	90
12	20	75	85

FIGURE E.4 Chemical Yield Study: Central Composite Design

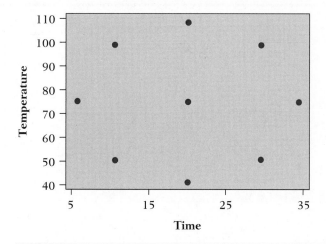

where

$$x_1 = \frac{\text{time} - 20}{15}$$

and

$$x_2 = \frac{\text{temp} - 75}{35}$$

was fit to the data (see Table E.12) and used to construct the contour plot shown in Figure E.5. Contour plots show those combinations of time and temperature that will give the same yield. Contour plots are similar to weather maps that show

TABLE E.12   Chemical Yield Study Regression Analysis

Response Surface Regression: YIELD versus TIME, TEMP

Estimated Regression Coefficients for YIELD

Term	Coef	SE Coef	T	P
Constant	87.530	1.055	82.930	0.000
Time	−2.118	1.087	−1.949	0.099
Temperature	0.636	1.051	0.606	0.567
Time × Time	−4.943	1.712	−2.887	0.028
Temperature × Temperature	−6.356	1.671	−3.804	0.009
Time × Temperature	−9.975	2.218	−4.498	0.004

S = 2.112 R^2 = 87.8% R^2 (adj) = 77.7%

FIGURE E.5   Response Surface Contour Plot: Yield as a Function of Time and Temperature

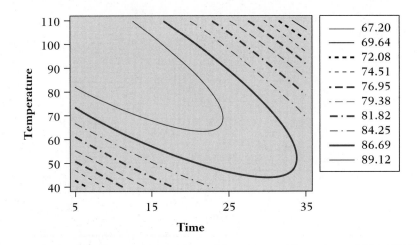

areas with the same temperature, and topographical maps that show areas with the same elevation. Figure E.5 shows that yield will be maximized at short reaction times and high reaction temperatures.

Another example of a response surface experiment is discussed in Table 7.5. Further details on the design, analysis, and interpretation of response surface designs can be found in the references summarized below.

REFERENCES

Box, G. E. P., Hunter, W. G., & Hunter, J. S. (1978). *Statistics for experimenters.* New York: Wiley Interscience, John Wiley.

Moen, R. D., Nolan, T. W., & Provost, L. P. (1991). *Improving quality through planned experimentation.* New York: McGraw-Hill.

Montgomery, D. C. (1991). *Design & analysis of experiments* (3rd ed.). New York: John Wiley.

Myers, R. H., & Montgomery D. C. (1995). *Response surface methodology.* New York: Wiley-Interscience, John Wiley.

Snee, R. D., Hare, L. B., & Trout, J. R. (1985). *Experiments in industry.* Milwaukee: Quality Press, American Society for Quality, pp. 25–35.

APPENDIX F

More on Inference Tools

To understand God's thoughts, we must study statistics, for these are the measure of His purpose.

Florence Nightingale

Chapter 8 presented inference tools, focusing on confidence intervals and hypothesis tests. Formulas that illustrated key concepts were included in the text; in other cases, no formula was given. This appendix provides one concise reference for the most common confidence intervals and hypothesis tests by listing the hypotheses being tested as well as the appropriate formulas and "degrees of freedom." Of course, you typically will perform all of these tests using statistical software, such as Minitab or JMP. For more information on inference tools, see Hogg and Ledolter (1987) or Walpole and Myers (1993).

As can be seen in Table F.1, the tests of averages involve the t distribution. This is because we are estimating the population standard deviation σ from the sample standard deviation s. Using a sample estimate adds additional uncertainty, or variation, which is accounted for in the t distribution. If we knew what σ was, we would utilize the z distribution. Using 95% confidence, the appropriate z value is 1.96. Note that $t_{.025}$ is the appropriate t value for 95% confidence. The .025 means that the area above this value in the t distribution is .025. There is another .025 below $-t_{.025}$, giving a total of .05, for 95% confidence. For cases where we have a large sample size, say 30 or greater, we can ignore the uncertainty in s and utilize the z distribution. The constant μ_0 is intended to represent some hypothesized value for the population average, such as 12 oz if we are trying to fill soft drink cans with exactly 12 oz in each can. The pooled estimate of the standard deviation is used to obtain a better estimate of variation. Recall from Chapter 8 that the typical t test for comparing two averages assumes that the populations have the same variance. It was also noted in Chapter 8 that most statistical software, such as Minitab or JMP, allows a version of the t test that does not make this assumption.

TABLE F.1
Inference Tools

Parameter of Interest	95% Confidence Interval	Null Hypothesis	Test Statistic	Degrees of Freedom
Averages				
μ	$\bar{x} \pm t_{.025}\dfrac{s}{\sqrt{n}}$	$\mu = \mu_0$	$t = \dfrac{\bar{x}-\mu_0}{s/\sqrt{n}}$	$n-1$
$\mu_1 - \mu_2$	$\bar{x}_1 - \bar{x}_2 \pm t_{.025}s_p\sqrt{\dfrac{1}{n_1}+\dfrac{1}{n_2}}$	$\mu_1 = \mu_2$	$t = \dfrac{\bar{x}_1-\bar{x}_2}{s_p\sqrt{\dfrac{1}{n_1}+\dfrac{1}{n_2}}}$	$n_1 + n_2 - 2$
Variances				
σ^2	$\dfrac{(n-1)s^2}{\chi^2_{.025}}$ to $\dfrac{(n-1)s^2}{\chi^2_{.975}}$	$\sigma^2 = \sigma_0^2$	$\chi^2 = \dfrac{(n-1)s^2}{\sigma_0^2}$	$n-1$
σ_1^2/σ_2^2	$\dfrac{s_1^2}{s_2^2}F_{.975}$ to $\dfrac{s_1^2}{s_2^2}F_{.025}$	$\sigma_1^2 = \sigma_2^2$	$F = \dfrac{s_2^2}{s_1^2}$	$n_2 - 1, n_1 - 1$
Proportions				
π	$p \pm 1.96\sqrt{\dfrac{p(1-p)}{n}}$	$\pi = \pi_0$	$z = \dfrac{p-\pi_0}{\sqrt{\dfrac{\pi_0(1-\pi_0)}{n}}}$	n/a
$\pi_1 - \pi_2$	$p_1 - p_2 \pm 1.96\sqrt{\dfrac{p_1(1-p_1)}{n_1}+\dfrac{p_2(1-p_2)}{n_2}}$	$\pi_1 = \pi_2$	$z = \dfrac{p_1 - p_2}{\sqrt{P(1-P)\left(\dfrac{1}{n_1}+\dfrac{1}{n_2}\right)}}$	n/a

where:

$s_p = \sqrt{\dfrac{(n_1-1)s_1^2 + (n_2-1)s_2^2}{n_1+n_2-2}}$ is called the pooled standard deviation

$P = \dfrac{n_1 p_1 + n_2 p_2}{n_1 + n_2}$ is a pooled estimate of the overall proportion.

For tests involving more than two averages, variances, or proportions, use ANOVA, homogeneity of variance (Levine's method), or the chi-squared test, respectively.

When testing a single variance we utilize the chi-squared distribution. We typically construct confidence intervals and test hypotheses for variances rather than for standard deviations because variances are easier to deal with mathematically. However, one can easily convert a confidence interval for the variance to one for the standard deviation by taking the square root of the interval. Similarly, we construct confidence intervals or test hypotheses for ratios of variances rather than differences because this is easier to do mathematically. This is done using the *F* distribution.

The tests and confidence intervals for proportions listed on Table F.1 utilize the normal approximation to the binomial and are therefore based on the *z* (standard normal) distribution. They use the 95% confidence *z* value of 1.96 instead of a *t* value. As explained in Chapter 9, the central limit theorem proves that as the sample size increases, the binomial will become approximately normal. In practice, this phenomenon occurs fairly quickly, and these formulas are reasonably accurate whenever the number of "successes" and the number of "failures" are both at least five. Alternatively, one can use the chi-squared test (discussed in Chapter 8), which has the same requirement for sample size. For smaller samples, we are required to use the binomial distribution directly, which is much more cumbersome. Note that the *z* distribution does not have any "degrees of freedom" associated with it.

REFERENCES

Hogg, R. V., & Ledolter, J. (1987). *Engineering statistics.* New York: Macmillan.
Walpole, R. E., & Myers, R. H. (1993). *Probability and statistics for engineers and scientists* (5th ed.). Englewood Cliffs, NJ: Prentice Hall.

APPENDIX G

More on Probability Distributions

Probabilities direct the conduct of the wise man

Cicero

This appendix will briefly review the concept of a probability distribution (presented in Chapter 9) and then present summary information (formulas, averages, and variances) on several common continuous and discrete distributions. Typical applications of these distributions will also be noted. For more information on probability distributions, see Walpole and Myers (1993).

G.1 Discrete Probability Distributions

Discrete probability distributions, also called "probability mass functions," quantify the probability of each possible outcome for a discrete random variable. For example, we might wish to calculate the probability of winning 0, 1, or 2 bids when submitting 2 bids for different municipal bonds.

Some rules about basic probability:

- Each individual probability must be somewhere between 0 and 1, where 0 means the outcome is not possible and 1 means that the outcome is certain. Lets suppose the probability of winning a given bid is .5.
- The sum of probabilities for all possible outcomes must equal 1. For example, if the probability of winning 0 bids is .25, and the probability of winning 1 is .5, then the probability of winning 2 must be .25.
- The probability of two events, A and B, both occurring (the intersection), is

$$P(A\&B) = P(A) \times P(B)$$

assuming A and B are independent of one another. For example, if A is the outcome of the first bid, and B is the outcome of the second, and we assume that the two bids are not related, then the probability of winning both bids is .5 × .5 = .25. If A and B are not independent events, the probability of their intersection must be calculated based on knowledge of their relationship. This is called conditional probability and is discussed in Walpole and Myers (1993).

- The probability of either of the two events A and B occurring (the union), is

$$P(A \cup B) = P(A) + P(B) - P(A\&B)$$

For example, the probability of winning at least one bid is .5 + .5 − .25 = .75.

- The probability of A occurring is 1 minus the probability that A does not occur. This is intuitively obvious, but in many cases it is easier to calculate the probability that A does not occur. For example, if we wish to calculate the probability that we win at least 1 bid in the next 10, it is much easier to calculate the probability that we don't.

Table G.1 depicts the formula, average, variance (standard deviation squared), and comments on Bernoulli, binomial, Poisson, geometric, and negative binomial distributions. These are five of the most commonly used discrete distributions in practice, but by no means constitute an exhaustive list. Note that all except the Poisson relate to situations where there are only two possible outcomes, a "success" and a "failure."

G.2   Continuous Probability Distributions

Continuous probability distributions, also called probability density functions, quantify the relative likelihood of potential values of a continuous random variable. Because an infinite number of outcomes are possible for a continuous variable, the probability function does not directly calculate the probability of specific outcomes. Rather, it is the area under the curve defined by the probability function that determines probability of specific outcomes. For example, for a standard normal (z) distribution, the probability of observing a value between −1 and +1 is the area under the normal curve between these two values. This is calculated using the branch of mathematics known as calculus, and is automated by standard statistical software.

Of course, values of the random variable x that have the largest probability densities are most likely to occur. For example, a standard normal variable is more likely to be around 0 than any other possible value, because 0 has the largest density (i.e., the density is highest at the average value of 0).

Again, Table G.1 lists several of the most commonly used—but not all—continuous probability distributions. Listed are the normal, standard normal, t, exponential, uniform, chi-squared, and F distributions. Each of these, other than the uniform distribution, was discussed in Chapters 8 and 9. The uniform is used when any value within a defined range is equally likely. Only a few real business situations produce this distribution, but it is useful in situations where we wish to document our uncertainty, but have little solid data to go by. One example might be forecasting the next several years' change in interest rates of adjustable rate mortgages (ARMs) in a volatile market. We may be reluctant to forecast a specific

TABLE G.1

Discrete and Continuous Probability Distributions

Discrete Probability Distributions

Distribution	Formula	Average	Variance	Comments
Bernoulli	$P(x) = \pi$ for $x = 1$ $= 1 - \pi$ for $x = 0$	π	$\pi(1 - \pi)$	Equivalent to one trial of the binomial. The binomial is the sum of n Bernoullis.
Binomial	$P(x) = \dfrac{n!}{(n-x)!\,x!}\,\pi^x(1-\pi)^{n-x}$	$n\pi$	$n\pi(1 - \pi)$	Often used for defectives data.
Poisson	$P(x) = \dfrac{e^{-\mu}\mu^x}{x!}$	μ	μ	Often used for defects data.
Geometric	$P(x) = \pi(1 - \pi)^{x-1}$	$\dfrac{1}{\pi}$	$\dfrac{1-\pi}{\pi^2}$	Used to calculate probabilities for the number of trials required to obtain the first "success" for a binomial.
Negative binomial	$P(x) = \dfrac{(x-1)!}{(r-1)!(x-r)!}\,\pi^r(1-\pi)^{x-r}$	$\dfrac{r}{\pi}$	$\dfrac{r(1-\pi)}{\pi^2}$	Used to calculate probabilities for the number of trials (x) required to obtain r successes for a binomial. Same as geometric for $r = 1$.

Note: These discrete distributions are only defined for positive (or 0) values of x. For the negative binomial, x must be $\geq r$.

(continued on next page)

TABLE G.1
Discrete and Continuous Probability Distributions *(continued)*

Continuous Probability Distributions

Distribution	Formula	Average	Variance	Comments
Normal	$f(x) = \dfrac{1}{\sqrt{2\pi\sigma^2}} \exp\left(\dfrac{-(x-\mu)^2}{2\sigma^2}\right)$	μ	σ^2	Most commonly used continuous distribution. The bell-shaped, or Gaussian curve.
Standard normal	$f(x) = \dfrac{\exp\left(-\dfrac{x^2}{2}\right)}{\sqrt{2\pi}}$	0	1	Normal curve with average of zero and standard deviation of 1. Also called the z distribution. Used extensively in the theory of statistical inference.
t	$f(x) = \dfrac{c}{(1+x^2/d)^{(d+1)/2}}$	0	$\dfrac{d}{d-2}$	Used in place of z when estimating the population standard deviation with the sample standard deviation. The constant c is chosen so that the total area under the curve is 1.0.
Exponential*	$f(x) = \dfrac{1}{\mu} e^{-x/\mu}$	μ	μ^2	Some application in reliability.
Uniform	$f(x) = \dfrac{1}{b-a}$	$\dfrac{a+b}{2}$	$\dfrac{(b-a)^2}{12}$	The uniform is a flat distribution, where any value between a and b is equally likely.

(continued on next page)

TABLE G.1
Discrete and Continuous Probability Distributions (*continued*)

Continuous Probability Distributions

Distribution	Formula	Average	Variance	Comments
Chi-squared*	$f(x) = cx^{(d/2)-1}e^{-x/2}$	d	$2d$	Used for chi-squared test of proportions and also for sample variances. The variable d is the degrees of freedom, which must be a positive integer. The constant c is chosen so that the total area under the curve is 1.0.
F^*	$f(x) = \dfrac{cx^{(d_1/2)-1}}{\left[1 + (d_1/d_2)x\right]^{(d_1+d_2)/2}}$	$\dfrac{d_2}{d_2-2}$	$\dfrac{2d_2^2\left(d_1+d_2-2\right)}{d_1(d_2-2)^2\left(d_2-4\right)}$	Used for comparing two variances. Note that the F distribution has 2 degrees of freedom parameters, d_1 and d_2. The constant c is chosen so that the total area under the curve is 1.0.

*Note: These continuous distributions are defined only for positive values of x.

value or even a distribution, but by contract the change is limited to a predefined range with a minimum and maximum.

If we subtract the population average from a normally distributed random variable and divide by the population standard deviation, we obtain a standard normal—that is, $z = (x - \mu)/\sigma$ has a standard normal distribution. In most practical situations, we don't actually know the population standard deviation, but are estimating it with s. This produces the t distribution: $t = (x - \mu)/s$ has a t distribution with $n - 1$ degrees of freedom (where n was the sample size used to calculate s). For sample sizes of 30 or more, the t and standard normal are almost the same. In most statistical inference situations, not knowing the population average (μ) is not important, because we are typically assuming that it is equal to some hypothesized value, such as the average of another population. Based on the process of hypothesis testing, we replace μ with the hypothesized value. Recall that in hypothesis testing, we begin with the assumption that the hypothesis is true.

In many practical situations, the random variable x is actually a sample average, in which case we need to understand the sampling distribution (see Chapter 9). In this case, the formula for t is:

$$t = \frac{\bar{x} - \mu}{s/\sqrt{n}}$$

since $s/\sqrt{n}$ is the sample standard deviation of $\bar{x}$.

If we take a sample variance, multiply it by $n - 1$, and divide by the hypothesized population variance, we obtain a chi-squared variable. In other words, $\chi^2 = (n - 1)s^2/\sigma^2$ has a chi-squared distribution with $n - 1$ degrees of freedom (i.e., the d in Table G.1 is $n - 1$). This can then be used to obtain a confidence interval for σ.

If we take the ratio of two chi-squareds, dividing each by its degrees of freedom, we obtain an F distribution—that is,

$$F = \frac{\chi_1^2/(n_1 - 1)}{\chi_2^2/(n_2 - 1)} = \frac{s_1^2/\sigma_1^2}{s_2^2/\sigma_2^2}$$

has an F distribution with $n_1 - 1$ and $n_2 - 1$ degrees of freedom. If we are testing the hypothesis that $\sigma_1^2 = \sigma_2^2$, then we assume them equal, and this simplifies to $F = s_1^2/s_2^2$, or just the ratio of two variances.

REFERENCE

Walpole, R. E., & Myers, R. H. (1993). *Probability and statistics for engineers and scientists* (5th ed.). Englewood Cliffs, NJ: Prentice Hall.

APPENDIX H

Process Design (Reengineering)

Reengineering a company's business processes ultimately changes practically everything *about the company.*

Michael Hammer and James Champy

H.1   Why Design?

The most common application of statistical thinking is to an existing process that needs improvement. In some cases, however, designing a completely new process is required. For example, an organization may wish to offer a new service, such as Web page design. Rather than improving an existing marketing process, we have to design a completely new one. In other cases, we may need to scrap an existing process and start over again, rather than improve the current one. For example, a financial system may be running on several old "legacy" computer systems and needs to be modernized to run on an Internet-based system. In such cases, we do not spend a lot of time studying the old process, because we plan to create a completely new one. We just need to understand what about the old system is limiting it, and what needs to be continued in the new system.

The term *reengineering* refers to situations wherein the old system is scrapped and we start with a "clean sheet of paper" in designing the new one. A very popular reengineering initiative swept through U.S. business in the 1990s, largely as a result of the book *Reengineering the Corporation,* by Hammer and Champy (1993). (This general business initiative was discussed briefly in Chapter 1.) Some of the principles discussed below are based on this groundbreaking book and on subsequent references, such as Champy (1995) and Hammer (1996). Both contain more details on process design and reengineering.

Given an existing process, the first step in reengineering is to decide whether improving it will achieve sufficient improvement, or if we must start over again with a completely new design. This difficult decision is analogous to a cost-conscious person who is driving an old car that is frequently in the repair shop. At some point it will become more economical to buy a new car, rather than continue to repair the old one—but it is hard to know whether now is the time. In business settings, this decision is often made after analyzing and determining the inherent capability of the process and the cost of achieving its maximum capability. A project may begin with the process improvement strategy and halfway through indicate that the process must be totally redesigned in order to achieve objectives. In other cases, such as that of the legacy computer system, the decision may be obvious from the beginning. Of course, if no process exists currently, then we begin with process design.

Improvement efforts may be classified in a continuum:

- Problem solving is the most basic and typically least time consuming and expensive.
- Process improvement typically results in more substantial improvement, but is also more time consuming and expensive.
- Process design, or reengineering, is capable of providing the most radical improvements, but also will likely take the longest and be most expensive.

Planners must carefully choose the most appropriate approach for the particular situation.

H.2 Measures of Success

Once the decision to design a new process has been made, determine the key output measures. These will be the "measures of success" for the process—without which it will be very hard to agree on the "best" design. Then design a process to meet these output measures most effectively and efficiently. Later in the design process, identify key process variables and, ultimately, key input variables. Document any constraints, including time limitations, allowable budgets, or restrictions on the design (e.g., "Must be compatible with legacy systems").

These measures of success must obviously be measurable, rather than fuzzy (subjective) concepts. For example, a customer may say that they want "knowledgeable sales reps." This is a rational need, but to determine if we were delivering it we need some means of measuring it, such as a technical test for the sales reps, surveys for the customers to fill out, or other means. Tools such as Quality Function Deployment (QFD) can be extremely helpful in converting fuzzy concepts into tangible measures.

H.3 Conceptual Design

Designers in a variety of fields have found it useful to split the design process into two major components. The first is conceptual design, which uses imagination, benchmarking, brainstorming, or other methods to think "outside the box," and identify the best conceptual design. Conceptual design is an idea for which we may not have worked out all the details. An example for designing a new consumer loan would be: "An electronic application with a rapid decision and electronic transfer of funds." At this stage, think about all possible alternatives to avoid copying inefficient aspects of the current design.

Think about the conceptual design holistically, although at a conceptual level. That is, think about each of the following potential elements of the design:

- Process design (via a flowchart)
- Product design (the output from the process)
- People design (skills required, organizational chart, and so on)
- Material design (the inputs needed)
- The information technology (IT) design (the computer technology required)
- Facility design (if a new or remodeled facility is needed)
- Equipment design (if equipment other than computers is needed)

Of course, even though a new design may not need each of these elements, they should always be considered.

In many cases, human nature leads people to want to begin the process with the detailed design stage. For example, some members of the design team may have a great idea for wireless connectivity to the Internet, and want to jump into the details of the wireless technology immediately. Clearly, teams should only proceed to detailed implementation after agreeing on the measures of success for the design and the overall conceptual design.

Once the conceptual design elements are agreed upon, critically analyze it to avoid spending time and money developing a design that will not work. Various methods can be used to evaluate the design, including computer simulation, failure mode and effects analysis (FMEA), benchmarking of similar designs, simple spreadsheet analysis, development of prototypes, and so on. Do enough work to be sure that the conceptual design selected will meet the measures of success defined in the first stages of the project.

H.4 Detailed Design

Once a conceptual design has been selected and evaluated, all of the details that were skipped over in conceptual design need to be worked out. For example, if we develop an "electronic application" for a loan, what would the Web page actually look like? How many servers will be required? What download speed will be required? These details in each of the design elements (IT design, people design, and so on) need to be worked out. In typical designs, most of the total cycle time is spent in this phase.

Make the detailed design as analytical as is feasible. Make decisions about the details of the design using the analytical methods discussed in this text. For example, to design an electronic loan application, we need to determine the appropriate number of servers to purchase. This could be decided using gut feel and intuition or using predicted volume to the Web site to model the relationship between number of servers and the response time of the site. Clearly an accurate model will enable the team to make a more informed business decision.

Another example requiring a quantitative approach is designing a new deal-making process for a service business that finances large business deals, such as leveraged buy-outs, acquisitions, and initial public offerings (IPOs). Once the design team has identified the key process steps and determined how they will be carried out, the team should consider how the cycle time of each step would relate to the overall process cycle time—a key output measure of success. Analyzing a sequential process would utilize linear combinations (discussed in Chapter 9) to predict the distribution of overall cycle time based on our estimates of the individual step cycle times. Similarly, the team can determine what individual step cycle time distributions are required to achieve the desired overall cycle time distribution. In this way, the requirements can flow from overall output measures of success to process and input needs. Such an approach makes the entire design process more analytical and rigorous.

After developing the detailed design, perform an overall evaluation of the design. This is similar to the evaluation of the conceptual design, but now there is much more detail to evaluate. It may be necessary to loop back through the detailed design process to address any serious deficiencies.

H.5   Trust but Verify!

"Data often destroy a good argument"—often we can make a strong argument that something will work, but obtaining actual data may disprove this argument. We no doubt believe that our new design will perform fabulously. However, we only have *predictions* of performance: Given that we haven't yet implemented the design, we have no actual data. It is almost always a good idea to pilot the design on a limited scale to obtain actual data on performance and minimize the risk of implementing the design full scale.

A pilot is a limited implementation of the entire system under realistic conditions. A pilot might involve identifying a target population of potential customers, making them aware of the Web site, and monitoring the system as this targeted set of customers access the prototype. The customers would access the system whenever they chose and would not already be familiar with the system.

In contrast, prototypes are tests of the system typically done under ideal conditions. For example, a prototype of the electronic loan application could be developed on an experimental server that is accessed only during off-business hours and only by people knowledgeable about the system. This evaluation can provide valuable proof-of-concept information, but does not provide information on how the system would perform under real conditions, hence it is not a pilot.

After obtaining real performance data on key output, process, and input measures from the pilot, the team is now in a position to utilize the problem-solving or process improvement strategies to improve the design prior to full implementation. This is an example of how the three improvement methodologies of process design, process improvement, and problem solving can be effectively integrated into an overall business improvement framework.

H.6 Implementation and Follow-up

Now that we have proven our design—not only by analytical prediction, but also in actual application—we are in a position to implement it fully. This may be done in several steps via a sequential rollout or in a one-step grand opening. Once we implement the design, we are back to the familiar situation of having an existing process. If our business has a continuous improvement mindset, we will realize that even with a successful implementation, there is plenty of room for improvement. The improvement focus will now logically shift to process improvement and problem solving.

H.7 Summary

Some key success factors to keep in mind for process design are:

- Carefully scope the design project prior to beginning to work on it.
- Ensure that quantitative output measures of success have been identified early.
- Consider options for the conceptual design before jumping into the details.
- Think about each of the design elements (IT, people, process, and so on).
- Make detailed design as quantitative as feasible; don't rely on intuition.
- Resist the temptation to implement prior to conducting a real pilot.
- Transition from design to process improvement and problem solving.

REFERENCES

Champy, J. (1995). *Reengineering management.* New York: HarperCollins.

Hammer, M. (1996). *Beyond reengineering.* New York: HarperCollins.

Hammer, M., & Champy, J. (1993). *Reengineering the corporation.* New York: HarperCollins.

APPENDIX I

t Critical Values

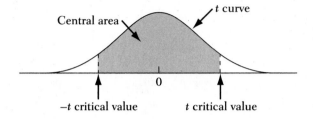

Central area captured:		.80	.90	.95	.98	.99	.998	.999
Confidence level:		80%	90%	95%	98%	99%	99.8%	99.9%
	1	3.08	6.31	12.71	31.82	63.66	318.31	636.62
	2	1.89	2.92	4.30	6.97	9.93	23.33	31.60
	3	1.64	2.35	3.18	4.54	5.84	10.21	12.92
	4	1.53	2.13	2.78	3.75	4.60	7.17	8.61
	5	1.48	2.02	2.57	3.37	4.03	5.89	6.86
	6	1.44	1.94	2.45	3.14	3.71	5.21	5.96
	7	1.42	1.90	2.37	3.00	3.50	4.79	5.41
	8	1.40	1.86	2.31	2.90	3.36	4.50	5.04
	9	1.38	1.83	2.26	2.82	3.25	4.30	4.78
	10	1.37	1.81	2.23	2.76	3.17	4.14	4.59
	11	1.36	1.80	2.20	2.72	3.11	4.03	4.44
	12	1.36	1.78	2.18	2.68	3.06	3.93	4.32
	13	1.35	1.77	2.16	2.65	3.01	3.85	4.22
	14	1.35	1.76	2.15	2.62	2.98	3.79	4.14
	15	1.34	1.75	2.13	2.60	2.95	3.73	4.07
	16	1.34	1.75	2.12	2.58	2.92	3.69	4.02
	17	1.33	1.74	2.11	2.57	2.90	3.65	3.97
Degrees of	18	1.33	1.73	2.10	2.55	2.88	3.61	3.92
freedom	19	1.33	1.73	2.09	2.54	2.86	3.58	3.88
	20	1.33	1.73	2.09	2.53	2.85	3.55	3.85
	21	1.32	1.72	2.08	2.52	2.83	3.53	3.82
	22	1.32	1.72	2.07	2.51	2.82	3.51	3.79
	23	1.32	1.71	2.07	2.50	2.81	3.49	3.77
	24	1.32	1.71	2.06	2.49	2.80	3.47	3.75
	25	1.32	1.71	2.06	2.49	2.79	3.45	3.73
	26	1.32	1.71	2.06	2.48	2.78	3.44	3.71
	27	1.31	1.70	2.05	2.47	2.77	3.42	3.69
	28	1.31	1.70	2.05	2.47	2.76	3.41	3.67
	29	1.31	1.70	2.05	2.46	2.76	3.40	3.66
	30	1.31	1.70	2.04	2.46	2.75	3.39	3.65
	40	1.30	1.68	2.02	2.42	2.70	3.31	3.55
	60	1.30	1.67	2.00	2.39	2.66	3.23	3.46
	120	1.29	1.66	1.98	2.36	2.62	3.16	3.37
z critical values	∞	1.28	1.645	1.96	2.33	2.58	3.09	3.29

APPENDIX J

Standard Normal Probabilities (Cumulative z Curve Areas)

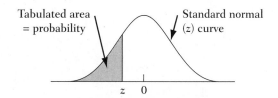

Tabulated area = probability

Standard normal (z) curve

Second Decimal of z Value

z	.00	.01	.02	.03	.04	.05	.06	.07	.08	.09
−3.8	.0000	.0000	.0000	.0000	.0000	.0000	.0000	.0000	.0000	.0000
−3.7	.0001	.0001	.0001	.0001	.0001	.0001	.0001	.0001	.0001	.0001
−3.6	.0002	.0002	.0001	.0001	.0001	.0001	.0001	.0001	.0001	.0001
−3.5	.0002	.0002	.0002	.0002	.0002	.0002	.0002	.0002	.0002	.0002
−3.4	.0003	.0003	.0003	.0003	.0003	.0003	.0003	.0003	.0003	.0002
−3.3	.0005	.0005	.0005	.0004	.0004	.0004	.0004	.0004	.0004	.0003
−3.2	.0007	.0007	.0006	.0006	.0006	.0006	.0006	.0005	.0005	.0005
−3.1	.0010	.0009	.0009	.0009	.0008	.0008	.0008	.0008	.0007	.0007
−3.0	.0013	.0013	.0013	.0012	.0012	.0011	.0011	.0011	.0010	.0010
−2.9	.0019	.0018	.0018	.0017	.0016	.0016	.0015	.0015	.0014	.0014
−2.8	.0026	.0025	.0024	.0023	.0023	.0022	.0021	.0021	.0020	.0019
−2.7	.0035	.0034	.0033	.0032	.0031	.0030	.0029	.0028	.0027	.0026
−2.6	.0047	.0045	.0044	.0043	.0041	.0040	.0039	.0038	.0037	.0036
−2.5	.0062	.0060	.0059	.0057	.0055	.0054	.0052	.0051	.0049	.0048
−2.4	.0082	.0080	.0078	.0075	.0073	.0071	.0069	.0068	.0066	.0064
−2.3	.0107	.0104	.0102	.0099	.0096	.0094	.0091	.0089	.0087	.0084
−2.2	.0139	.0136	.0132	.0129	.0125	.0122	.0119	.0116	.0113	.0110
−2.1	.0179	.0174	.0170	.0166	.0162	.0158	.0154	.0150	.0146	.0143
−2.0	.0228	.0222	.0217	.0212	.0207	.0202	.0197	.0192	.0188	.0183
−1.9	.0287	.0281	.0274	.0268	.0262	.0256	.0250	.0244	.0239	.0233
−1.8	.0359	.0351	.0344	.0336	.0329	.0322	.0314	.0307	.0301	.0294
−1.7	.0446	.0436	.0427	.0418	.0409	.0401	.0392	.0384	.0375	.0367
−1.6	.0548	.0537	.0526	.0516	.0505	.0495	.0485	.0475	.0465	.0455
−1.5	.0668	.0655	.0643	.0630	.0618	.0606	.0594	.0582	.0571	.0559
−1.4	.0808	.0793	.0778	.0764	.0749	.0735	.0721	.0708	.0694	.0681
−1.3	.0968	.0951	.0934	.0918	.0901	.0885	.0869	.0853	.0838	.0823
−1.2	.1151	.1131	.1112	.1093	.1075	.1056	.1038	.1020	.1003	.0985
−1.1	.1357	.1335	.1314	.1292	.1271	.1251	.1230	.1210	.1190	.1170
−1.0	.1587	.1562	.1539	.1515	.1492	.1469	.1446	.1423	.1401	.1379
−0.9	.1841	.1814	.1788	.1762	.1736	.1711	.1685	.1660	.1635	.1611
−0.8	.2119	.2090	.2061	.2033	.2005	.1977	.1949	.1922	.1894	.1867
−0.7	.2420	.2389	.2358	.2327	.2296	.2266	.2236	.2206	.2177	.2148
−0.6	.2743	.2709	.2676	.2643	.2611	.2578	.2546	.2514	.2483	.2451
−0.5	.3085	.3050	.3015	.2981	.2946	.2912	.2877	.2843	.2810	.2776
−0.4	.3446	.3409	.3372	.3336	.3300	.3264	.3228	.3192	.3156	.3121
−0.3	.3821	.3783	.3745	.3707	.3669	.3632	.3594	.3557	.3520	.3483
−0.2	.4207	.4168	.4129	.4090	.4052	.4013	.3974	.3936	.3897	.3859
−0.1	.4602	.4562	.4522	.4483	.4443	.4404	.4364	.4325	.4286	.4247
−0.0	.5000	.4960	.4920	.4880	.4840	.4801	.4761	.4721	.4681	.4641

(continued)

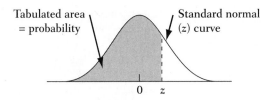

Tabulated area = probability   Standard normal (z) curve

Second Decimal of z Value

z	.00	.01	.02	.03	.04	.05	.06	.07	.08	.09
0.0	.5000	.5040	.5080	.5120	.5160	.5199	.5239	.5279	.5319	.5359
0.1	.5398	.5438	.5478	.5517	.5557	.5596	.5636	.5675	.5714	.5753
0.2	.5793	.5832	.5871	.5910	.5948	.5987	.6026	.6064	.6103	.6141
0.3	.6179	.6217	.6255	.6293	.6331	.6368	.6406	.6443	.6480	.6517
0.4	.6554	.6591	.6628	.6664	.6700	.6736	.6772	.6808	.6844	.6879
0.5	.6915	.6950	.6985	.7019	.7054	.7088	.7123	.7157	.7190	.7224
0.6	.7257	.7291	.7324	.7357	.7389	.7422	.7454	.7486	.7517	.7549
0.7	.7580	.7611	.7642	.7673	.7704	.7734	.7764	.7794	.7823	.7852
0.8	.7881	.7910	.7939	.7967	.7995	.8023	.8051	.8078	.8106	.8133
0.9	.8159	.8186	.8212	.8238	.8264	.8289	.8315	.8340	.8365	.8389
1.0	.8413	.8438	.8461	.8485	.8508	.8531	.8554	.8577	.8599	.8621
1.1	.8643	.8665	.8686	.8708	.8729	.8749	.8770	.8790	.8810	.8830
1.2	.8849	.8869	.8888	.8907	.8925	.8944	.8962	.8980	.8997	.9015
1.3	.9032	.9049	.9066	.9082	.9099	.9115	.9131	.9147	.9162	.9177
1.4	.9192	.9207	.9222	.9236	.9251	.9265	.9279	.9292	.9306	.9319
1.5	.9332	.9345	.9357	.9370	.9382	.9394	.9406	.9418	.9429	.9441
1.6	.9452	.9463	.9474	.9484	.9495	.9505	.9515	.9525	.9535	.9545
1.7	.9554	.9564	.9573	.9582	.9591	.9599	.9608	.9616	.9625	.9633
1.8	.9641	.9649	.9656	.9664	.9671	.9678	.9686	.9693	.9699	.9706
1.9	.9713	.9719	.9726	.9732	.9738	.9744	.9750	.9756	.9761	.9767
2.0	.9772	.9778	.9783	.9788	.9793	.9798	.9803	.9808	.9812	.9817
2.1	.9821	.9826	.9830	.9834	.9838	.9842	.9846	.9850	.9854	.9857
2.2	.9861	.9864	.9868	.9871	.9875	.9878	.9881	.9884	.9887	.9890
2.3	.9893	.9896	.9898	.9901	.9904	.9906	.9909	.9911	.9913	.9916
2.4	.9918	.9920	.9922	.9925	.9927	.9929	.9931	.9932	.9934	.9936
2.5	.9938	.9940	.9941	.9943	.9945	.9946	.9948	.9949	.9951	.9952
2.6	.9953	.9955	.9956	.9957	.9959	.9960	.9961	.9962	.9963	.9964
2.7	.9965	.9966	.9967	.9968	.9969	.9970	.9971	.9972	.9973	.9974
2.8	.9974	.9975	.9976	.9977	.9977	.9978	.9979	.9979	.9980	.9981
2.9	.9981	.9982	.9982	.9983	.9984	.9984	.9985	.9985	.9986	.9986
3.0	.9987	.9987	.9987	.9988	.9988	.9989	.9989	.9989	.9990	.9990
3.1	.9990	.9991	.9991	.9991	.9992	.9992	.9992	.9992	.9993	.9993
3.2	.9993	.9993	.9994	.9994	.9994	.9994	.9994	.9995	.9995	.9995
3.3	.9995	.9995	.9995	.9996	.9996	.9996	.9996	.9996	.9996	.9997
3.4	.9997	.9997	.9997	.9997	.9997	.9997	.9997	.9997	.9997	.9998
3.5	.9998	.9998	.9998	.9998	.9998	.9998	.9998	.9998	.9998	.9998
3.6	.9998	.9998	.9999	.9999	.9999	.9999	.9999	.9999	.9999	.9999
3.7	.9999	.9999	.9999	.9999	.9999	.9999	.9999	.9999	.9999	.9999
3.8	.9999	.9999	.9999	.9999	.9999	.9999	.9999	.9999	.9999	1.0000

Index

Regression coefficient
    confidence interval, 343–344
    test, 360–361
Regression equation, prediction interval with, 344–345
Regression modeling process, 328. *See also* Model
    check model fit, 246–251
    coefficients, 329
    extrapolation, 252
    in general, 236–237
    least squares calculations, 238–243
    method, 238
    multiple predictor variables, 237–238
    report and use model, 251–252
Reports. *See also* Presentations
    in general, 467
    graphics, 467–468
    statistical, 468
Residual analysis
    model building, 246–251, 259
    normal probability scale, 249–250
    observation sequence, 250–251
    outlier residuals, 250–251
    predicted values, 247–248
    predictor variables, 248–249
Resin variation, case study, 96–102
Response surface experiments, 492–494. *See also* Experimental design
Rework, 66. *See also* Work
Reynard, S., 464
Ricoh, Numazu plant, 96–102
Ritter, D., 172, 185
Roosevelt, Franklin, 150
Root sum of squares formula, 407
Rummler, G. A., 57, 82
Run chart, 108, 113, 119, 125, 126. *See also* Control chart
    compared to control chart, 179
    discussed, 178–179
    residual analysis, 250–251
Run rules, 172

S

Salamone, 195
Sample. *See also* Sampling
    frame selection, 146
    random, 41, 148–149
Sample size, 353. *See also* Sample size formulas
    determining, 146–148

    for hypothesis test, 365–366
    key drivers, 365
    survey, 470
Sample size formulas. *See also* Statistical inference tools
    in general, 361
    sampling from finite population, 364–365
    sampling from infinite population, 361–364
Sample variance, 403–404. *See also* Variance
    formula, 403
Sampling. *See also* Data collection
    data quality, 148–151
    data quantity, 151–152
    from infinite population, 361–365
    practical concerns, 41, 170
    time dimension in, 335–336
Sampling distribution
    central limit theorem, 400–402
    in general, 399
    sample average, 399–400
    sample variance, 403–404
    t distribution, 404–406
SAS Incorporated, 224
Scatter plot, 100, 125
    defined, 109
    discussed, 180–183
    "labeled," 181
Scholtes, P. R., 11, 67, 460
Schroeder, R., 11
Scientific method
    experiment, 42
    hypothesis, 42
    observation, 43
    Plan-Do-Check-Act cycle, 43
Scrap, 66
Scribe, 190
    meetings, 462
Self-managed work teams, 11. *See also* Teams
Senge, P., 10, 11
Service, defined, 61
Shade, J. E., 132, 197
Shewhart cycle. *See* Plan-Do-Check-Act cycle
Shewhart, Walter, 43, 169
SIPOC model, business process, 56, 61–64, 75, 83
Six Sigma, 95, 284. *See also* Problem solving; Process improvement
    discussed, 11–12, 476–479
    evolution, 479–480

Statistical Thinking Model

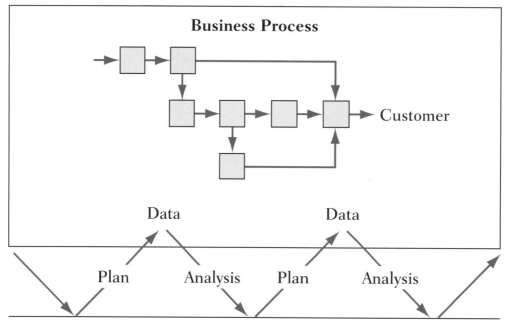

Source: Adapted from G. E. P. Box, W. G. Hunter, & J. S. Hunter, *Statistics for Experimenters.* New York: Wiley Interscience, 1978.